Advances and Applications in Structural Vibration Control

Advances and Applications in Structural Vibration Control

Guest Editors

Ruisheng Ma
Haoran Zuo
Kunjie Rong
Siyuan Wu

MDPI

Basel • Beijing • Wuhan • Barcelona • Belgrade • Novi Sad • Cluj • Manchester

Guest Editors

Ruisheng Ma
State Key Laboratory of Bridge Safety and Resilience
Beijing University of Technology
Beijing
China

Haoran Zuo
College of Civil Engineering
Tongji University
Shanghai
China

Kunjie Rong
Department of Mechanical Engineering
North China Electric Power University
Baoding
China

Siyuan Wu
China Construction Eighth Engineering Division Co., Ltd.,
Shanghai
China

Editorial Office
MDPI AG
Grosspeteranlage 5
4052 Basel, Switzerland

This is a reprint of the Special Issue, published open access by the journal *Buildings* (ISSN 2075-5309), freely accessible at: https://www.mdpi.com/journal/buildings/special_issues/1O4X676L7D.

For citation purposes, cite each article independently as indicated on the article page online and as indicated below:

Lastname, A.A.; Lastname, B.B. Article Title. *Journal Name* **Year**, *Volume Number*, Page Range.

ISBN 978-3-7258-5743-2 (Hbk)
ISBN 978-3-7258-5744-9 (PDF)
https://doi.org/10.3390/books978-3-7258-5744-9

Contents

About the Editors

Ruisheng Ma

Ruisheng Ma is a Professor in the College of Architecture and Civil Engineering at Beijing University of Technology (BJUT). He earned his Ph.D. from Curtin University in 2021 and joined BJUT in 2022. His research focuses on Structural Vibration Control and Structural Dynamics. Dr. Ma has authored over 60 peer-reviewed papers in well-known international journals, which have garnered more than 1,800 citations and an H-index of 22 (Google Scholar, as of October 2025). His work has received quite a few highly competitive research grants, including grants from the National Natural Science Foundation of China and the China Postdoctoral Science Foundation. In 2025, he was ranked among the world's top 2% most-cited scientists by Stanford University.

Haoran Zuo

Haoran Zuo is an Associate Professor in the College of Civil Engineering at Tongji University, appointed in 2025. Prior to this, he held the prestigious Australian Research Council (ARC) Discovery Early Career Researcher Award (DECRA) Fellowship in the School of Civil and Mechanical Engineering at Curtin University. He received his Ph.D. in Civil Engineering from Curtin University in 2019 and worked as a Postdoctoral Fellow in the Department of Civil and Environmental Engineering at The Hong Kong Polytechnic University from 2020 to 2022. Dr. Zuo's research centers on the dynamic response analysis and vibration control of offshore structures. His outstanding contributions have been recognized through several prestigious honors, including the ARC DECRA Fellowship, the Marie Skłodowska-Curie Actions Fellowship, China's National Excellent Young Scientists Fund (Overseas), and the Best Young Researcher Award at the 18th East Asia-Pacific Conference on Structural Engineering and Construction. He has published over 30 SCI-indexed journal papers, including Cover Papers, Editor's Featured Papers, and ESI Highly Cited Papers, which have collectively received more than 1,360 citations and an H-index of 17 on Google Scholar as of October 2025. Dr. Zuo currently serves as an Editor and Guest Editor for several international journals and actively contributes to the academic community through his roles on scientific and organizing committees. In addition, Dr. Zuo was recognized in the Top 2% Scientists list by Elsevier and Stanford University.

Kunjie Rong

Kunjie Rong received his Ph.D. degree from Tongji University. He now works as an Associate Professor at the Department of Mechanical Engineering, North China Electric Power University. His research focuses on disaster prevention and mitigation for transmission line systems. He has presided over three projects from the Provincial and National Foundations, including the National Natural Science Foundation of China, China Postdoctoral Science Foundation, and Shandong Postdoctoral Science Foundation. Based on these projects, he has published more than 30 papers in peer-reviewed journals and has been awarded 25 invention patents.

Siyuan Wu

Siyuan Wu received his Ph.D. degree from Tongji University and is currently conducting postdoctoral research at Dalian University of Technology. His research focuses on vibration control and optimal design. He has presided over five projects at the provincial and enterprise levels. He has obtained 31 patents and published over 40 technical papers. Moreover, he has also been selected by the Jiangsu Province Innovative Talent Program.

Preface

The evolving paradigm of modern civil engineering is characterized by the proliferation of structural systems of increasing geometrical extremity: heightened elevation, pronounced slenderness, and significant spatial expanse. This class of infrastructure—encompassing tall buildings, long-span bridges, wind turbine towers, and offshore floating structures—is frequently deployed in geographies susceptible to multiple hazards. Consequently, these structures are subjected to a multifaceted spectrum of external environmental loads, including, but not limited to, wind, seismic ground motions, and waves.

The dynamic responses of such structures to these environmental loads present a fundamental challenge to their serviceability, safety, and long-term durability. The mitigation of such dynamic responses necessitates moving beyond strength-based design principles to embrace performance-based engineering and resilience. In this context, structural vibration control has emerged as a critical and indispensable discipline, providing a systematic methodology for ensuring structural integrity and performance under demanding loading conditions.

This reprint offers a comprehensive treatise on the evolution and application of novel vibration control technologies, with a specific focus on the development of new materials and devices. Its examination is underpinned by the critical premise that advanced control strategies are indispensable for ensuring the resilience and reliability of next-generation engineering structures—a necessity for safeguarding both significant economic investment and, most fundamentally, human life.

Ruisheng Ma, Haoran Zuo, Kunjie Rong, and Siyuan Wu
Guest Editors

buildings

MDPI

Article

Study on the Influence and Optimization Design of Viscous Damper Parameters on the Damping Efficiency of Frame Shear Wall Structure

Xiang Lan [1,2,*], Guanglan Wei [3] and Xingxian Zhang [4]

1 College of Architecture and Civil Engineering, Kunming University, Kunming 650214, China
2 Institute of Engineering Earthquake Resistance and Disaster Reduction, Kunming University, Kunming 650214, China
3 College of Architecture and Civil Engineering, Dianchi College of Yunnan University, Kunming 650228, China
4 College of Urban Construction, Yunnan Open University & Yunnan Vocational and Technical College of National Defense Industry, Kunming 650032, China
* Correspondence: lx523947015@kmu.edu.cn

Abstract: This study investigates the influence of viscous damper parameters on the damping efficiency of frame shear wall structures. Taking a specific frame shear wall structure as the background, a three-dimensional finite element model is established using a nonlinear dynamic time–history analysis method. The damping ratio, reduction in vertex displacement, reduction in base shear, and inter-story drift utilization rate are selected as the damping performance indicators. Firstly, a sensitivity analysis is conducted to study the influence of different viscous damper parameters on these indicators. Then, the relationship models between the viscous damper parameters and the indicators are fitted using the response surface method, and the fitting effect is evaluated through an F-test and determination coefficient R^2. Finally, an objective function based on key damping performance indicators is established to solve for the optimal parameters. The results show that the traditional sensitivity analysis method is unable to comprehensively consider the combined effects of different damping efficiency indicators. The response surface method has high fitting accuracy and good predictability and can serve as an optimization model. Considering the stiffness of supporting components matched with the viscous damper parameters, the feasibility of the optimal damping parameters is demonstrated from an engineering application perspective. A simple and easy-to-operate damping design flowchart is proposed, providing important guidance and reference for designers in frame shear wall structure damping design in the future.

Keywords: energy dissipation structure; frame shear wall structure; viscous damper parameter; damping efficiency; response surface method

1. Introduction

On 1 September 2021, with the implementation of the "Regulations on Seismic Management of Construction Engineering" in China, the application of energy dissipation and isolation technology has sparked a new wave nationwide. Energy dissipation structures [1,2] refer to the installation of energy dissipation devices in building structures to dissipate a portion of the seismic energy input, thereby reducing the response of the upper structure under seismic excitation and mitigating or even avoiding damage to the main structure. The control types of energy dissipation systems mainly include active control, semi-active control, passive control, and intelligent control [3–5]. The common passive energy dissipation devices include velocity-dependent dampers, displacement-dependent dampers, and hybrid dampers. Viscous dampers [6] are the most common velocity-based dissipative devices used to reduce the vibration response of structures by utilizing the damping characteristics of viscous materials. When a structure is subjected to external forces, the viscous material inside the damper generates damping forces, allowing the structure to absorb and dissipate energy. Viscous dampers have several advantages compared to other types of dampers. Firstly, they can provide higher additional damping ratios, effectively reducing the vibration amplitude and response of the structure. Secondly, they exhibit stable and reliable damping characteristics within the frequency response range. Lastly, viscous dampers have a relatively simple construction, resulting in lower installation and maintenance costs. As a result, they have been widely applied in engineering practice.

Lin et al. [7] conducted a comprehensive study on the earthquake response of elastic single-degree-of-freedom systems with attached nonlinear viscous dampers. And Gherbi et al. [8] researched the simplified design strategy of nonlinear fluid viscous dampers for MDOF structures. A computational framework is proposed for the design of earthquake-resistant buildings using genetic algorithms to optimize robust designs with passive dampers [9]. Rayegani et al. [10] conducted a study on retrofitting a steel moment frame structure with four-joint rotational friction dampers to improve seismic behavior during mainshock–aftershock sequences. Ding et al. [11] presented a seismic performance analysis of viscous damping outrigger technology for super high-rise buildings, demonstrating its effectiveness in enhancing structural security and reducing construction costs. The application of viscous dampers in adjacent buildings is also widespread. Scholars have conducted in-depth research on the dynamic response [12–15] and seismic performance [16,17] of structures under seismic excitation. It can be observed from the literature mentioned above that passive energy dissipation devices, particularly viscous dampers, are widely utilized in various structural systems. However, their application in reinforced concrete frame shear walls is almost nonexistent.

Numerous scholars have conducted comprehensive research on the calculation methodology for additional damping ratio and the effective evaluation of the added dampers. Xu et al. [18] proposed a method for calculating the additional damping ratio considering the influence of excitation frequency. Diotallevi et al. [19] conducted a simplified method for assessing the supplemental damping ratio using a new dimensionless parameter. This was verified through numerical investigations on single- and multi-degree of freedom systems under harmonic excitations and ground motions. Hwang et al. [20] investigated how to incorporate the consideration of additional damping ratio in the practical engineering application of the normative design methods (FEMA273 or FEMA356) and related research report methods. Rama et al. [21], Xie et al. [22], and Occhiuzzi [23] conducted a comprehensive study on the enhancement of structural seismic performance and effectiveness assessment of dampers installed in buildings. However, there has been no specific research on the influence of viscous damper parameters on the additional damping ratio and the effectiveness of the damper.

The purpose of attaching viscous dampers is to reduce the structural response during an earthquake. Different scholars have conducted detailed research on optimal seismic response control and optimal damper placement. Singh et al. [24] and Aguirre et al. [25] investigated the optimal seismic response of structures with additional viscous dampers. The optimal placement of dampers has been investigated in terms of the simplified methods of multi-degree-of-freedom systems [26], transfer functions [27], critical excitation [28], and distribution of damping coefficients [29]. Pollini et al. [30] conducted research on minimum-cost optimization of viscous dampers and their supporting members for seismic retrofitting. The optimal design of frame structures [31], steel frame structures [32], and seismic retrofit structures [33] with added viscous dampers has also been studied. Seismic response is essentially a random vibration, which needs to be understood from a probabilistic perspective. The researchers have also studied the random vibration response of structures with additional viscous dampers under seismic excitation [34], as well as the research on their optimized design [35,36], and conducted probabilistic seismic response assessment studies [37,38]. However, they did not consider the influence of different damper parameters on the optimal seismic response control and optimal damper placement.

In summary, there is still limited research detailing the influence of viscous dampers' parameters on the seismic performance of damping structures, especially in the application research of frame shear wall structures. Therefore, this paper aims to investigate the influence of damper parameters in detail on the seismic efficiency of frame shear wall structures with attached viscous dampers through various damping indicators. The study proposes to use a response surface-based method, which has been applied in bridge structures [39,40]. In previous conventional structural design optimization methods, optimization was often performed using a large number of finite element analyses, which is computationally expensive for multi-degree-of-freedom systems such as those used in practical structural engineering. By introducing the response surface method and conducting sensitivity analysis on a limited number of damper parameters that affect the damping efficiency, an explicit objective function is established through response surface fitting based on the obtained results. This enables the prediction of various damping indicators of structural response under other parameter combinations. The optimization problem of damper parameters is then solved by optimizing the objective function. This approach avoids frequent repetitive finite element calculations, reduces computational time and resource consumption, and improves the efficiency of optimizing damper parameters for damping structures. In order to provide a clearer design approach for the damping design of frame shear walls with additional viscous dampers, a simplified and easy-to-operate "damping design flowchart considering the influence of damper parameters on damping efficiency" is proposed, aiming to provide important guidance and reference for designers in the damping design of frame shear wall structures in the future.

2. Methodology

2.1. Optimization Method for Damper Parameters Based on Response Surface Method

2.1.1. Establishment of Response Surface Model

First, we establish a finite element analysis model to analyze the key indicators of structural response under common damper parameters and their neighboring parameters. Then, fitting is performed using the response surface method, which approximates a specific polynomial explicit function suitable for solving problems related to nonlinear data processing. The basic expression of the response surface model is as follows.

$$y = \beta_0 + \sum_{i=1}^{n} \beta_i x_i + \sum_{i=1}^{n} \beta_{n+i} x_i^2 + \sum_{i=1}^{n} \beta_{2n+i} x_i^3 + \ldots + \sum_{i \neq j}^{n} \beta_{ij} x_i x_j \quad (1)$$

In the equation, where x_i and y are the parameter variables and calculated results of the response surface model, respectively; β_0, β_i, β_{n+i}, β_{2n+i}, …, β_{ij} are unknown coefficients, and the coefficient vector formed by them is β.

$$\beta = (X^T X)^{-1} X^T Y \quad (2)$$

In the equation, vector X is composed of the x values at the sample points of finite element calculations, $X = (x_1, x_2, \ldots, x_n)^T$; vector Y is composed of the y values at the sample points of finite element calculations, $Y = (y_1, y_2, \ldots, y_n)^T$.

After determining the appropriate order of the function and obtaining the unknown coefficients, the response surface model is obtained. To verify the accuracy of the model, the coefficient of determination R^2 and the F-test are used to perform accuracy testing and significance analysis on the response surface model. The calculation expressions are as follows.

$$R^2 = 1 - SSE/SST \quad (3)$$

$$F = \frac{SSR/p}{SSE/(n-p-1)} \quad (4)$$

$$SSE = \sum_{i=1}^{n} (y_i - \hat{y}_i)^2, SST = \sum_{i=1}^{n} (y_i - \overline{y}_i)^2,$$
$$SSR = \sum_{i=1}^{n} (\hat{y}_i - \overline{y})^2 \quad (5)$$

In the equation, SSE, SST, and SSR represent the sum of squares of errors, total sum of squares, and sum of squares of regression, respectively, for function y. p and $n-p-1$ represent the degrees of freedom for regression and error, respectively, for SSR and SSE. y_{i} denotes the finite element value at sample i, $\hat{y}_i$ represents the response surface model value at sample i, and $\overline{y}$ denotes the average value of the finite element.

2.1.2. Solution of Extremum for Multivariate Function

The response surface model established in the previous section is a multivariate function. The optimization of the damper parameters mainly involves two parameters: the damping coefficient C and the damping exponent α. Therefore, the response surface model is a bi-variate function. The extremum of the bi-variate function can be obtained by applying the method of finding extremum for bi-variate functions. The specific steps are as follows:

1. Assume the bi-variate function is $z = f(x,y)$. Calculate the first-order partial derivatives and set them equal to zero to find the stationary point p_0, as shown in Equation (6).
2. Calculate the second-order partial derivatives of p_0 and obtain the coefficients A, B, and C, as shown in Equation (7).
3. Determine the extremum situation obtained, as shown in Equation (8).

$$f_{\mathrm{x}}(p_0) = 0,\ f_{\mathrm{y}}(p_0) = 0 \quad (6)$$

$$f_{\mathrm{xx}}(p_0) = A,\ f_{\mathrm{xy}}(p_0) = B,\ f_{\mathrm{yy}}(p_0) = C \quad (7)$$

$$\begin{array}{ll} AC - B^2 > 0 & \begin{cases} A > 0, \text{ minimum at point } p_0 \\ A < 0, \text{ maximum at point } p_0 \end{cases} \\ AC - B^2 < 0 & \text{No extremum} \\ AC - B^2 = 0 & \text{Uncertain value} \end{array} \quad (8)$$

2.2. Engineering Case on the Optimization of Damper Parameters in Frame Shear wall Structure

Frame shear wall structure is one of the most commonly used structural systems, and attaching viscous dampers at appropriate locations can achieve a good damping effect. When the placement and quantity of dampers are fixed, the influence of common parameters of the viscous damper (damping coefficient C and damping exponent α) on the damping efficiency of the structure deserves further research. In this paper, an actual engineering case study is conducted for analysis and research.

2.2.1. Engineering Overview

Taking a senior high school dormitory building as an example, the structure system adopts a reinforced concrete frame shear wall system, consisting of 6 floors with a total height of 21.7 m and a floor height of 3.7 m for the first floor and 3.6 m for the other floors, and the total building area is 6716 m^2. According to Chinese standards, the seismic design intensity is set at level 8. That is to say, the peak ground acceleration is 0.20 g. The site category is class II, the design seismic grouping is group 3, the site characteristic period is T_{g} = 0.45 s, and the inherent damping ratio value for reinforced concrete frame shear wall structure is 5% according to Chinese standards.

Shear walls are arranged according to the principles of uniformity, dispersion, and periphery, as shown in the shaded part of Figure 1a in the architectural plan. The walls are all made of 200 mm reinforced concrete with an axial compressive strength design value of 14.3 MPa and the Young's modulus is 30,000 MPa; this is referred to as C30 concrete in

Chinese standards. According to the above standards, the reinforcement of the shear walls is as follows: (1) The vertical distribution reinforcement ratio is considered according to the structural requirement of 0.25%. (2) The horizontal distribution reinforcement ratio is configured based on the calculated results according to the magnitude of shear force on the shear walls. The reinforcement ratio on the first floor is mainly between 0.34% and 0.64%, and the reinforcement ratio on the second floor is between 0.28% and 0.32% for some parts. The horizontal distribution reinforcement ratio of the remaining shear walls is also considered to be 0.25%.

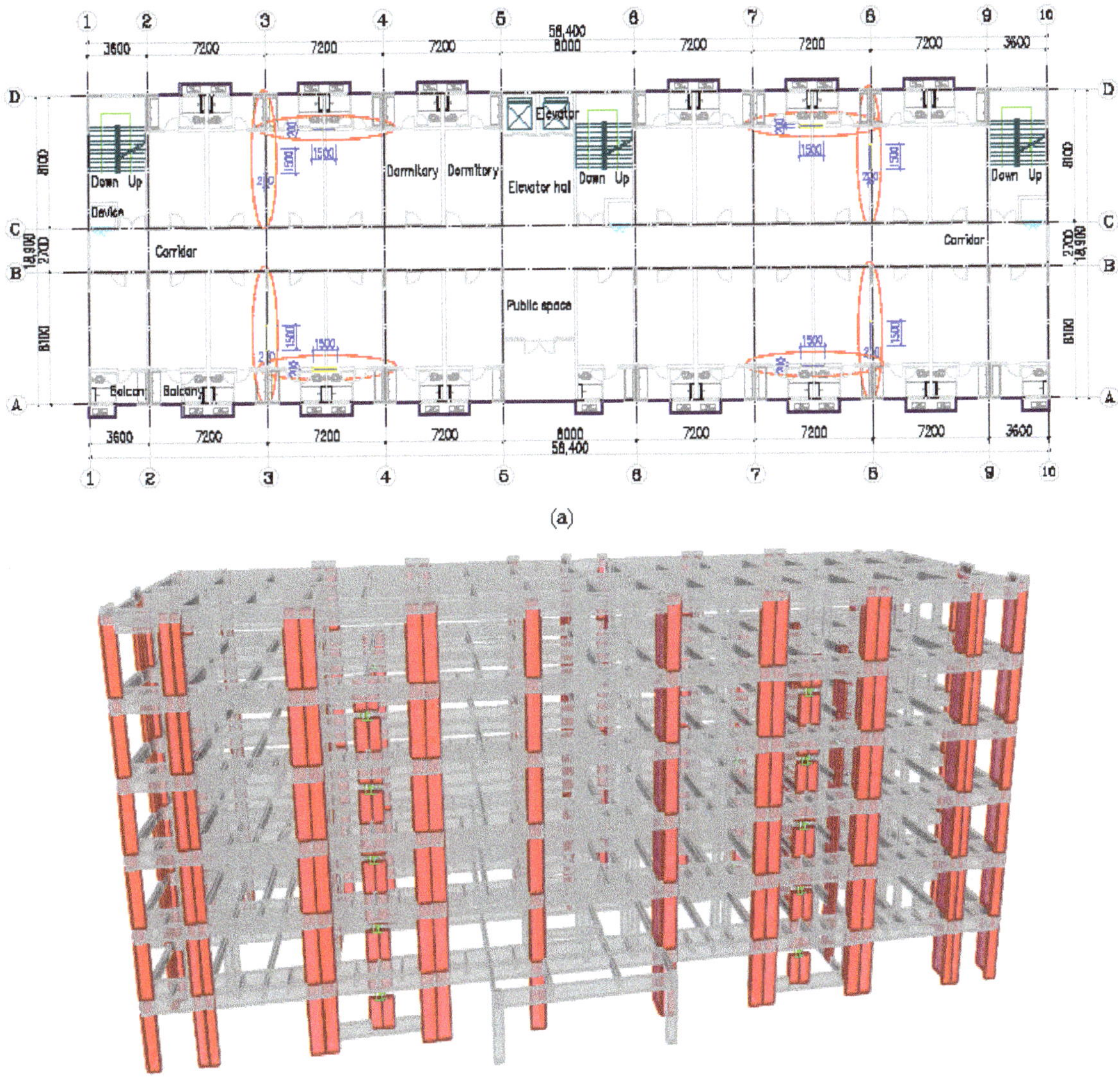

Figure 1. Layout diagram of dampers and calculation model of damping structure: (a) the layout diagram of dampers; (b) the calculation model of damping structure.

The columns are arranged at the intersection of axis A with axis 3, 4, 7, and 8, axis B and D with axes 2 to 9, and axis C with axes 1 to 10, as shown in Figure 1a. The column sections mainly consist of two types: 500×500 mm^2 and 400×400 mm^2. The column sections are the same for floors 1 to 6, while the column section at the intersection of axis D with axis 2, 5, 6 and 9 is 400×400 mm^2, and the remaining column sections are 500×500 mm^2. In addition, there are only two special columns with dimensions of 400×600 mm^2 used to support a canopy of a gate on one floor. The main beam sections have dimensions of 200×750 mm^2 in the X direction and 200×600 mm^2 in the Y direction. The secondary beam sections have dimensions of 200×600 mm^2 for larger spans in the Y direction, and there are two other types of secondary beam sections: 200×500 mm^2 and 200×400 mm^2. The concrete axial compressive strength design value for beams and columns is also 14.3 MPa, with a Young's modulus of 30,000 MPa. The longitudinal and transverse reinforcement for beams and columns needs to be configured according to the calculation results.

Considering that the structure has a total of 6 floors and the top floor has a smaller inter-story drift angle, it is customary to place the dampers in the layers with larger inter-story displacement angles, which are the 1st to 5th floors. Next, the number, placement, and parameters of dampers will be determined primarily according to the following steps [41,42]:

1. Determine the target additional damping ratio added to the structure by the viscous dampers. In this study, a value of 3.0% is considered.
2. The formula for calculating the additional damping ratio through the energy method is used to deduce the total damping coefficient and the selected damping exponent required for the structure under the target additional damping ratio. The detailed procedure of this derivation can be referred to in reference [41].
3. Divide the total damping coefficient by the damping coefficient of a single damper to obtain the number of dampers. The damping coefficient and exponent of a single damper are generally selected based on engineering experience. In this study, the damping coefficient is taken as C = 60 kN/(mm/s)$^{0.25}$, and the damping exponent α is 0.25.
4. Determine the number and placement of dampers in each floor. The dampers are typically arranged in a simplified manner with uniform distribution in height, and the placement is the same so as not to affect the functionality of the building. Therefore, in this study, the number and placement of viscous dampers are the same for the 1st to 5th floors. Four sets of dampers are arranged in the X and Y directions on each floor, resulting in a total of 40 sets of dampers. The arrangement diagram of dampers is shown in Figure 1a.

2.2.2. Finite Element Model Establishment and Earthquake Waves Selection

A three-dimensional finite element damping analysis model shown in Figure 1b is established using SAP2000 (version 15.2.1). The beam-column elements are simulated using rod elements, while the shear walls are simulated using shell elements. All three types of elements are considered linearly in this study. If readers need to consider the nonlinear behavior of the structure, it is recommended to simulate the beam-column members using plastic hinges or fiber hinges and simulate the mechanical behavior of concrete and reinforcement separately in the shear walls using layered shells. During the modeling process, rigid connections are adopted between beams and columns to ensure the consistency of linear and angular displacements of beams and columns at the nodes.

The viscous damper utilizes a nonlinear model that is based on the damper unit to achieve realistic simulation, and the viscous dampers are positioned as indicated by the blue rectangle within the ellipse in Figure 1a. The Maxwell constitutive model is employed in the viscous damper, and its restoring force model is depicted in Figure 2, as expressed by Equation (9).

$$F = C\,v^{\alpha} \tag{9}$$

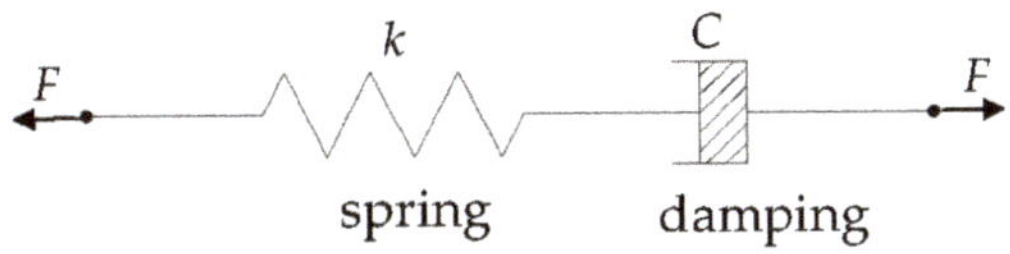

Figure 2. Maxwell mechanical model.

In the equation, F is the damping force, C is the damping coefficient, v is the relative velocity of the damper element, and α is the damping exponent. $\alpha = 1.0$ is a linear damper and $\alpha \neq 1.0$ is a nonlinear damper. The commonly used values in structural engineering are between 0.15 and 1.0.

The upper and lower braces connected to the damper are designed according to the specifications [2], which need to meet the stiffness requirements. The reinforcement is generally designed according to the strength requirements for components that remain elastic under rare earthquakes. The upper and lower braces connected in this study adopt cantilever walls with a thickness of 200 mm and a width of 1500 mm. The height of the upper and lower supports is half of the floor height minus the beam height and the damper height. The design value of the concrete axial compressive strength is 14.3 MPa and the Young's modulus is 30,000 MPa.

Elastic time–history analysis based on Ritz vectors (fast nonlinear analysis, i.e., FNA) is used, with 144 vibration modes selected. According to the requirements of the "Code for Seismic Design of Buildings" and the "Technical Specification for Seismic Energy Dissipation of Buildings" [1,2], 5 natural waves and 2 artificial waves are selected for time–history analysis, and the time–history curve, response spectrum curve, and comparison with the normative response spectrum are shown in Figure 3. Seismic wave records are normalized to peak amplitude, and when used, they are amplified to the corresponding peak amplitude according to different seismic levels using a scaling factor. However, the time axis of seismic wave records is not adjusted for scaling. In the numerical simulation analysis, the additional damping ratio of the structure, the force and displacement of the damper, the shear force and displacement of the floor, and other relevant indicators are selected based on the average value of the seven seismic waves.

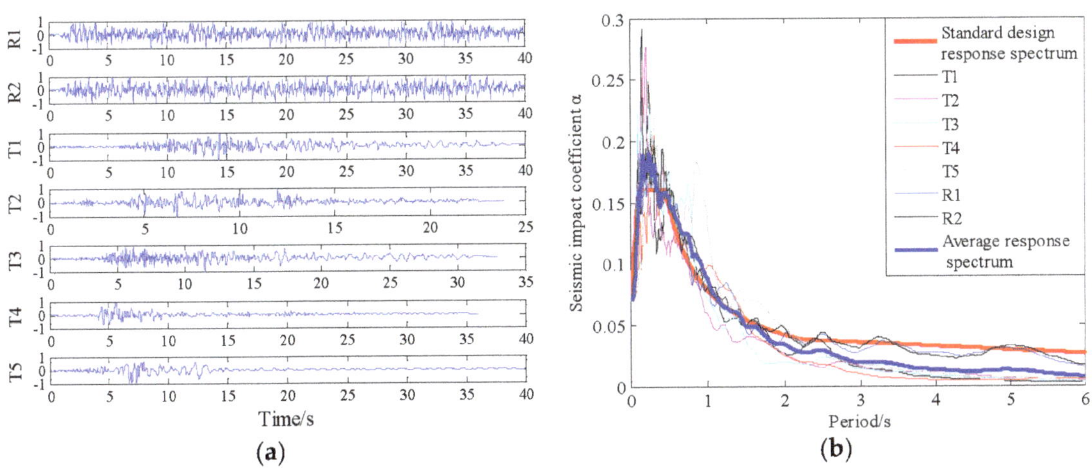

Figure 3. Seismic wave time–history curves and response spectrum curves: (**a**) the time–history curves; (**b**) comparison of response spectrum curves.

3. Results and Discussion

3.1. Sensitivity Analysis of the Impact of Damping Parameters

In order to investigate the influence of parameters of the viscous damper (mainly damping coefficient C and damping exponent α) on the damping efficiency in frame shear wall structure, a comparative finite element analysis of two sets of parameters is conducted based on the information in Section 2.2.1 regarding damper parameters. The first set consists of damping coefficients of 60 $kN/(mm/s)^{\alpha}$ with varying damping exponents. The second set consists of a damping exponent of 0.25 with varying damping coefficients, as detailed in Table 1.

Table 1. The damper parameters.

The First Set of Parameters			The Second Set of Parameters		
Serial Number	Damping Coefficient C	Damping Exponent α	Serial Number	Damping Coefficient C	Damping Exponent α
1	60	0.15	1	20	0.25
2	60	0.25	2	40	0.25
3	60	0.35	3	60	0.25
4	60	0.45	4	80	0.25
5	60	0.55	5	100	0.25
6	60	0.65	6	120	0.25

After attaching viscous dampers, the damping efficiency of damper is usually measured by the magnitude of the additional damping ratio. The reduction ratio of the vertex displacement and base shear of the damping structure relative to the non-damped structure (the structure without viscous dampers) are studied, and the force and displacement of the damper themselves are also considered. The following analysis will be conducted in detail based on these aspects.

3.1.1. The Additional Damping Ratio

The additional damping ratio is one of the important indicators for measuring the damping efficiency in a structural system. Viscous dampers belong to the typical velocity-dependent damper. According to the current standards in China [1,2], the effective additional damping ratio provided by dampers to the structural system can be calculated using the following equation.

$$\xi_d = \frac{\sum_{j=1}^{n} W_{cj}}{4\pi W_s} \tag{10}$$

where ξ_d represents the additional effective damping ratio dampers provide to the structure; W_{cj} is the energy dissipated by the jth dissipative component in the structure during one cycle of reciprocating motion with an inter-story displacement Δu_j; n is the number of dampers; W_s is the total strain energy of the energy dissipation and damping structure under the expected displacement.

The additional damping ratios in the X and Y directions under two sets of parameters are calculated based on Equation (10), and the results are shown in Figure 4.

From the figure, it can be observed that the damping coefficient C remains constant at 60 $kN/(mm/s)^{\alpha}$ and the damping exponent α varies between 0.15 and 0.65. The additional damping ratio of the structure increases initially and then decreases, ranging from 1.75% to 4.05%. The maximum additional damping ratio is achieved when α is between 0.25 and 0.35. When the damping exponent α remains constant at 0.25 and the damping coefficient C varies between 20 and 120 $kN/(mm/s)^{0.25}$, the additional damping ratio of the structure also initially increases and then decreases, ranging from 1.63% to 4.05%. The maximum additional damping ratio is obtained when C is 60 $kN/(mm/s)^{0.25}$.

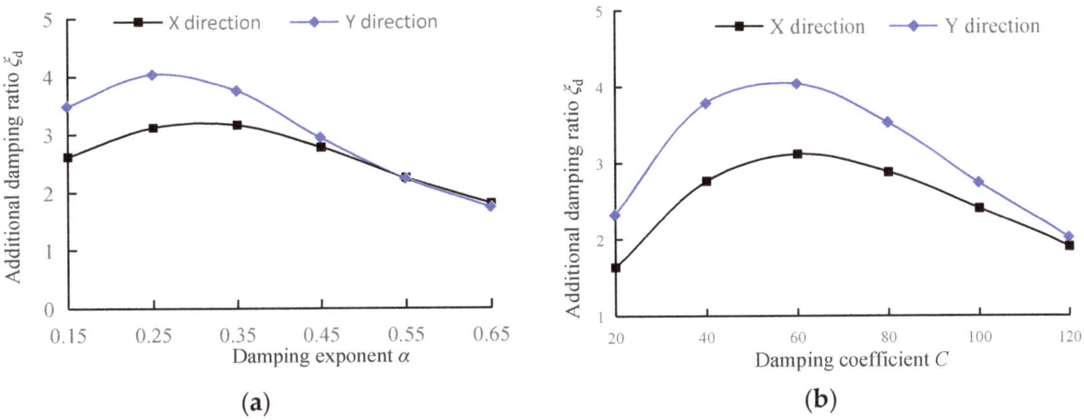

Figure 4. The curves of additional damping ratio: (**a**) the first set of parameters; (**b**) the second set of parameters.

3.1.2. The Reduction Rate of Vertex Displacement

The reduction rate of vertex displacement is also one of the important indicators for measuring the damping efficiency in a structural system. It is represented by the reduction rate of vertex displacement in the damping structure relative to that in the non-damped structure. The calculation expression is as follows.

$$\mu_u = \frac{u_0 - u}{u_0} \times 100\% \tag{11}$$

where μ_u represents the reduction rate of vertex displacement, u_0 represents the vertex displacement in the non-damped structure, and u represents the vertex displacement in the damping structure.

The reduction rates of vertex displacement in the X and Y directions of the structure are calculated for two sets of parameters according to Equation (11), as shown in Figure 5.

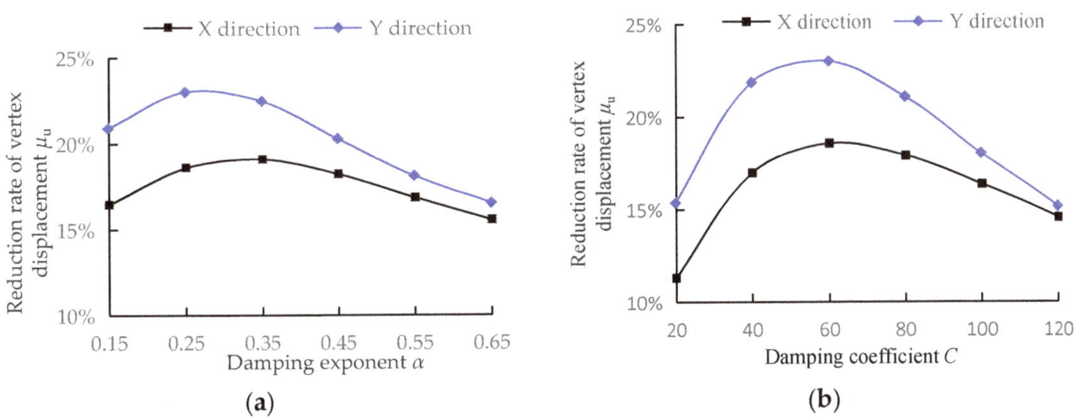

Figure 5. The curves of vertex displacement reduction rate: (**a**) the first set of parameters; (**b**) the second set of parameters.

From the figure, it can be observed that when the damping coefficient C remains constant and the damping exponent α varies between 0.15 and 0.65, the reduction rate of vertex displacement increases initially and then decreases. Overall, there is not a significant difference in the reduction rate of vertex displacement, which ranges from 16% to 23%. The maximum reduction rate of vertex displacement is achieved when α is between 0.25 and 0.35. When the damping exponent α is fixed at 0.25 and the damping coefficient C varies between 20 and 120 kN/(mm/s)$^{0.25}$, the reduction rate of vertex displacement first increases rapidly and then decreases slowly, ranging from 11% to 23%. The maximum reduction rate of vertex displacement is achieved when C is 60 kN/(mm/s)$^{0.25}$.

3.1.3. The Reduction Rate of Base Shear

The base shear reduction rate is another important indicator for measuring the damping efficiency in a structural system. It is expressed by the ratio of the reduction in base shear for a structure with damping measure to the base shear without such measure. The calculation expression is as follows.

$$\mu_V = \frac{V_0 - V}{V_0} \times 100\% \quad (12)$$

In the equation, μ_V represents the base shear reduction rate, V_0 represents the base shear of the non-damped structure, and V represents the base shear of the damping structure.

The base shear reduction rates in the X and Y directions are calculated according to Equation (12) for two sets of parameters, as shown in Figure 6.

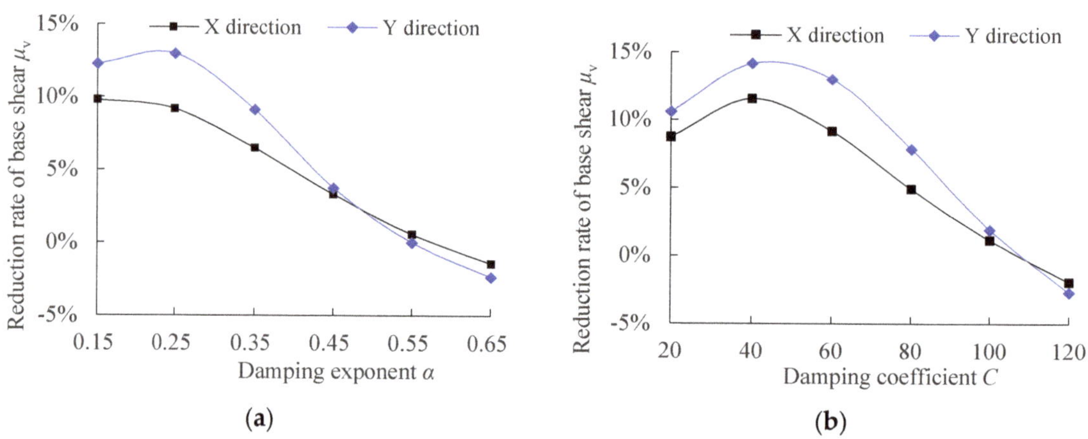

Figure 6. The curves of base shear reduction rate: (**a**) the first set of parameters; (**b**) the second set of parameters.

From the graph, it can be observed that when the damping coefficient C remains constant at 60 kN/(mm/s)$^{\alpha}$, and the damping exponent α is adjusted between 0.15 and 0.65, the base shear reduction rate initially increases and then decreases, varying between −2% and 13%. The maximum base shear reduction rate is achieved when α = 0.25. When the damping exponent α is fixed at 0.25, and the damping coefficient C is adjusted between 20 and 120 kN/(mm/s)$^{0.25}$, the base shear reduction rate also initially increases and then decreases, varying between −3% and 14%. The maximum base shear reduction rate is achieved when C = 40 kN/(mm/s)$^{0.25}$.

The aforementioned analysis of base shear includes the force of the bottom dampers. Therefore, negative values appear when calculating the base shear reduction rate, indicating an increase in base shear after the installation of additional dampers. However, the base shear of the structure itself does not change in this way. By deducting the force of the bottom dampers from the original base shear, a new base shear is obtained. The base shear reduction rates in the *X* and *Y* directions for two sets of parameters are calculated according to Equation (12), as shown in Figure 7.

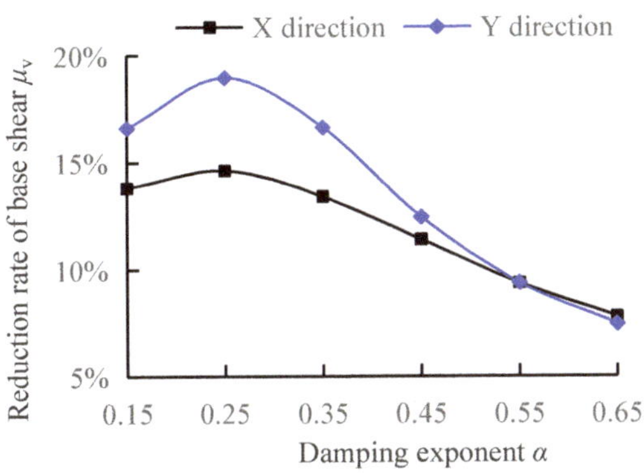

(a)

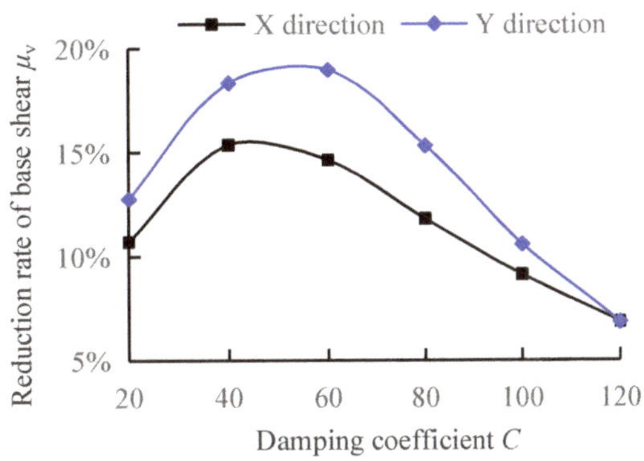

(b)

Figure 7. The curves of base shear reduction rate (excluding damper output force): (**a**) the first set of parameters; (**b**) the second set of parameters.

From the graph, it can be observed that when the damping coefficient *C* remains constant at 60 kN/(mm/s)$^{\alpha}$, and the damping exponent α is adjusted between 0.15 and 0.65, the base shear reduction rate initially increases and then decreases, ranging from 7% to 19%. The maximum base shear reduction rate is achieved when α = 0.25. When the damping exponent α is fixed at 0.25, and the damping coefficient *C* is adjusted between 20 and 120 kN/(mm/s)$^{0.25}$, the base shear reduction rate initially increases rapidly and then decreases slowly, ranging from 7% to 19%. The maximum base shear reduction rate is achieved when *C* = 40~60 kN/(mm/s)$^{0.25}$.

Comparing Figures 6 and 7, it can be seen that the base shear reduction rate significantly increases after deducting the force of the bottom dampers from the original base shear, and the pattern becomes more reasonable. Therefore, this method should be adopted to calculate the base shear reduction rate. Comparing Figures 5 and 7, it can be observed that the base shear reduction rate is slightly smaller than the vertex displacement reduction rate.

3.1.4. The Inter-Story Displacement Utilization Rate

The inter-story displacement utilization ratio is commonly used to measure the damping efficiency in a structural system. It is expressed as the ratio between the displacement of the damping device in the damping structure and the displacement of the corresponding floor. The calculation expression is as follows.

$$\eta = \frac{u_{\mathrm{d}}}{u} \times 100\% \tag{13}$$

where η represents the inter-story displacement utilization rate; u represents the displacement of the floor where the damping device is located; u_{d} represents the displacement of the damping device.

The inter-story displacement utilization rates in the *X* and *Y* directions of the structure are calculated based on Equation (13), as shown in Figure 8.

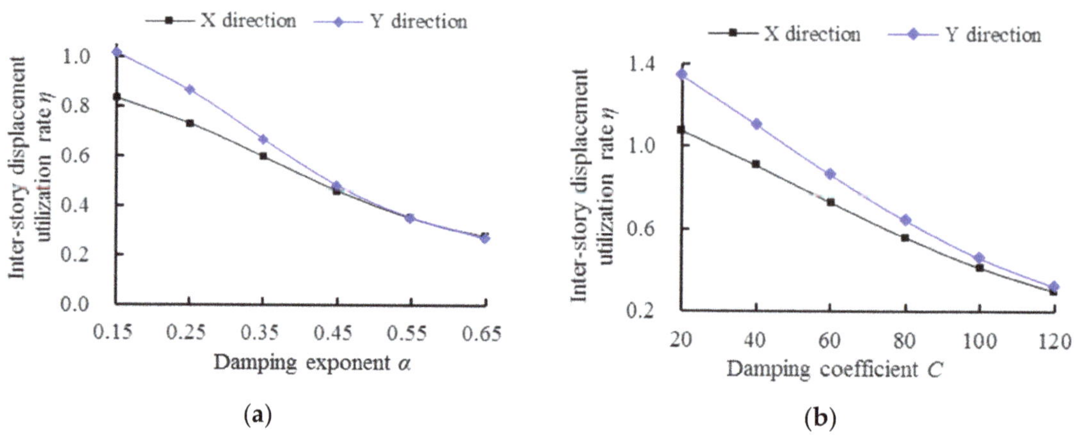

Figure 8. The curves of inter-story displacement utilization rate: (**a**) the first set of parameters; (**b**) the second set of parameters.

From the figure, it can be observed that when the damping coefficient C remains constant at 60 kN/(mm/s)$^{\alpha}$, the inter-story displacement utilization rate rapidly decreases from 1.02 to 0.27 as the damping exponent α is adjusted between 0.15 and 0.65. Similarly, when the damping exponent α remains constant at 0.25, the inter-story displacement utilization rate decreases from 1.35 to 0.30 as the damping coefficient C is adjusted between 20 and 120 kN/(mm/s)$^{0.25}$. Both the damping coefficient and damping exponent have significant effects on the inter-story displacement utilization rate, with the damping coefficient having a more pronounced effect. The smaller inter-story displacement utilization rate indicates that the displacement of the damping device can make less use of the floor displacement, resulting in a poorer energy dissipation performance.

3.1.5. Damping Force and Displacement

The force and displacement response of dampers allow it to dissipate energy, which is the reason for its ability to achieve a damping effect. The output force and displacement response of dampers vary greatly under different damper parameters. A higher output force and displacement result in more energy dissipation. However, the magnitude of the damper force has a significant impact on the damping structure, especially on the components connected to the damper. Therefore, a reasonable balance between the damper output force and displacement is required during the design of the damping structure. The damper output force and displacement for the X and Y directions of the structure under two sets of parameters are shown in Figures 9 and 10, respectively.

From Figures 9 and 10, it can be seen that the output force of dampers increases with the increase in the damping exponent and damping coefficient. The displacement of dampers decreases with the increase in the damping exponent and the damping coefficient. When the dynamic characteristics of structure X and Y directions are similar, the output force and displacement responses of the dampers are also similar. Therefore, in practical engineering applications, it is particularly important to balance the output of the damper. Excessive output of the damper will increase the burden on the supporting components of the damper, making the design of the supporting components difficult. On the other hand, insufficient output of the damper will result in lower energy dissipation, seriously affecting the damping effect of the structure.

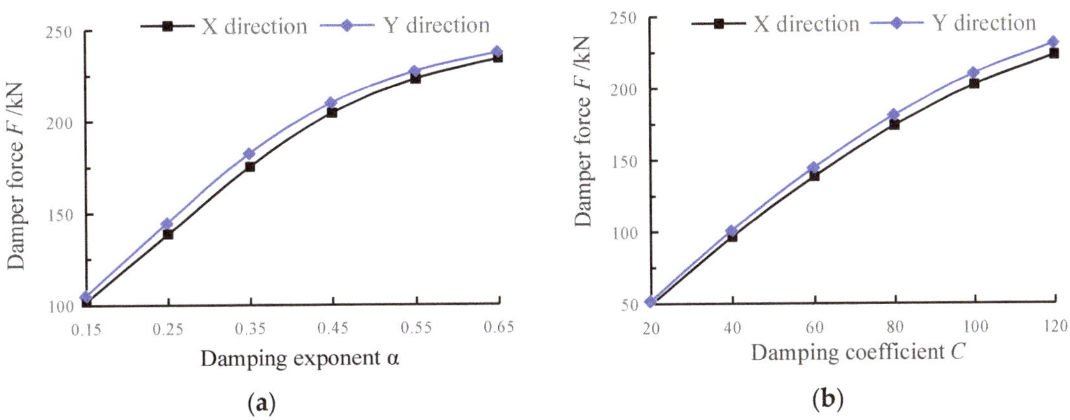

Figure 9. The curves of damper output force: (**a**) the first set of parameters; (**b**) the second set of parameters.

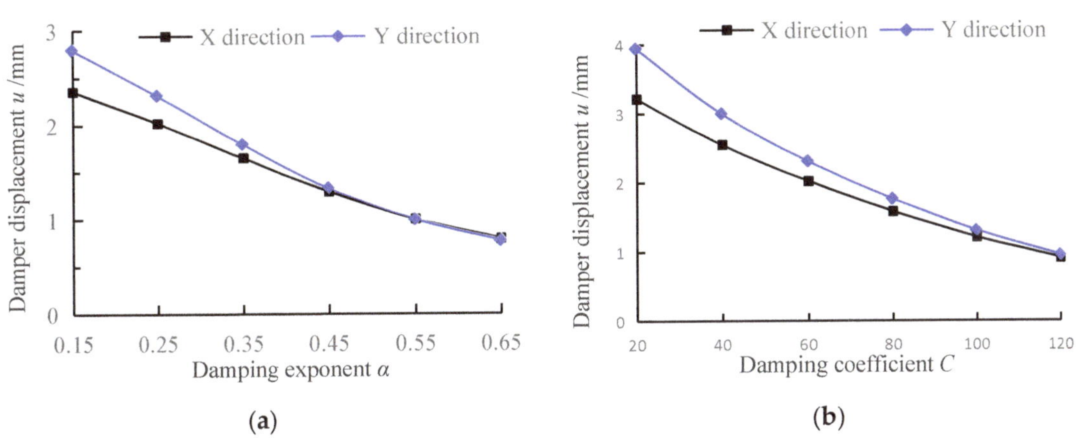

Figure 10. The curves of damper displacement: (**a**) the first set of parameters; (**b**) the second set of parameters.

3.2. *Optimization of Parameters for Viscous Dampers and Flowchart for Damping Design*

Optimization design of viscous dampers' parameters for damping of frame shear wall structure based on response surface methodology and multivariate function extremum.

3.2.1. Response Surface Model

In the response surface model, the polynomial order directly affects the fitting accuracy of the model. When the order is too high, it is prone to "Runge's phenomenon", which reduces the accuracy of the response surface. On the other hand, a too low order cannot correctly reflect the nonlinear characteristics with implicit functional functions. Therefore, different basic expressions of response surface models are selected based on the curve laws of various indicators in the previous section.

In the precision test of the response surface model, the determination coefficient R^2 exhibits values between 0 and 1, and the closer it is to 1, the higher the fitting accuracy. The significance level p value is calculated using the F-test method, and if $p < 0.01$, it indicates that the parameters of the response surface model are significant.

For convenience of simulation and expression, the average values of the corresponding indicators X-axis and Y-axis responses in the previous section are taken, and then the relationship between the damping coefficient and the damping exponent is examined. The surface fitting results of the seismic response for various indicators are shown in Table 2.

Table 2. Seismic response surface fitting results.

Indicators	Response Surface Types	p	R^2
Additional damping ratio ξ_d	Poly2D	<0.0001	0.917
Reduction rate of vertex displacement μ_u	Poly2D	<0.0001	0.900
Reduction rate of base shear μ_V	Poly2D	<0.0001	0.901
Inter-story displacement utilization rate η	Plane	<0.0001	0.988
Damper force F/kN	Plane	<0.0001	0.970
Damper displacement u/mm	Plane	<0.0001	0.973

Note: The response surface types are the types of nonlinear surface fitting in Origin2017, where Poly2D represents a second-order polynomial for two variables, and Plane represents a first-order polynomial for two variables.

Referring to the fitting types of response surfaces corresponding to various damping indicators in Table 2, the fitting curves for the six indicators shown in Figure 11 are obtained. The red dots in the figure represent the target working condition sample points in Section 3.1. It can be seen from Table 2 that the response surface fitting surfaces of various indicators are in excellent agreement with the calculated results at the sample points, indicating excellent fitting effect.

From Figures 4 and 11, it can be observed that the additional damping ratio, reduction rate of vertex displacement, and reduction rate of the base shear of the structure first increase and then decrease with the increase in damping coefficient C, and also first increase and then decrease with the increase in damping exponent α, but the rate of change is not consistent. The inter-story displacement utilization ratio and damper displacement decrease with the increase in both the damping coefficient and the damping exponent. The damper force increases with the increase in both the damping coefficient and the damping exponent. Therefore, it can be seen that there is no consistent or monotonic variation pattern in the response of different damping efficiency indicators of the structure, which is related to the values of the damper parameters C and α, showing a nonlinear surface variation pattern: response surface of additional damping ratio, response surface of vertex displacement reduction rate, and response surface of base shear reduction rate. As for the inter-story displacement utilization ratio, damper displacement, and damper output force, they show relatively simple linear surface variation patterns. Therefore, the influence pattern of different combinations of viscous damper parameters on the response of damping efficiency indicators is complex, and traditional sensitivity analysis methods struggle to comprehensively consider the combined effects of different damping efficiency indicators and obtain the optimal parameters.

3.2.2. Objective Function

In the above indicators of the damping efficiency, the focus is on analyzing the indicators that first increase and then decrease with the variation in the damper parameters (damping coefficient and damping exponent), i.e., the indicators that have extremum values, mainly including the additional damping ratio, the reduction rate of vertex displacement, and the reduction rate of base shear. Regarding the utilization ratio of inter-story displacement, it can be seen from the curve in Figure 8 that the smaller the damping coefficient and damping exponent, the greater the utilization ratio of inter-story displacement. The output force and displacement curves of the damper in Figure 9 to Figure 10 also approximately monotonically increase or decrease with the relationship with the damper parameters.

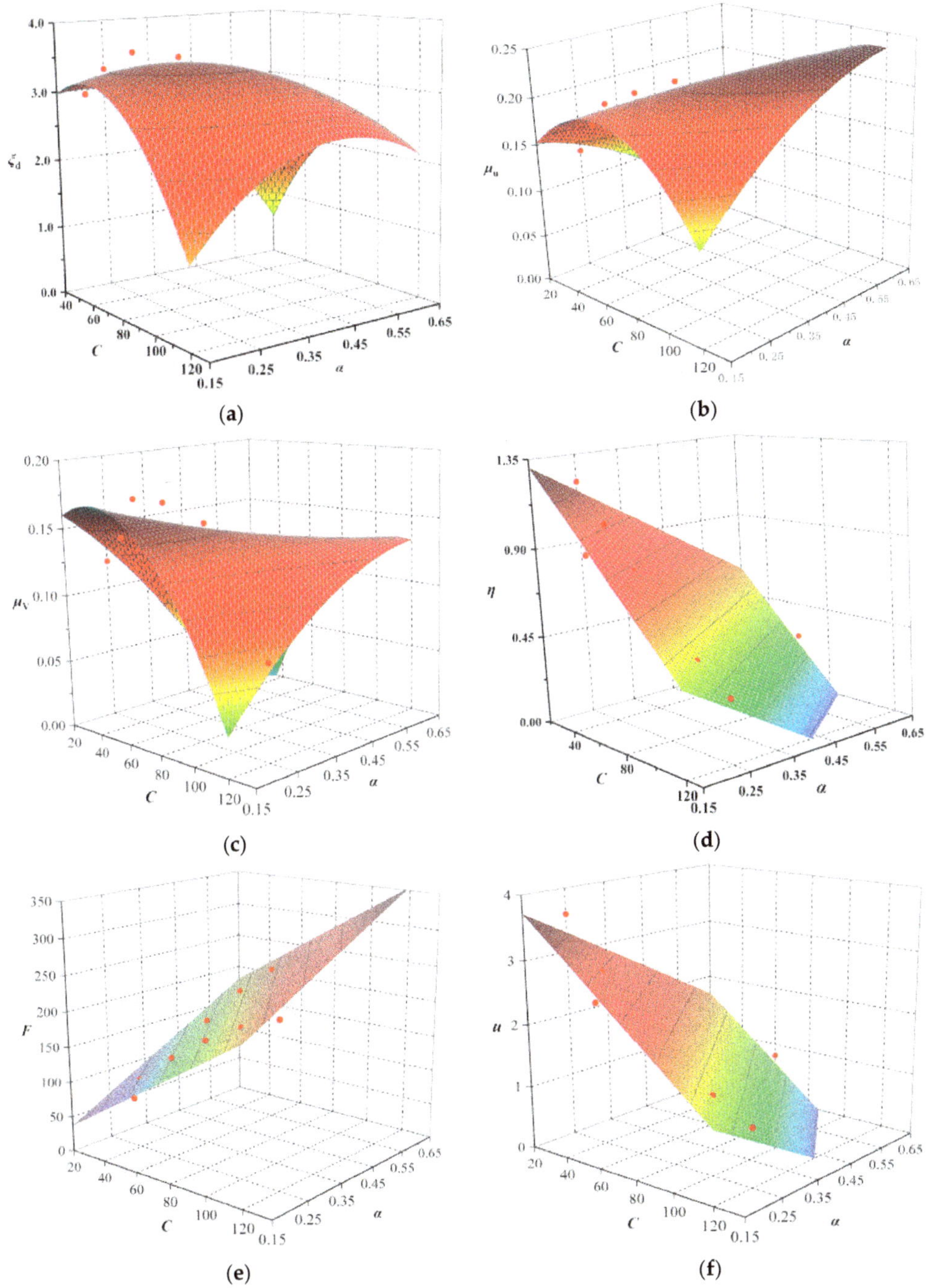

Figure 11. The response surface fitting of each indicator: (**a**) additional damping ratio; (**b**) reduction rate of vertex displacement; (**c**) reduction rate of base shear; (**d**) inter-story displacement utilization rate; (**e**) damper force; (**f**) damper displacement.

By using the linear weighting method, the three indicators with extremum values are linearly superimposed to obtain a response evaluation function that considers the combined influence of different damping efficiency indicators, i.e., the objective function. The weight coefficients of each indicator are determined based on the degree of influence of the damping efficiency indicator on the structural response. In this study, it is considered that the influence evaluations of the additional damping ratio, the reduction rate of vertex displacement, and the reduction rate of base shear on the structural response are equivalent, so they are assigned the same weight coefficients a, b, c, respectively. Therefore, the response evaluation function is expressed as shown in the equation.

$$S = a\xi_{\mathrm{d}} + b\mu_{\mathrm{u}} + c\mu_{\mathrm{V}} \tag{14}$$

where S is the response evaluation function (objective function) of the structure under the combined influence of different damping efficiency indicators. The weight coefficients, a, b, and c, are all set to 1.0. ξ_{d} is the additional effective damping ratio provided by the damper; μ_{u} is the reduction rate of vertex displacement of the damping structure; μ_{V} is the reduction rate of the base shear of the damping structure. By optimizing Equation (14), the maximum value of the response S is obtained, which corresponds to the parameters of the viscous damper after the optimization design of the damping structure.

3.2.3. Analysis of Optimized Results

The objective function of Equation (14) is optimized with respect to the damping coefficient and the damping exponent as a bi-variate function. The specific steps follow those described in Equations (6)–(8) in Section 2.1.2. It is found that $A < 0$ and $AC-B^2 > 0$, indicating the existence of a maximum value. Based on the previous analysis, this maximum value is the maximum within the given range. The optimized values for the damping parameters are found to be $(C, \alpha) = (72, 0.32)$.

Substituting these optimal parameters into the actual engineering case, the additional damping ratio, vertex displacement reduction ratio, and base shear reduction ratio of the structure are obtained as shown in Table 3.

Table 3. Summary of damping indicators under optimal parameters.

Indicators	X Direction	Y Direction
Additional damping ratio ξ_d	2.91%	3.38%
Reduction rate of vertex displacement μ_u	18%	21%
Reduction rate of base shear μ_V	12%	14%
Inter-story displacement utilization rate η	0.53	0.59
Damper force F/kN	186	194
Damper displacement u/mm	1.48	1.61

From Table 3, it can be seen that the optimal parameters $(C, \alpha) = (72, 0.32)$ are used for calculations in the actual engineering project in Section 2.2. The obtained damping indicator differs from the results calculated in Section 3.1 with damping parameters (60, 0.25), for example, the additional damping ratio index: $\xi_{\mathrm{d}}(60, 0.25) = 3.58\%$, $\xi_{\mathrm{d}}(72, 0.32) = 3.14\%$ (taking the average value in the X and Y directions). It seems that the result of the optimal damping parameters (72, 0.32) is not as expected. The main reason for this is that the stiffness of the supporting damper components has a certain influence on the effectiveness of the damper parameters, which will be discussed in detail in the following content. In addition, the utilization rate of inter-story displacement is relatively low, indicating that the displacement of the damper is not fully utilized by the floor displacement, which is also related to the stiffness of the supporting damper components.

In fact, the effectiveness of the damper is to some extent influenced by the stiffness of the supporting damper components. Specifications [1,2] specifically stipulate that the stiffness of the supporting components of the linear viscous damper in the direction of energy dissipation should comply with Equation (15). Subsequently, using Equation (15) as a criterion, the support stiffness level is matched with the corresponding optimal damping parameters (72, 0.32) with the support stiffness level of the given damping parameters (60, 0.25).

$$K_b \geq 6\pi C_D/T_1 \tag{15}$$

where K_b is the stiffness of the supporting components in the direction of energy dissipation (kN/m); T_1 is the fundamental natural period of the damping structure (s); C_D is the linear damping coefficient of the damper ($kN \cdot m^{-1} \cdot s^{-1}$), and for the nonlinear viscous damping coefficient, it needs to be converted into the equivalent linear viscous damping coefficient according to Equation (16) and satisfy the requirement of Equation (15).

$$C_D = C_{eq} = \frac{\lambda}{\pi(\omega A)^{1-\alpha}} C_\alpha \tag{16}$$

In the equation, C_{eq} represents the equivalent linear viscous damping coefficient, which replaces C_D in Equation (15); C_α represents the nonlinear viscous damping coefficient; ω is the circular frequency of the damping structure system; A is the displacement amplitude of the damper in the damping structure; α is the damping exponent; λ is the gamma function related to the damping exponent α, given by $2^{2+\alpha}\Gamma^2(1 + \alpha/2)/\Gamma(2 + \alpha)$.

According to Equations (15) and (16), the support component stiffness of the optimal damper parameters (72, 0.32) should be adjusted to a level equivalent to the parameters (60, 0.25). The support component stiffness needs to be amplified by a factor of 1.80, and its width should also be increased from 1500 mm to 2090 mm, while the thickness remains unchanged at 200 mm. Then, further analysis will be conducted. The results are presented in Table 4.

Table 4. Summary of optimal parameters damping indicators after stiffness matching.

Indicators	X Direction	Y Direction
Additional damping ratio ξ_d	4.50%	5.70%
Reduction rate of vertex displacement μ_u	25%	30%
Reduction rate of base shear μ_V	19%	24%
Inter-story displacement utilization rate η	0.70	0.81
Damper force F/kN	199	209
Damper displacement u/mm	1.79	1.97

From Table 4, it can be seen that after considering the matching problem of the supported component stiffness under the optimal damper parameters, various damping performance indicators, especially the additional damping ratio, vertex displacement reduction rate, and base shear reduction rate, all significantly increase. The output force and displacement of the damper also increase significantly, leading to a significant increase in energy dissipation and a significant improvement in the damping efficiency of the structure. The establishment of the response surface model and the optimization of the objective function based on the above theoretically prove the existence of the optimal damping parameters. The calculation results in this table, based on the practical engineering case, demonstrate the feasibility of the optimal damping parameters from a practical perspective.

3.2.4. The Damping Design Flowchart

The application of viscous dampers in frame shear wall structures is investigated through traditional sensitivity analysis and response surface-based methods. The influence of damper coefficients and damper exponents on the structural seismic performance is studied in detail. A simple and easy-to-operate damping design flowchart is proposed, which can be used to complete the damping design of other additional viscous dampers in frame shear wall structures according to the idea and method presented in the flowchart.

Therefore, it can provide important guidance and reference for designers in the future when conducting damping design of frame shear wall structures, as shown in Figure 12.

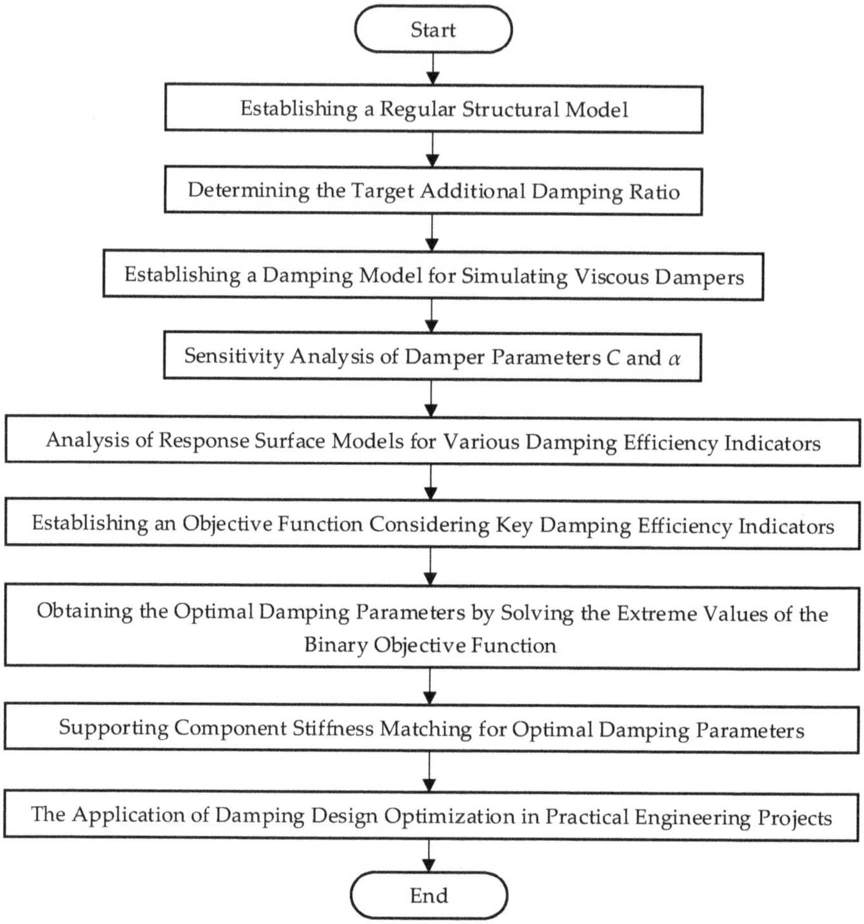

Figure 12. Damping design flowchart considering the impact of damper parameters on damping efficiency.

4. Conclusions

This paper focuses on the application of viscous dampers in frame shear wall structures and extensively investigates the influence of damper parameters on the damping efficiency of frame shear wall structures. It also proposes a damping design flowchart that takes into account the impact of damper parameters on the damping efficiency. The main conclusions are as follows:

(1) The parameters within the range of damping coefficient 20~120 kN/(mm/s)$^{\alpha}$ and damping exponent 0.15~0.65 are combined, and the traditional sensitivity analysis method is used to calculate various damping efficiency indicators. When the damping coefficient and damping exponent are approximately (60, 0.25), the additional damping ratio, reduction rate of vertex displacement, and reduction rate of base shear achieve relatively large values of 4.05%, 23%, and 19%, respectively. As the damping coefficient and damping exponent increase, the inter-story displacement utilization rate decreases from significantly greater than 1.0 to far less than 1.0, and the

displacement of the damper decreases while the damping force increases. The study shows that the influence of different combinations of viscous damper parameters on the response of damping efficiency indicators is complex, and it is difficult for the traditional sensitivity analysis method to comprehensively consider the combined effects of different damping efficiency indicators to obtain the optimal parameters.

(2) By explicitly formulating the relationship between viscous damper parameters and various damping efficiency indicators of the structural system using response surface methodology, and combining the *F*-test and coefficient of determination R^2 to evaluate the fitting effect of the response surface function, high fitting accuracy and good predictability are achieved, making our model suitable as an optimization model.

(3) The influence of support component stiffness on the damping efficiency indicators of the structure is significant. After the variation of damper parameters, it is advisable to match the corresponding support component stiffness according to the specifications in order to obtain the true results of various damping efficiency indicators under optimal parameters. After matching the stiffness of the supporting members, the additional damping ratios in the case increased from 2.91% and 3.38% to 4.50% and 5.70%, respectively, representing an approximately 55% increase. Other damping efficiency indicators also showed significant improvement. This pattern is applicable to other frame shear wall structures with additional viscous dampers.

(4) A simple and easy-to-use "damping design flowchart considering the impact of damper parameters on the damping efficiency" is proposed. By following the idea and method presented in this flowchart, designers can complete the damping design of other frame shear wall structures with additional viscous dampers. This flowchart can provide important guidance and reference for future designers in conducting damping design for frame shear wall structures.

Author Contributions: Conceptualization, X.L. and X.Z.; methodology, X.L.; software, X.L.; validation, X.L., G.W. and X.Z.; formal analysis, X.L.; investigation, G.W.; resources, X.L.; data curation, X.L.; writing—original draft preparation, X.L.; writing—review and editing, X.L.; visualization, X.Z.; supervision, G.W.; project administration, X.L.; funding acquisition, X.L. All authors have read and agreed to the published version of the manuscript.

Funding: This research was funded by the Yunnan Provincial Basic Research Special Project (Grant No. 202101AU070059), the Foundation of Yunnan Provincial Educational Committee (No. 2022J0596) and the Research Project of Introducing Talents in Kunming University (No. XJ20210006).

Data Availability Statement: All the data utilized in the research has been fully presented in the paper. No additional data has been employed.

Conflicts of Interest: The authors declare no conflicts of interest.

References

1. *GB 50011-2016;* Chinese Standard, Code for Seismic Design of Buildings. Chinese Architecture and Building Press: Beijing, China, 2016.
2. *JGJ 297-2013;* Chinese Standard, Technical Specification for Seismic Energy Dissipation of Buildings. Chinese Architecture and Building Press: Beijing, China, 2013.
3. Titirla, M.D. A state-of-the-art review of passive energy dissipation systems in steel braces. *Buildings* **2023**, *13*, 851. [CrossRef]
4. Rayegani, A.; Nouri, G. Application of smart dampers for preventing of seismic pounding in isolated structures subjected to near-fault earthquakes. *J. Earthq. Eng.* **2020**, *26*, 4069–4084. [CrossRef]
5. Rayegani, A.; Nouri, G. Seismic Collapse Probability and Life Cycle Cost Assessment of Isolated Structures Subjected to Pounding with Smart Hybrid Isolation Systems Using a Modified Fuzzy Based Controller. *Structures* **2022**, *44*, 30–41. [CrossRef]
6. Zoccolini, L.; Bruschi, E.; Cattaneo, S.; Quaglini, V. Current trends in fluid viscous dampers with semi-active and adaptive behavior. *Appl. Sci.* **2023**, *13*, 10358. [CrossRef]
7. Lin, W.H.; Chopra, A.K. Earthquake response of elastic SDF systems with nonlinear fluid viscous dampers. *Earthq. Eng. Struct. D* **2002**, *31*, 1623–1642. [CrossRef]
8. Gherbi, A.; Belgasmia, M. A simplified design strategy of nonlinear fluid viscous dampers for MDOF structures. *Iran. J. Sci. Technol. Trans. Civ. Eng.* **2022**, *46*, 857–864. [CrossRef]

9. Dargush, G.F.; Sant, R.S. Evolutionary aseismic design and retrofit of structures with passive energy dissipation. *Earthq. Eng. Struct. D* **2005**, *34*, 1601–1626. [CrossRef]
10. Rayegani, A.; Soureshjani, O.K.; Alaee, S.A.M.; Mualla, I.H.; Nemati, F. Seismic performance of buildings equipped with four-joint rotational friction dampers in mainshock-aftershock sequences. *J. Struct. Eng.* **2024**, *150*, 12980. [CrossRef]
11. Ding, J.; Wang, S.; Wu, H. Seismic performance analysis of viscous damping outrigger in super high-rise buildings. *Struct. Des. Tall Spec. Build.* **2018**, *27*, e1486. [CrossRef]
12. Chen, P.; Wu, X. Investigations on the dynamic response of adjacent buildings connected by viscous dampers. *Buildings* **2022**, *12*, 1480. [CrossRef]
13. De Souza Pippi, A.; Avila, S.M.; Doz, G. A review on the use of the inerter device in the structural coupling technique for adjacent building vibration control. *Structures* **2022**, *42*, 480–501. [CrossRef]
14. Tubaldi, E. Dynamic behavior of adjacent buildings connected by linear viscous/viscoelastic dampers. *Struct. Control Health Monit.* **2015**, *22*, 1086–1102. [CrossRef]
15. Palermo, M.; Silvestri, S. Damping reduction factors for adjacent buildings connected by viscous-viscous dampers. *Soil. Dyn. Earthq. Eng.* **2020**, *138*, 106323. [CrossRef]
16. Kazemi, F.; Miari, M.; Jankowski, R. Investigating the effects of structural pounding on the seismic performance of adjacent RC and steel MRFs. *Bull. Earthq. Eng.* **2021**, *19*, 317–343. [CrossRef]
17. Zhai, C.; Jiang, S.; Li, S.; Xie, L. Dimensional analysis of earthquake-induced pounding between adjacent inelastic MDOF buildings. *Earthq. Eng. Eng. Vib.* **2015**, *14*, 295–313. [CrossRef]
18. Xu, W.; Du, D.; Wang, S.; Li, W. A new method to calculate additional damping ratio considering the effect of excitation frequency. *Adv. Civ. Eng.* **2020**, *2020*, 3172982. [CrossRef]
19. Diotallevi, P.P.; Landi, L.; Dellavalle, A. A methodology for the direct assessment of the damping ratio of structures equipped with nonlinear viscous dampers. *J. Earthq. Eng.* **2012**, *16*, 350–373. [CrossRef]
20. Hwang, J.S.; Huang, Y.N.; Yi, S.L.; Song, Y.H. Design formulations for supplemental viscous dampers to building structures. *J. Struct. Eng.* **2008**, *134*, 22–31. [CrossRef]
21. Rama, R.K.; Ansu, M.; Iyer, N.R. A methodology of design for seismic performance enhancement of buildings using viscous fluid dampers. *Struct. Control Health Monit.* **2014**, *21*, 342–355. [CrossRef]
22. Xie, Y.; Zhang, J.; Xi, W. Effectiveness evaluation and optimal design of nonlinear viscous dampers for inelastic structures under pulse-type ground motions. *Earthq. Eng. Struct. D* **2018**, *47*, 2802–2820. [CrossRef]
23. Occhiuzzi, A. Additional viscous dampers for civil structures: Analysis of design methods based on effective evaluation of modal damping ratios. *Eng. Struct.* **2009**, *31*, 1093–1101. [CrossRef]
24. Singh, M.P.; Moreschi, L.M. Optimal seismic response control with dampers. *Earthq. Eng. Struct. D* **2001**, *30*, 553–572. [CrossRef]
25. Aguirre, J.J.; Almazan, J.L.; Paul, C.J. Optimal control of linear and nonlinear asymmetric structures by means of passive energy dampers. *Earthq. Eng. Struct. D* **2013**, *42*, 377–395. [CrossRef]
26. Lopez, G.D. A simple method for the design of optimal damper configurations in MDOF structures. *Earthq. Spectra* **2001**, *17*, 387–398.
27. Takewaki, I. Optimal damper placement for minimum transfer functions. *Earthq. Eng. Struct. D* **1997**, *26*, 1113–1124. [CrossRef]
28. Takewaki, I. Optimal damper placement for critical excitation. *Probabilist. Eng. Mech.* **2000**, *15*, 317–325. [CrossRef]
29. Lin, T.K.; Hwang, J.S.; Chen, K.H. Optimal distribution of damping coefficients for viscous dampers in buildings. *Int. J. Struct. Stab. Dyn.* **2017**, *17*, 1750054. [CrossRef]
30. Pollini, N.; Lavan, O.; Amir, O. Minimum-cost optimization of nonlinear fluid viscous dampers and their supporting members for seismic retrofitting. *Earthq. Eng. Struct. D* **2017**, *46*, 1941–1961. [CrossRef]
31. Parcianello, E.; Chisari, C.; Amadio, C. Optimal design of nonlinear viscous dampers for frame structures. *Soil. Dyn. Earthq. Eng.* **2017**, *100*, 257–260. [CrossRef]
32. Domenico, D.D.; Hajirasouliha, I. Multi-level performance-based design optimisation of steel frames with nonlinear viscous dampers. *Bull. Earthq. Eng.* **2021**, *19*, 5015–5049. [CrossRef]
33. Lavan, O.; Amir, O. Simultaneous topology and sizing optimization of viscous dampers in seismic retrofitting of 3D irregular frame structures. *Earthq. Eng. Struct. D* **2014**, *43*, 1325–1342. [CrossRef]
34. Tubaldi, E.; Kougioumtzoglou, I.A. Nonstationary stochastic response of structural systems equipped with nonlinear viscous dampers under seismic excitation. *Earthq. Eng. Struct. D* **2015**, *44*, 121–138. [CrossRef]
35. Zhang, Y.; Xu, W.; Du, D.; Wang, S. Stochastic optimization of dissipation structures based on lyapunov differential equations and the full stress design method. *Buildings* **2023**, *13*, 665. [CrossRef]
36. Su, C.; Li, B.; Chen, T.; Dai, X. Stochastic optimal design of nonlinear viscous dampers for large-scale structures subjected to non-stationary seismic excitations based on dimension-reduced explicit method. *Eng. Struct.* **2018**, *175*, 217–230. [CrossRef]
37. Tubaldi, E.; Ragni, L.; Dall'Asta, A. Probabilistic seismic response assessment of linear systems equipped with nonlinear viscous dampers. *Earthq. Eng. Struct. D* **2015**, *44*, 101–120. [CrossRef]
38. Rodolfo, J.T.G.; Yeudy, F.V.A.; Ramon, G.D. Probabilistic estimation of the dynamic response of high-rise buildings via transfer functions. *Eng. Struct.* **2024**, *302*, 117299.
39. Worden, K.; Cross, E.J. On switching response surface models, with applications to the structural health monitoring of bridges. *Mech. Syst. Signal Process.* **2018**, *98*, 139–156. [CrossRef]

40. Deng, L.; Cai, C.S. Bridge model updating using response surface method and genetic algorithm. *J. Bridge Eng.* **2010**, *15*, 553–564. [CrossRef]
41. Lan, X.; Pan, W.; Zhang, L.; Yu, W.; Wu, K. Application and research of viscous fluid damper in frame structure. *Sichuan Build. Sci.* **2022**, *48*, 10–17. (In Chinese) [CrossRef]
42. Aydin, E.; Boduroglu, M.H.; Guney, D. Optimal damper distribution for seismic rehabilitation of planar building structures. *Eng. Struct.* **2007**, *29*, 176–185. [CrossRef]

buildings MDPI

Article

Seismic Isolation of Fragile Pole-Type Structures by Rocking with Base Restraints

Sheng Li [1,2,3,*], Yao Hu [4], Zhicheng Lu [3], Bo Song [1] and Guozhong Huang [1,2]

[1] School of Civil and Resource Engineering, University of Science and Technology Beijing, Beijing 100083, China
[2] Research Institute of Macro-Safety Science, University of Science and Technology Beijing, Beijing 100083, China
[3] China Electric Power Research Institute, Beijing 100055, China
[4] Department of Infrastructure Engineering, The University of Melbourne, Melbourne, VIC 3010, Australia
* Correspondence: shengli@ustb.edu.cn

Abstract: Pole-type structures are vulnerable to earthquake events due to their slender shapes, particularly porcelain cylindrical equipment in electrical substations, which has inherent fragility and low strength in its materials. Traditional base isolation designs configure the bottom of the pole-type equipment as hinges with restraints. It fully relies on the restrainers to re-center the pole-type equipment, posing a risk of tilting and functionality failure after earthquakes. This study proposes a solution to this challenge by introducing a restrained rocking mechanism at the base of the structure. The design leverages the self-centering nature of rocking motion and uses restrainers to control the amplitude of rotation. Hence, it can effectively avoid tilting of the pole-type structures after earthquakes. Experimental investigations conducted on a 1:1 full-scale specimen revealed that the proposed restrained rocking design can achieve a reduction in seismic internal forces of over 50% while maintaining equipment in an upright position. Furthermore, an analytical model for the proposed isolation system of pole structures was developed and validated through comparison with experimental results. This paper introduces a novel solution for seismic isolation of pole-type structures through restrained rocking, specifically addressing the research gap regarding a reliable self-centering mechanism under seismic excitation. This advancement significantly enhances the seismic resilience of fragile pole-type structures and provides practical design methodologies for the seismic isolation of slender structures.

Keywords: restrained rocking; seismic isolation; pole-type structure; shaking table test; porcelain electrical equipment

1. Introduction

Pole-type structures are commonly engineered to withstand moderate loads at considerable heights. This category of structures encompasses a variety of utility poles, such as electricity poles, flagpoles, communication facility poles, and overhead antennas. Typically, these structures are constructed using materials such as timber, steel tubes, or steel-reinforced concrete. Despite the supported loads being significantly lower than their own weight, the inherent self-weight and slender profile render these pole-type structures susceptible to lateral forces induced by seismic events [1,2] or wind loads [3,4]. Consequently, various efforts have been made to enhance the flexural strength of the pole-type structures [5,6].

As illustrated in Figure 1a, cylindrical equipment within electric power substations represents an example of pole-type structures. The cylindrical electrical equipment comprises an array of insulators that support and house electrical components and conductors [7]. To ensure an adequate ground insulation distance, pole-type equipment in high-voltage substations can be more than ten meters in height, with length-to-diameter ratios of up to 1:20 [7]. Porcelain materials have excellent electrical insulation performance and are resistant to environmental corrosion, making them the preferred choice for forming the structural elements of cylindrical electrical equipment. As depicted in Figure 1a, hollow core insulators are formed by porcelain bushings, with both ends connected to metal flanges, and multiple insulators are assembled to create a piece of equipment. The electrical components housed within the hollow core insulators serve as non-structural parts.

Porcelain materials are characterized by low strength and brittleness, rendering them susceptible to brittle fracture even under low tensile stress [8,9]. Additionally, porcelain materials have a relatively high density, slightly exceeding that of concrete, and the electrical components housed within add considerable weight to the equipment. Given their brittle nature, substantial weight, and tall, slender profiles, the seismic resilience of pole-type cylindrical electrical equipment is compromised. Various types of pole-type electrical equipment, such as transformer bushings [10,11], surge arresters [12], capacitor voltage transformers [13], and post insulators [14,15], have been observed to be damaged during major earthquakes in recent decades [16–19], adversely affecting essential power supplies in the aftermath of these earthquakes [20–22]. Figure 1b illustrates the extensive damage to porcelain cylindrical electrical equipment during the 2008 Wenchuan earthquake. While glass fiber composite materials [23] can be employed as substitutes for porcelain in insulator tubes, challenges related to inadequate bending strength at connections persist [24]. Furthermore, replacing insulators necessitates repeated electrical design and verification efforts, incurring significant costs.

(a) (b)

Figure 1. Example of seismic fragile pole-type equipment in engineering (photo by the author) and damage of an electrical substation within the Sichuan electric power grid during the 2008 Wenchuan earthquake (cited from [25]). (**a**) Pole-type cylindrical equipment; (**b**) Earthquake damage.

Given the aforementioned considerations, the optimal approach for seismic enhancement of cylindrical equipment is seismic isolation at the base. In relevant studies, the base of pole-type equipment is engineered to be a rotational hinge in order to achieve isolation objectives [26,27]. Additionally, energy dissipation devices, such as metallic yielding dampers [28,29] or wire rope isolators [30–32], are integrated as restraints to control the rotation at the base hinge. This method, which mitigates base moments and prolongs the structural vibration period, has demonstrated effectiveness in enhancing seismic resistance. However, it is crucial to note that restoring moments at the bottom are provided by restraint elements. This approach necessitates the restraint elements to maintain a recoverable state; otherwise, post-seismic residual deformations may lead to structural tilting and subsequent functional or structural failures. The absence of a reliable self-centering mechanism for seismic isolation of pole-type equipment represents a notable research gap. Addressing this gap is the primary objective of this study, which has the potential to further enhance the seismic resilience of such equipment.

The rocking motion of structures under horizontal base excitation [33] inherently possesses self-centering capabilities owing to the gravitational effects after uplifting. Seismic protection by rocking has been exemplified by the stone pillars in ancient temple architecture, which have withstood earthquakes over hundreds of years [34,35]. In recent decades, leveraging rocking action for seismic isolation design has seen significant application in building and bridge engineering, such as the use of rocking walls (or rocking shallow foundations) in buildings [36–39] and rocking piers in bridges [40–43]. Various types of restraints [44–47] have been proposed to mitigate the overturning hazard of rocking systems.

Incorporating restrained rocking mechanisms to achieve reliable rocking isolation designs for pole-type electrical equipment is the primary focus of this study. Equally important is the development of an analytical model as an analysis method for the restrained rocking design of such facilities. Previous studies have proposed analytical models for the rocking response of flexible structures [48–52], which serve as the basis for the models presented in this paper. Additionally, a significant contribution of this paper lies in the validation of the isolation design through shaking table tests of full-scale electrical equipment, the results of which will be extensively reported.

The novelty of the paper lies in the restrained rocking mechanism for seismic isolation of pole-type cylindrical electrical equipment. While previous studies have explored seismic isolation methods involving rotational hinges and energy dissipation devices, the proposed design integrates rocking interfaces, rotational restraint devices, and stopples to allow for a controlled rocking motion of the structure under seismic excitation while limiting excessive rotation. This innovative design addresses the lack of a reliable self-centering mechanism in existing seismic isolation methods for pole-type equipment, representing a significant advancement in enhancing the seismic resilience of such structures.

In this paper, Section 2 will introduce the design concept of restrained rocking isolation for pole-type structures. The analytical model of the proposed system will be developed in Section 3. The validation of the restrained rocking design for a full-scale high-voltage surge arrester will be presented in Section 4, and Section 5 will conduct a seismic response analysis to explore the seismic response patterns of the proposed design.

2. Conceptual Development

The aspect ratio of pole-type structures typically ranges from 1:10 to 1:20. According to previous experimental research [12], pole-type electrical equipment primarily experiences bending loads during seismic events, leading to significant tensile stress at the base. Consequently, implementing base shear isolation becomes challenging. The high tensile stress at the base necessitates horizontal shear isolation devices with a robust capacity for tensile resistance. However, common solutions for base shear isolation, such as friction pendulums or lead-core rubber bearings, lack sufficient tensile resistance in the vertical direction. Therefore, a rocking motion damping approach is considered more suitable for pole-type structures.

The design of restrained rocking isolation for pole-type structures in this study encompasses three main parts: rocking interfaces set at the base, rotational restraint devices circularly arranged at the base, and stopples set to prevent excessive rotation. These parts are compactly located between the base of the structure and the foundation (or elevation podium). The design is shown in Figure 2.

The isolation design in this study introduces a rocking mechanism at the base that allows the structure to undergo rocking motions under horizontal seismic excitation, as depicted in Figure 3. Unlike setting a hinge at the base in which the supporting block functions as a part of the base hinge [29], the rocking pads are set to provide an adjustable contact surface of width W (the diameter of the contact circle). Reducing the width W tends to make the structure more conducive to uplift, while increasing W enables the structure to enter a rocking state under larger base excitation. Another parameter of the rocking pad is the height h_p, which should be coordinated with the initial length of the restrainers and its expected deformation during an earthquake event.

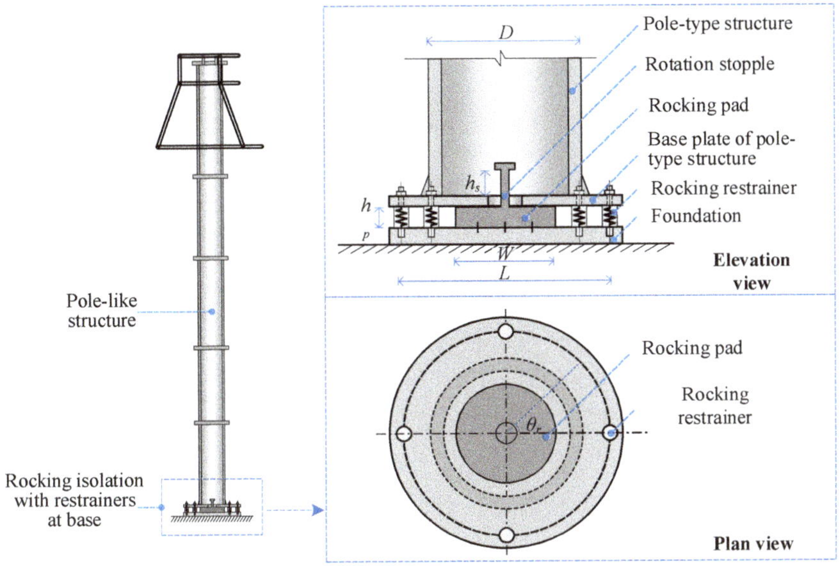

Figure 2. Proposed design of restrained rocking isolation of pole-type cylindrical electrical equipment.

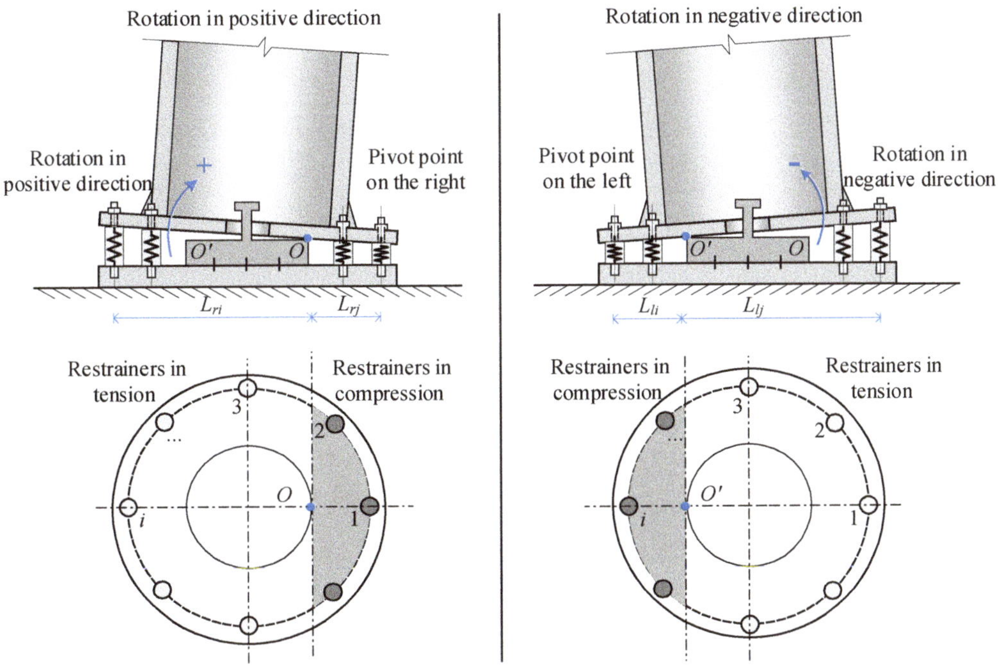

Figure 3. Rocking motion at the isolation device.

The rocking restrainers are arranged around the base of the structure, as shown in Figure 2, with the lower end fixed to the base and the upper end connected to the bottom plate of the pole-type structure. The example in Figure 2 shows eight evenly spaced restrainers, and its number can vary in different cases. The diameter D of the installation position needs to be selected carefully. A larger diameter D will result in a larger moment arm for restraining rocking motion and larger deformations in a certain rotation. When D is small, the deformation of the restrainer decreases, but larger force is required for restraining rocking.

The level arm of the circularly arranged restrainers relative to the pivot point is not constant during the rocking motion. For a given restrainer, the level arm is larger in the tension phases as compared to the compression phases. Additionally, among the restrainers, the majority are under tension, with only a few experiencing compression, as depicted in Figure 3. This force distribution characteristic of rocking restrainers is beneficial in reducing the demand on the height of rocking pads. Moreover, it facilitates the utilization of common springs as restrainers, which typically exhibit a large tensile stroke and a small compressive stroke. Figure 4 outlines the hysteresis relationships of various restrainers, such as linear springs, elasto-plastic elements, and yielding elements using the Wen model. Although not explicitly listed, restrainers with viscous damping can also be adopted.

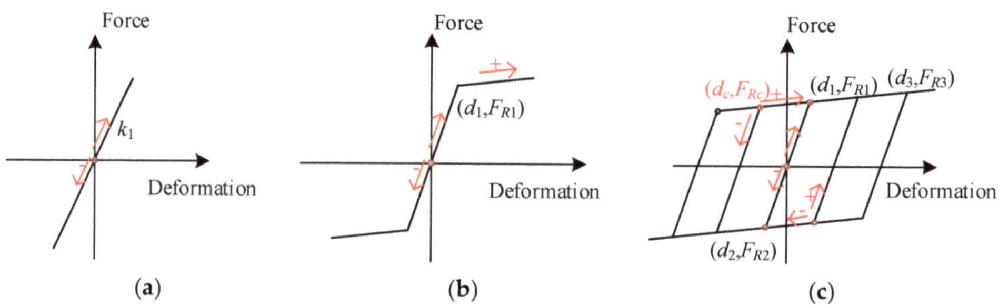

Figure 4. Mechanical behaviours of rocking restrainers.

In the design of restrained rocking isolation, it is necessary to implement stopples to limit the maximum rotation of the pole-structure within a certain range. The rotation stopples can be integrated with the rocking pad; for instance, it can be placed at the midpoint of the rocking pad, as shown in Figure 2. The length of the stopple (i.e., h_s) restricts rotation within the range of arctan ($2h_s/W$). The rotation limit can be set according to the maximum allowable inclination of the structure during earthquakes, and, in normal cases, the restrainers, rather than stopples, can keep the rotation within the limit so that the stopples become an additional safeguard against overturning.

Figure 3 specifies the positive direction of rotation in the analysis. The structure rotates around the right-side point O of the rocking pad at a given time, t, and the rotation is represented by $\theta(t)$, which is a positive value. The level arm of ith the restrainer is expressed as:

$$L_{Ri} = L_{ri}, (i = 1, 2, \ldots, n) \tag{1}$$

in which L_{Ri} represents the lever arm of the ith restrainer relative to the rotation point and L_{ri} denotes the lever arm when the rotation point is on the right. When the structure rotates around the left-side point O', $\theta(t)$ is negative and the level arm of the restrainer is:

$$L_{Ri} = L_{li}, (i = 1, 2, \ldots, n) \tag{2}$$

in which L_{Ri} represents the lever arm of the ith rocking restrainer relative to the rotation point O' on the left. The values of L_{ri} and L_{li} are contingent upon the distance of the restrainer relative to the rotation point, as depicted in Figure 3. The deformation of the restrainer is expressed as:

$$\Delta_{Ri}(t) = L_{Ri}\tan\theta(t) \tag{3}$$

The corresponding restraining forces are determined based on the constitutive relationship of the restrainers. When the loading and unloading paths of the restrainers are consistent, as illustrated in Figure 4a,b, they can be represented as:

$$F_{Ri}(t) = f[\Delta_{Ri}(t)] \tag{4}$$

In cases where the loading and unloading paths of the restrainers are exhibiting hysteretic behaviour, $F_{Ri}(t)$ is dependent on both the current deformation and the loading history. It also necessitates analysis combined with hysteresis relationships, as in the case shown in Figure 4c.

The total restoring moment of the restrainers, $M_R(t)$, is given by:

$$M_R(t) = \sum_{i=1}^{n} F_{Ri} L_{Ri} \tag{5}$$

In addition to the restoring force from the restrainers, it is important to note that additional restoring moments can be generated from the rocking structure under the gravitational effects, $M_G(t)$,

$$M_G(t) = f[\theta(t), u(h,t)] \tag{6}$$

in which $u(h,t)$ denotes the elastic displacement of the structure at height h.

3. Dynamics Modeling

The pole-type structure is simplified as a beam with a uniform cross-section, considering bending deformation as dominant and neglecting shear deformation. The sectional bending modulus of the column is denoted by EI, and the height of the column is H. The column is assumed to have a uniformly distributed mass with a mass density of m_c/H. An additional mass m is added to the top to represent the supported load.

The fundamental period of pole-type equipment in electrical substations is typically within 0.8 s, placing it within or near the resonant period range of earthquake excitations. Consequently, the seismic response of pole-type equipment is predominantly governed by its first vibration mode. In former studies by the author on the shaking table test of pole-type electrical equipment with fixed bases, including the testing of surge arresters, post insulators and capacitor voltage transformers [7], it was demonstrated that the seismic response of the equipment primarily exhibited the first-order bending mode [12]. The shear deformation of the slender pole-type equipment is negligible considering the small cross-sectional dimensions compared to their length; therefore, it could be modelled by an Euler–Bernoulli beam with a damping coefficient c. The first-order mode shape of the cantilever column can be described by the following equation [53]:

$$\psi(h) = \frac{-h^3}{2H^3} + \frac{3h^2}{2H^2} \tag{7}$$

With a fixed base, the frequency of the structure is given by:

$$\omega_1 = \sqrt{\frac{3EI}{\left(\frac{33}{140}m_c + m\right)H^3}} \tag{8}$$

It is important to highlight that, for pole-type structures with fundamental periods significantly longer than the resonant period range of earthquake excitations, the effectiveness of base isolation might be compromised due to the contribution of higher vibration modes [54,55]. While the exploration of seismic isolation methods for such flexural pole-type structures holds significance, it falls beyond the scope of this paper and warrants further research.

For the restrained rocking design in Section 2, the width of the rocking pad at the bottom of the structure is denoted as W, and the characteristic angle of the rocking structure is α, which equals arctan (W/H). There are n restrainers at the base, evenly distributed on a circle with diameter D. When the structure undergoes rocking vibrations, the system has two degrees of freedom: the rocking rotation degree of freedom (DOF) θ around the pivot point (O or O') and the translational DOF representing the deformation at top u, as shown in Figure 5. The system forms a dynamic equilibrium in the rotational and horizontal directions under the action of gravity, inertia forces, and forces of restrainers.

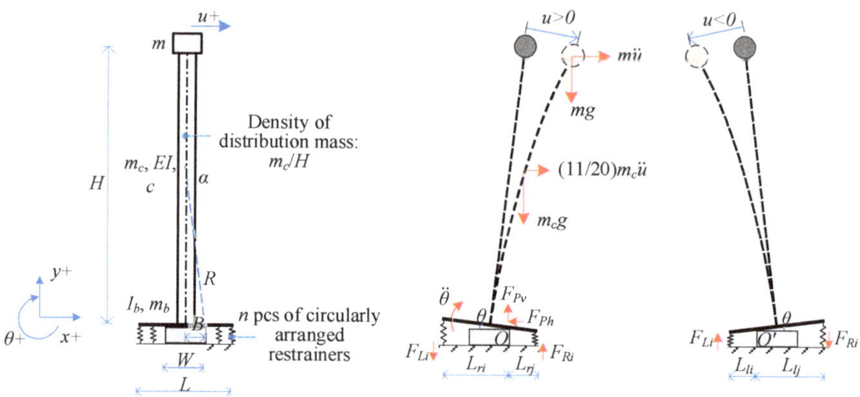

Figure 5. Dynamics modeling of the pole-type structure with restrained rocking isolation.

The process of the rocking vibration of the pole-type structure, as shown in Figure 6, is analyzed here. In the initial state of rest, both rotation θ and translational deformation u are zero. Under seismic excitation, the structure undergoes lateral displacement first and responds like a fixed-base SDOF system. It does not rotate until the lateral displacement reaches a critical value u_{cr}. The u_{cr} can be obtained by referring to the state where an equivalent lateral force acting at the top of the cantilever structure produces the same moment as that of the gravity of the structure [56], which is:

$$u_{cr} = \frac{(m_c + m)gBH^2}{3EI} \tag{9}$$

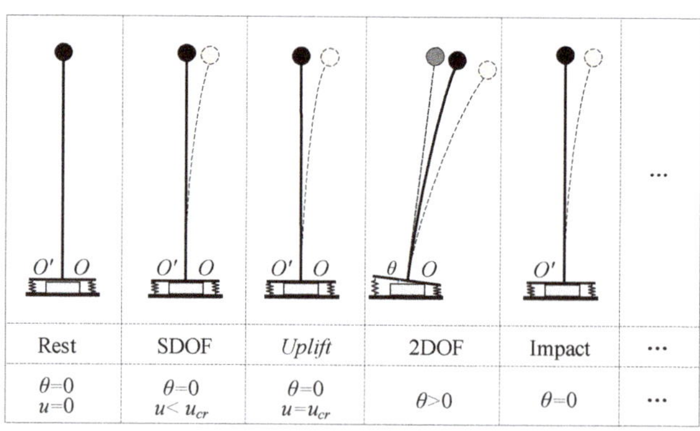

Figure 6. Sequence of rocking.

After that, the system undergoes rotation, resulting in a state of 2DOF. The occurrence of rotation produces a deformation of the rocking restrainers. The vibration frequency of translational deformation shifts due to the supporting conditions changing from a fixed base to pivoting rotation [52]. Under the combined restoring moment from gravity and the rocking restrainer, the rocking rotation of the structure will return to zero after a cycle, and rocking pounding will occur. In this study, the Vertical Velocity Energy Loss (VVEL) model [52] is adopted to update the translational velocity and rotational velocity after pounding. It assumes that the energy of the vertical motion component is completely consumed while the energy of the horizontal motion remains unchanged, which is:

$$\dot{\theta}_{post} = 0 \tag{10}$$

$$\dot{u}_{post} = sign\left(H\dot{\theta} + \dot{u}\right)\sqrt{\frac{m\left(H\dot{\theta} + \dot{u}\right)^2 + m_c\left(\frac{1}{3}H^2\dot{\theta}^2 + \frac{11}{20}H\dot{\theta}\dot{u} + \frac{33}{140}\dot{u}^2\right)}{m + \frac{33}{140}m_c}} \tag{11}$$

After the rocking pounding, the pole-type structure enters the SDOF state again and another cycle continues.

In a 2DOF state, both the angular acceleration $\ddot{\theta}$ and the translational acceleration $\ddot{u}$ will affect the dynamic balance in the translational direction and rotational direction. Analyzing from the perspective of dynamic equilibrium of forces and moments can be complex. Alternatively, analyzing the system dynamics from an energy perspective using the *Lagrange* equation is more straightforward.

The total kinetic energy E in the system is formed from kinetic energy generated by the motion of the added mass at the top and the column [52].

$$E = \frac{1}{2}m\left(\left(B^2 + H^2 + u^2 - 2\cdot sign(\theta)Bu\right)\dot{\theta}^2 + \dot{u}^2 + 2H\dot{u}\dot{\theta}\right)$$

$$+\frac{1}{2}m_c\left(\left(B^2 + \frac{H^2}{3} + \frac{33}{140}u^2 - \frac{3}{4}sign(\theta)Bu\right)\dot{\theta}^2 + \frac{33}{140}\dot{u}^2 + \frac{11}{20}H\dot{u}\dot{\theta}\right) \tag{12}$$

in which B equals 0.5 W as shown in Figure 5.

The potential energy V in the system comprises the potential energy V_G due to gravity acceleration g and horizontal ground motion $\ddot{u}_g$; the strain energy of the column V_E; and the energy developed in the rocking restrainer V_R:

$$V = V_G + V_E + V_R \tag{13}$$

The expressions' potential energies are expressed in Equations (14)–(16). These expressions are similar to that of an unrestrained rocking system [52], with the exception of the term rocking restrainers V_R.

$$V_G = Bmsign(\theta)(g\sin\theta - \ddot{u}_g\cos\theta) + Hm(g\cos\theta + \ddot{u}_g\sin\theta)$$

$$+um(-g\sin\theta + \ddot{u}_g\cos\theta) + Bm_c sign(\theta)(g\sin\theta - \ddot{u}_g\cos\theta)$$

$$+\frac{1}{2}Hm_c(g\cos\theta + \ddot{u}_g\sin\theta) + \frac{3}{8}um_c(-g\sin\theta + \ddot{u}_g\cos\theta) \tag{14}$$

$$V_E = \frac{1}{2}\cdot\frac{3EI}{H^3}u^2 \tag{15}$$

$$V_R = \sum_{i=1}^{n}\left(\int F_{Ri}L_{Ri}\tan\theta d\theta\right) \tag{16}$$

where L_{Ri} is the level arm of restraints and F_{Ri} is the force of restrainers obtained from hysteresis model (Figure 4).

By adopting *Lagrangian* operations [52,57], the governing equations of the system can be built based on considering transitions between potential and kinetic energy of the system. By the partial derivation operation in Equation (17), the governing equation for the rotation θ is obtained as Equation (18)

$$\frac{d}{dt}\frac{\partial(T-V)}{\partial\dot{\theta}}-\frac{\partial(T-V)}{\partial\theta}=0 \tag{17}$$

$$\begin{gathered}\left(m_c\left(B^2+\frac{H^2}{3}\right)+\frac{33}{140}m_c u^2-sgn(\theta)\frac{3}{4}m_c Bu+m\left(B^2+H^2\right)-sgn(\theta)2mBu+mu^2\right)\ddot{\theta}= \\ -\left(\frac{33}{70}m_c+2m\right)u\dot{u}\dot{\theta}-\left(\frac{11}{40}m_c+m\right)H\ddot{u}+sgn(\theta)\left(\frac{3}{4}m_c+2m\right)B\dot{u}\dot{\theta}+ \\ \ddot{u}_g\left(-sgn(\theta)(m+m_c+m_b)B\sin\theta-\left(\frac{1}{2}m_c+m\right)H\cos\theta+\left(\frac{3}{8}m_c+m\right)u\sin\theta\right)+ \\ g\left(-sgn(\theta)(m+m_c+m_b)B\cos\theta+\left(\frac{1}{2}m_c+m\right)H\sin\theta+\left(\frac{3}{8}m_c+m\right)u\cos\theta\right)- \\ sgn(\theta)\sum_{i=1}^{n}\left(F_{Ri}L_{Ri}\cos\theta\right)\end{gathered} \tag{18}$$

By the partial derivation operation in Equation (19), the governing equation for the translational deformation u is obtained as Equation (20) [52]:

$$\frac{d}{dt}\frac{\partial(T-V)}{\partial\dot{u}}-\frac{\partial(T-V)}{\partial u}=-c\dot{u} \tag{19}$$

$$\begin{gathered}\left(\frac{33}{140}m_c+m\right)\ddot{u}+\left(\frac{11}{40}m_c+m\right)H\ddot{\theta}=-\frac{3EI}{H^3}u+\left(\frac{3}{8}m_c+m\right)g\sin\theta+\left(\frac{33}{140}m_c+m\right)u\dot{\theta}^2 \\ -sgn(\theta)\left(\frac{3}{8}m_c+m\right)B\dot{\theta}^2-\left(\frac{3}{8}m_c+m\right)\ddot{u}_g\cos\theta-c\dot{u}\end{gathered} \tag{20}$$

A case study of a pole structure with a height H of 10 m is given to illustrate the analysis. The uniformly distributed mass of the column along height is 500 kg/m, and no lump mass is added at the top. The bending stiffness of the pole is 3.5×10^7 N.m^2. When the base is fixed, the first-order frequency and the damping ratio are 1.5 Hz and 0.025, respectively. A base isolation design with a rocking pad width of 300 mm, as shown in Figure 2, is adopted in this structure. A total of 10 elastic restrainer are arranged circularly with a diameter of 750 mm, and the stiffness of each restrainer is 5 kN/mm. The initial condition of the structure is zero rotation, zero translational deformation, and zero angular velocity, while the initial translational velocity is 3 m/s. The structure enters a state of free rocking attenuation. Figure 7 shows the numerical analysis results based on the model in this section, including rotation angle, angular velocity, translational deformation, and the energy of the system. It can be seen that the system has a relatively short residence time in the 1DOF state. This is because a small amount of deformation (u_{cr} = 5 mm) can cause the structure to enter a rocking state. The deformation of the structure is accompanied by high-frequency vibrations while maintaining the same pace as the rocking motion. This high-frequency part is the manifestation of the first-order deformation mode of the structure with a shifted frequency due to changes in supporting conditions after uplifting [48,58].

The energy curve represents the exchange of potential energy and kinetic energy, as well as the energy dissipation during the response. In this case study, the system exhibits initial velocity and experiences no base excitation. Therefore, monitoring changes in system energy provides insights into the damping characteristics of the system. As illustrated in Figure 7, the total energy dissipation of the system originates from two primary sources. Firstly, structural damping manifests as a gradual decrease in system energy, aligning with the inherent damping of the pole-type structure. Secondly, abrupt energy decreases signify dissipation due to rocking collisions, contributing to the system damping as the second source. If rocking restrainers with energy dissipation properties are utilized, their hysteresis energy dissipation constitutes the third source of system damping.

The structural damping of the pole-type structure can be quantified when it is with a fixed base. However, energy dissipation from rocking collisions and the hysteretic energy dissipation from restrainers depend on the vibration amplitude and cannot be accurately represented by a fixed equivalent damping coefficient. The influences of rocking impact and hysteretic behaviour of restrainers have been factored into the analytical model.

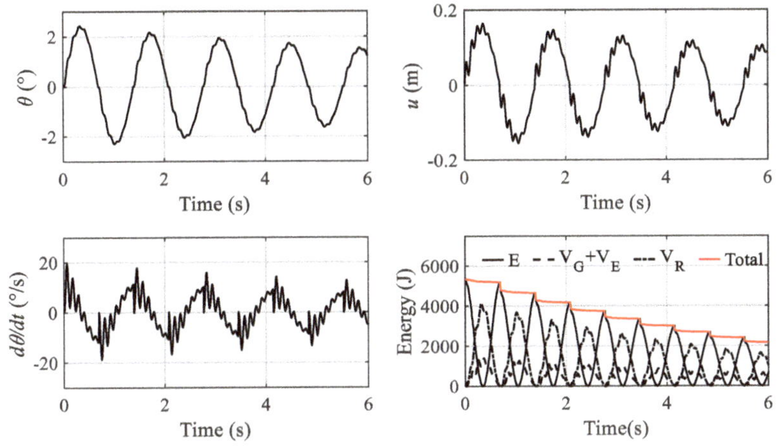

Figure 7. Energy of the system in rocking vibration.

4. Experimental Study

4.1. Testing Specimen, Setup and Schedule

The effectiveness of the restrained rocking isolation design for pole-type structures is validated by the experimental study in this section. Additionally, verifying the accuracy of the analytical model proposed in this paper is of equal significance. The shaking table test conducted in this section utilized a full-scale prototype as the specimen; therefore, the results obtained carry greater practical significance.

The specimen involved surge arrester equipment used in an Ultra High Voltage (UHV) electric power substation, as shown in Figure 8. The surge arrester comprised six porcelain insulators, with the upper five insulators being 2.12 m in height and 1350 kg in weight. The lower insulator was 0.65 m in height and 400 kg in weight, making the total weight of the pole-type structure 7150 kg. The main structural component of the insulator was a porcelain hollow core bushing with an outer diameter of 510 mm and an inner diameter of 400 mm. The elastic modulus of the porcelain material was 1.1×10^{11} N/m^2. The fundamental frequency of this pole-type structure was around 1.5 Hz, while the structural damping of the fixed based specimen was around 2%.

In the rocking isolation design, the width of the rocking pad was 375 mm. Ten restrainers were installed at the bottom of the pole-type structure, evenly distributed in a circle with a diameter of 770 mm. The restrainers exhibited typical metallic yielding hysteretic behaviour, as depicted in Figure 9. This device features an internally filled lead alloy core, which has typical metallic yielding behaviour under axial loading. Subject to shear forces, the core undergoes repetitive yielding of lead alloy. The recrystallization properties inherent in lead alloy ensure the stability of the mechanical performance of the core throughout the repeated shear-yielding cycles. The yielding load of the damper was 40 kN, the stiffness before yielding was 60 kN/mm, and the stiffness after yielding was 1% of the initial stiffness.

The setup of the shaking table test is shown in Figure 8, and the test was carried out according to the schedule outlined in Table 1. Test cases #1 to #5 were conducted on the fixed-base structure, whereas test cases #6 to #9 represented the rocking isolation condition. Due to the full-scale equipment specimen needing to be used in subsequent electrical performance tests following the seismic test, the input motion was chosen in accordance with the GB50260 standard [59], in which code-compatible artificial motion was employed. The time history of the artificial motion is depicted in Figure 10, exhibiting an acceleration response spectrum consistent with the code spectrum in the resonant segment.

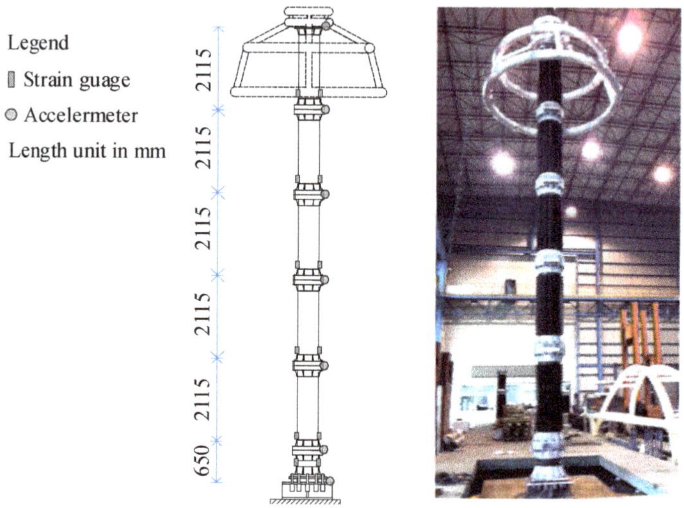

Figure 8. Test specimen of surge arrester equipment used in a Ultra High Voltage (UHV) electric power substation.

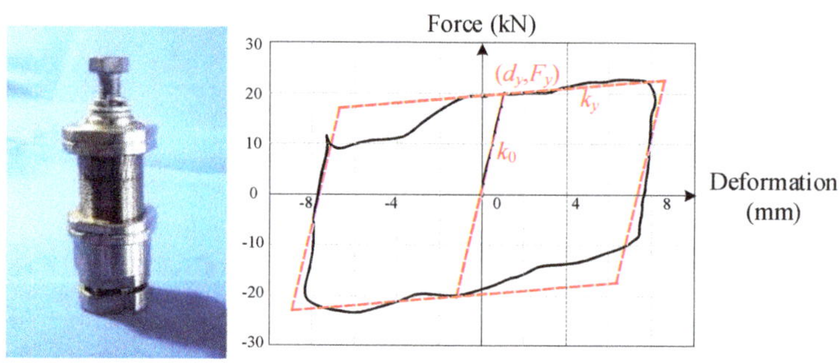

Figure 9. Rocking restrainer adopted in the experimental study.

Table 1. Test schedule.

No.	Setup	Input Motion	PGA	Direction
1		White noise	0.05 g	
2		Code compatible wave	* 1.4 × 0.15 g	
3	Fixed base	White noise	0.05 g	
4		Code compatible wave	1.4 × 0.20 g	
5		White noise	0.05 g	X direction
6		Code compatible wave	1.4 × 0.20 g	
7	Restrained rocking	White noise	0.05 g	
8		Code compatible wave	1.4 × 0.40 g	
9		White noise	0.05 g	

* The input motion was scaled by 1.4 times for considering the amplification effect of supporting podium or frame [12].

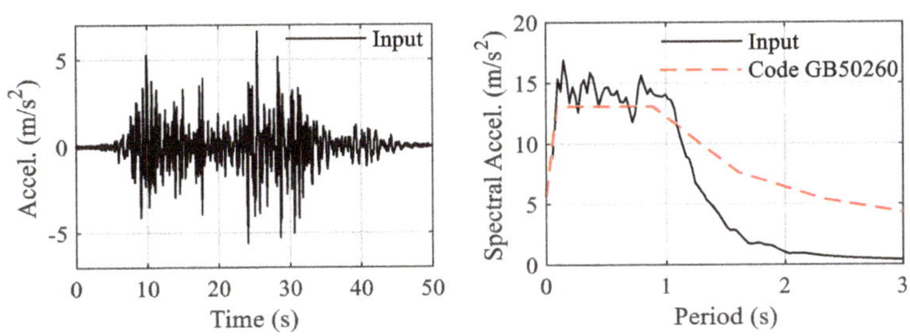

Figure 10. Input motion.

4.2. Testing Results

Comparisons were made between the seismic response results of the specimen in the conditions of rocking isolation and a fixed-base. In the comparison of acceleration response between test case #4 and #6, shown in Figure 11, it can be seen that the acceleration at the top decreased from 1 g to 0.45 g under the input with PGA of 1.4 × 0.2 g. At the same time, the Frequency Response Function (FRF) on the right shows that the predominant frequency of the equipment decreased from 1.4 Hz to an unfixed frequency of around 0.5–0.8 Hz. This was a result from the rocking motion in which the frequency was not fixed but related to the amplitude. The strain response indicated the seismic internal force at the porcelain component. In the fixed-base test shown in Figure 12, the strain at the base was about 195 με and the corresponding stress was 21.5 MPa, which was around 60% of the allowable stress of the porcelain material. In the rocking isolation test case, this value was lowered to 95 με or 10.4 MPa, with a decreasing rate of 55%. After the test, the specimen returned to an upright position without any residual tilting.

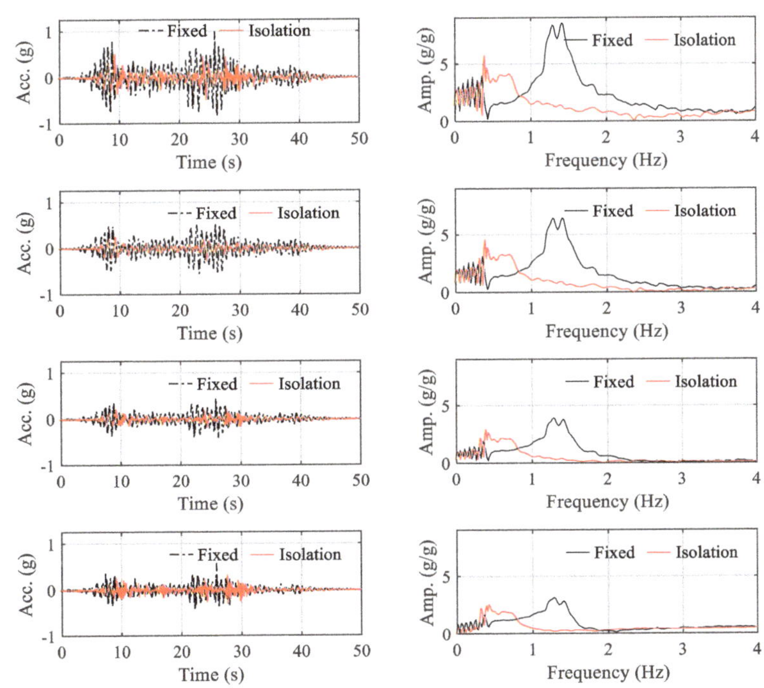

Figure 11. Acceleration responses in test case #4 and #6.

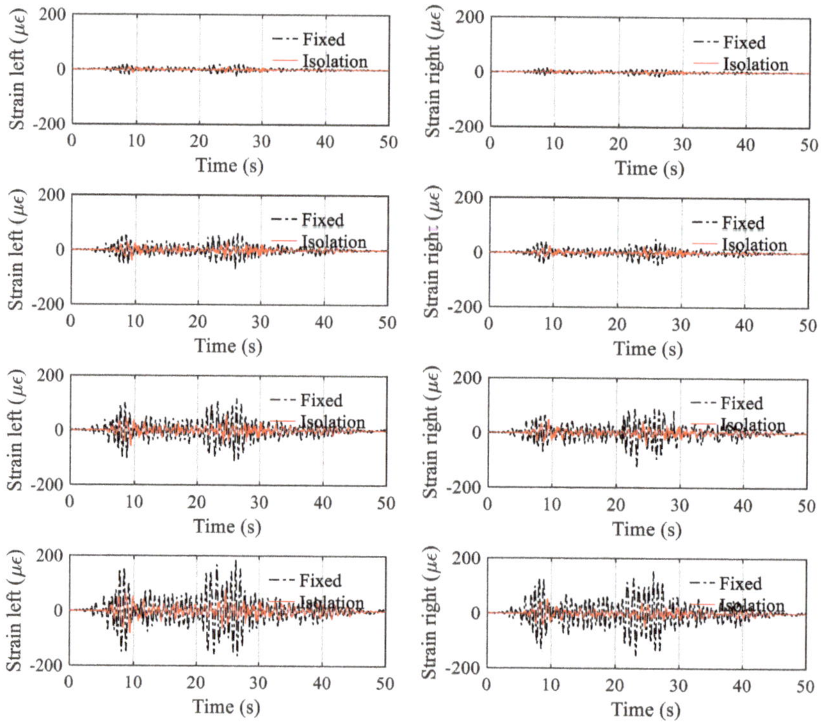

Figure 12. Strain responses in test case #4 and #6.

Similar experimental results were further observed in the test case of 1.4 × 0.4 g. Figure 13 shows a comparison of the acceleration at the top and the bending moment at the base (derived from strain). The results of the rocking isolation were recorded in test case #8, while the fixed-base case results were obtained from numerical analysis of a calibrated non-isolation model. It is evident that, by employing the restrained rocking design, the bending moment was reduced from 485 kN·m to 202 kN·m, achieving a reduction rate of 58%. Importantly, there was no residual tilting observed after the earthquake, indicating the effectiveness of the self-centering capability of the rock mechanism.

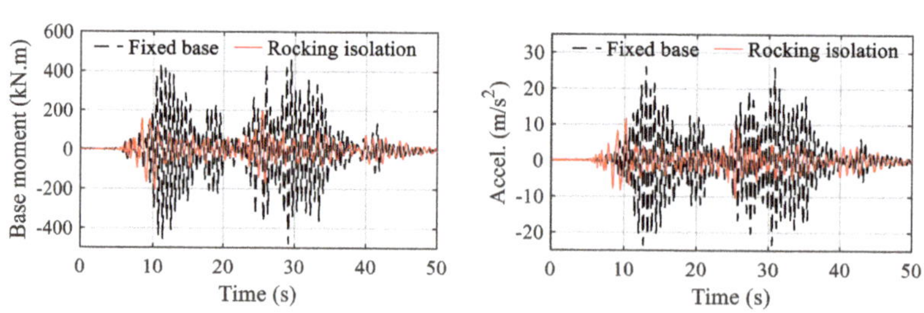

Figure 13. Acceleration and bending moment in test case #8.

4.3. *Validation of the Restrained Rocking Model*

Table 2 presents a summary of the parameters utilized in the seismic response analysis, based on the analytical model developed in Section 3. Figures 14 and 15 illustrate the comparison between experimental results and analysis results for test case #6 and test case #8, respectively. The differences between the predicted values and experimental values of amplitudes of bending moment and acceleration were less than 12%, validating the effectiveness of the analytical model in Section 3. The rotational behaviour depicted in the figures indicates that the structure underwent a continuous rocking motion throughout the process, thereby cutting down the bending moment at the base. Moreover, the rotation remained below two degrees in the 1.4 × 0.4 g test case, underscoring the effectiveness of the rocking restrainers.

Table 2. Parameters of testing specimen.

Pole-type structure					
Height	11.2 m	Diameter (outside)	510 mm	Diameter (inside)	400 mm
Distributive mass	632 kg/m	Added mass at top	50 kg	Elastic modulus of porcelain	110 GPa
Allowable stress of porcelain	36 MPa	Natural Frequency	1.6 Hz	Equivalent sect. stiffness	8.5×10^7 N.m^2
Damping	3.5%				
Design of rocking isolation at base					
Width of rocking base	375 mm	Number of restrainers	10	Arrangement circle diameter	770 mm
Yielding force of restrainer	40 kN	Initial stiffness of restrainer	60 kN/mm	Post yielding stiffness	0.6 kN/mm

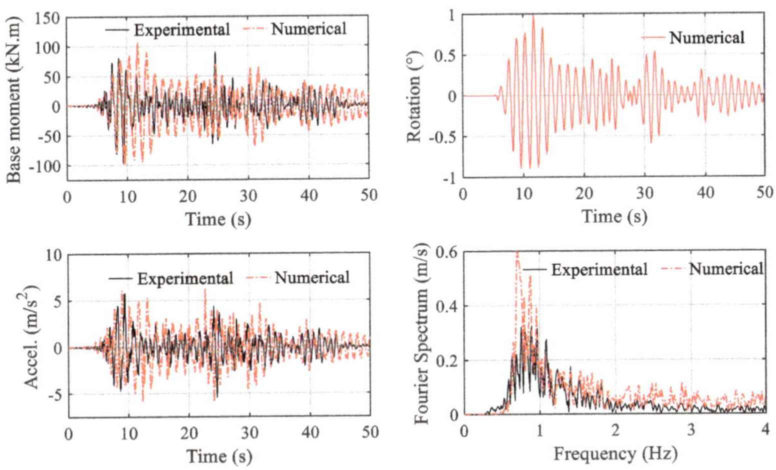

Figure 14. Comparison between experiment results and analysis results in test case #6.

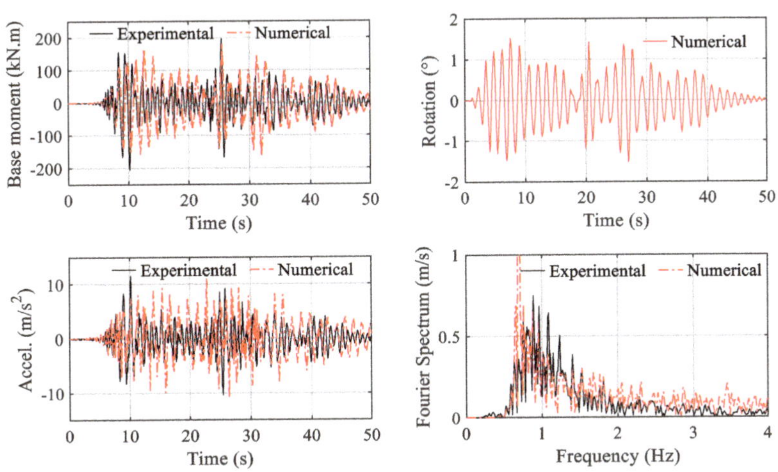

Figure 15. Comparison between experiment results and analysis results in test case #8.

5. Discussion on the Intensity Measure Governing the Response

This section uses the validated model to study the characteristics of the responses of pole-type structures under different seismic excitations. A total number of 21 earthquake records with magnitudes of 6.5 or above, as well as PGAs ranging from 0.1 g to 0.8 g, were selected from the PEER strong ground motion database [60]. The structural model with a restrained rocking design, as described in Table 2, was adopted in the numerical study. The analysis results were visualized in Figures 16 and 17, including the Intensity Measures (IM) of the input motion and the response of the structure.

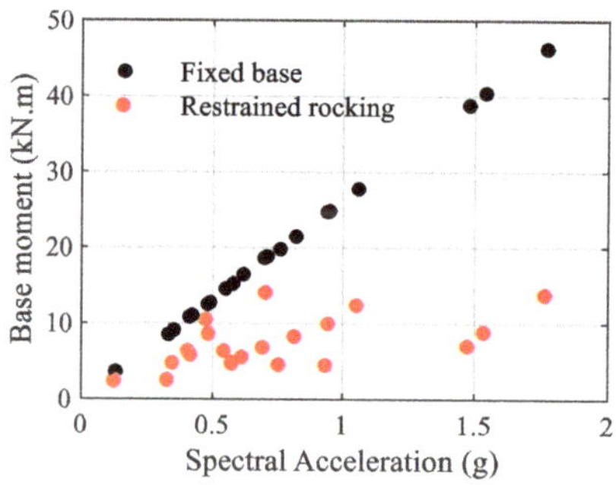

Figure 16. A comparison of base moment demand of the pole-type structure with and without rocking isolation under various input excitations.

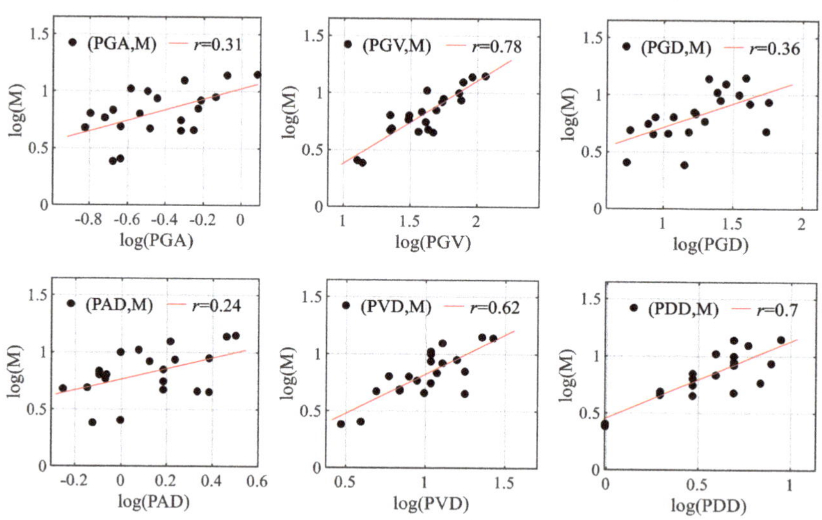

Figure 17. Relationship between intensity measures of input motion and base moment response (kN.m) of the restrained rocking pole-type structure. The units for PGA, PGV, PGD, PAD, PVD, and PDD are g, cm/s, cm, g, m/s, and m, respectively.

Figure 16 compares the base moment of the two structures, with the horizontal axis representing the spectral acceleration at the fundamental period. It is evident that the restrained rocking design achieved an average of 57% reduction in seismic internal forces (ranging from 16% to 82%). Additionally, there is a trend indicating that, the greater the seismic internal force in the fixed base structure, the more significant the isolation effect. For the fixed base structure, the seismic demand is primarily controlled by spectral acceleration.

However, no significant correlation is observed between spectral acceleration and base moment for restrained rocking structures.

This study also investigated the seismic demand on pole-type structures with restrained rocking, considering intensity measures (IM) such as peak ground acceleration (PGA), peak ground velocity (PGV), peak ground displacement (PGD), peak spectral acceleration demand (PAD), peak spectral velocity demand (PVD), and peak spectral displacement demand (PDD). The regression analysis results are shown in Figure 17. The PGA and PAD exhibit the lowest correlation with the seismic responses, whereas the PGV and PDD demonstrate the strongest correlation.

A seismic demand model taking base moment M as the Engineering Demand Parameter (EDP) can be expressed as:

$$log_{10}(M) = a + b \cdot log_{10}(PGV) \text{ or } log_{10}(M) = c + d \cdot log_{10}(PDD) \tag{21}$$

In this example, the values for a and b are -0.36 and 0.73, respectively; and the values for c and d are 0.46 and 0.67, respectively. This observation aligns with findings from other types of rocking structures [61–63], indicating a sensitivity to major velocity pulses. Consequently, this result underscores the importance of considering the velocity spectrum characteristics of the construction site when defining input motion in the design of pole-type structures with restrained rocking.

6. Conclusions

The traditional seismic isolation designs for pole-type structures typically involve engineering the base of the structure into a rotational hinge with restraints. However, this approach lacks a re-centring capability in the event of failure or yielding of the restrainer, thereby posing a risk of tilting and functionality failure following earthquakes. To address this challenge, a novel isolation design incorporating a restrained rocking mechanism at the base of the structure was proposed. The main outcomes of the paper are outlined below:

(1) The proposed design features a width-adjustable rocking interface at the base of the pole-type structure, surrounded by circularly arranged rocking restrainers. This configuration disrupts the transmission path of earthquake energy while introducing a dual mechanism for re-centring, restoring moments by gravitational force and restraining force.

(2) Experimental studies were conducted on full-scale porcelain surge arresters in high-voltage substations. The results demonstrated a reduction in internal force by more than 50%, with rotation controlled within two degrees under both 0.2 g and 0.4 g level excitation. Importantly, the specimen could return to an upright position after the test.

(3) The analytical model for the pole-type structures with restrained rocking isolation was developed and validated in this study. The establishment of analytical models provides an effective tool for the parameter design of the isolation system.

(4) This study explored the seismic demand model for the pole-type structure with the restrained rocking design. It was found that the structure's response exhibited a minor correlation with spectral acceleration but was sensitive to peak ground velocity and peak displacement demand. This finding underscores the importance of considering the velocity spectrum when selecting input motions in seismic design.

Author Contributions: Conceptualization, S.L. and Z.L.; methodology S.L., Y.H. and B.S.; validation, S.L.; formal analysis S.L., Y.H., B.S. and G.H.; writing—original draft, S.L.; writing—review and editing, Y.H., B.S. and G.H.; project administration, Z.L.; supervision, G.H.; funding acquisition, S.L. and Z.L. All authors have read and agreed to the published version of the manuscript.

Funding: This research was funded by the University of Science and Technology Beijing, project number: TZB20230101.

Data Availability Statement: Data are contained within the article.

Acknowledgments: The tests in the study were organized by the China Electric Power Research Institute.

Conflicts of Interest: The authors declare no conflicts of interest.

References

1. Siringoringo, D.M.; Fujino, Y.; Nagasaki, A.; Matsubara, T. Seismic Performance Evaluation of Existing Light Poles on Elevated Highway Bridges. *Struct. Infrastruct. Eng.* **2021**, *17*, 649–663. [CrossRef]
2. Baghmisheh, A.G.; Mahsuli, M. Seismic Performance and Fragility Analysis of Power Distribution Concrete Poles. *Soil Dyn. Earthq. Eng.* **2021**, *150*, 106909. [CrossRef]
3. Ibrahim, A.M.; El Damatty, A.A.; El Ansary, A.M. Finite Element Modelling of Pre-Stressed Concrete Poles under Downbursts and Tornadoes. *Eng. Struct.* **2017**, *153*, 370–382. [CrossRef]
4. Stephens, M.; Xu, Z.; Whittaker, C.; Wotherspoon, L. Vulnerability of Power Distribution Utility Poles to Tsunami Bore Impacts. *J. Coast. Hydraul. Struct.* **2023**, *3*. [CrossRef]
5. Ju, Y.; Zhao, J.; Wang, D.; Song, Y. Experimental Study on Flexural Behaviour of Reinforced Reactive Powder Concrete Pole. *Constr. Build. Mater.* **2021**, *312*, 125399. [CrossRef]
6. Sasaki, T.; Nozawa, S.; Tsukishima, D.; Kaneko, A. Earthquake Damage to Concrete Utility Poles for Shinkansen and Remedial Measure. *Concr. J.* **2015**, *53*, 622–628. [CrossRef]
7. Li, S.; Tsang, H.-H.; Cheng, Y.; Lu, Z. Effects of Sheds and Cemented Joints on Seismic Modelling of Cylindrical Porcelain Electrical Equipment in Substations. *Earthq. Struct.* **2017**, *12*, 55–65. [CrossRef]
8. Xue, Y.; Cheng, Y.; Zhu, Z.; Li, S.; Liu, Z.; Guo, H.; Zhang, S. Study on Seismic Performance of Porcelain Pillar Electrical Equipment Based on Nonlinear Dynamic Theory. *Adv. Civ. Eng.* **2021**, *2021*, 8816322. [CrossRef]
9. Takhirov, S.; Schiff, A.; Kempner, L.; Fujisaki, E. Breaking Strength of Porcelain Insulator Sections Subjected to Cyclic Loading. In *TCLEE 2009: Lifeline Earthquake Engineering in a Multihazard Environment*; American Society of Civil Engineers: Reston, VA, USA, 2009; pp. 1–12.
10. He, C.; Xie, Q.; Yang, Z.; Xue, S. Seismic Evaluation and Analysis of 1100-KVUHV Porcelain Transformer Bushings. *Soil Dyn. Earthq. Eng.* **2019**, *123*, 498–512. [CrossRef]
11. Ma, G.L.; Xie, Q.; Whittaker, A.S. Physical and Numerical Simulations of the Seismic Response of a 1,100 KV Power Transformer Bushing. *Earthq. Spectra* **2018**, *34*, 1515–1541. [CrossRef]
12. Li, S.; Tsang, H.-H.; Cheng, Y.; Lu, Z. Considering Seismic Interaction Effects in Designing Steel Supporting Structure for Surge Arrester. *J. Constr. Steel Res.* **2017**, *132*, 151–163. [CrossRef]
13. Li, S.; Cheng, Y.; Lu, Z.; Lam, N.; Xue, Y.; Wang, H. Full-Scale Testing on Seismic Performance of Surge Arrester with Retrofitted Composite Insulators. *Buildings* **2022**, *12*, 1720. [CrossRef]
14. Epackachi, S.; Dolatshahi, K.M.; Oliveto, N.D.; Reinhorn, A.M. Mechanical Behavior of Electrical Hollow Composite Post Insulators: Experimental and Analytical Study. *Eng. Struct.* **2015**, *93*, 129–141. [CrossRef]
15. Moustafa, M.A.; Mosalam, K.M. Structural Performance of Porcelain and Polymer Post Insulators in High Voltage Electrical Switches. *J. Perform. Constr. Facil.* **2016**, *30*, 04016002. [CrossRef]
16. Yu, Y.; Li, G.; Li, P.; Zhu, Q. Investigation and Analysis of Electric Equipment Damage in Sichuan Power Grid Caused by Wenchuan Earthquake. *Power Syst. Technol.* **2008**, *32*, T1.
17. Wang, Z. A Preliminary Report on the Great Wenchuan Earthquake. *Earthq. Eng. Eng. Vib.* **2008**, *7*, 225–234. [CrossRef]
18. Kwasinski, A.; Eidinger, J.; Tang, A.; Tudo-Bornarel, C. Performance of Electric Power Systems in the 2010-2011 Christchurch, New Zealand, Earthquake Sequence. *Earthq. Spectra* **2014**, *30*, 205–230. [CrossRef]
19. Mimura, N.; Yasuhara, K.; Kawagoe, S.; Yokoki, H.; Kazama, S. Damage from the Great East Japan Earthquake and Tsunami—A Quick Report. *Mitig. Adapt. Strateg. Glob. Chang.* **2011**, *16*, 803–818. [CrossRef]
20. Adachi, T.; Ellingwood, B.R. Serviceability of Earthquake-Damaged Water Systems: Effects of Electrical Power Availability and Power Backup Systems on System Vulnerability. *Reliab. Eng. Syst. Saf.* **2008**, *93*, 78–88. [CrossRef]
21. Salman, A.M.; Li, Y. A Probabilistic Framework for Seismic Risk Assessment of Electric Power Systems. *Procedia Eng.* **2017**, *199*, 1187–1192. [CrossRef]
22. Ang, H.S.; Pires, J.A.; Villaverde, R. A Model for the Seismic Reliability Assessment of Electric Power Transmission Systems. *Reliab. Eng. Syst. Saf.* **1996**, *51*, 7–22. [CrossRef]
23. Cheng, K.; Wang, Y.; Fang, H.; Qian, C.; Wu, P. Experimental Investigation and Prediction for Bending Creep of Glass Fiber-Reinforced Polymer Pultruded Tube. *Buildings* **2023**, *13*, 2714. [CrossRef]
24. Li, S.; Tsang, H.-H.; Cheng, Y.; Lu, Z. Seismic Testing and Modeling of Cylindrical Electrical Equipment with GFRP Composite Insulators. *Compos. Struct.* **2018**, *194*, 454–467. [CrossRef]
25. Xiao, J. Achievement of Electrical Technology in Sichuan Wenchuan Earthquake. *Electr. Age* **2008**, *12*, 32–34.
26. Gökçe, T.; Yüksel, E.; Orakdöğen, E. Seismic Performance Enhancement of High-Voltage Post Insulators by a Polyurethane Spring Isolation Device. *Bull. Earthq. Eng.* **2019**, *17*, 1739–1762. [CrossRef]
27. Xie, Q.; Shi, G.; Liu, Y. Influence of Lumped Mass and Rotary Inertia on Seismic Isolated Post Equipment. *J. Constr. Steel Res.* **2022**, *199*, 107604. [CrossRef]
28. Cheng, Y.; Li, S.; Lu, Z.; Liu, Z.; Zhu, Z. Seismic Risk Mitigation of Cylindrical Electrical Equipment with a Novel Isolation Device. *Soil Dyn. Earthq. Eng.* **2018**, *111*, 41–52. [CrossRef]

29. Liu, Z.; Zhang, L.; Cheng, Y.; Lu, Z.; Zhu, Z. Seismic Performance Improvement Using Bolt-on Isolators on Interconnected Slender Electrical Equipment. *Eng. Struct.* **2023**, *289*, 116238. [CrossRef]
30. Yang, Z.; Xie, Q.; He, C.; Xue, S. Numerical Investigation of the Seismic Response of a UHV Composite Bypass Switch Retrofitted with Wire Rope Isolators. *Earthq. Eng. Eng. Vib.* **2021**, *20*, 275–290. [CrossRef]
31. Alessandri, S.; Giannini, R.; Paolacci, F.; Malena, M. Seismic Retrofitting of an HV Circuit Breaker Using Base Isolation with Wire Ropes. Part 1: Preliminary Tests and Analyses. *Eng. Struct.* **2015**, *98*, 251–262. [CrossRef]
32. Alessandri, S.; Giannini, R.; Paolacci, F.; Amoretti, M.; Freddo, A. Seismic Retrofitting of an HV Circuit Breaker Using Base Isolation with Wire Ropes. Part 2: Shaking-Table Test Validation. *Eng. Struct.* **2015**, *98*, 263–274. [CrossRef]
33. Housner, G.W. The Behavior of Inverted Pendulum Structures during Earthquakes. *Bull. Seismol. Soc. Am.* **1963**, *53*, 403–417. [CrossRef]
34. Kavvadias, I.E.; Vasiliadis, L.K.; Elenas, A. Seismic Response Parametric Study of Ancient Rocking Columns. *Int. J. Archit. Herit.* **2017**, *11*, 791–804. [CrossRef]
35. Papadopoulos, K.; Vintzileou, E.; Psycharis, I.N. Finite Element Analysis of the Seismic Response of Ancient Columns. *Earthq. Eng. Struct. Dyn.* **2019**, *48*, 1432–1450. [CrossRef]
36. Grigorian, M.; Grigorian, C. An Introduction to the Structural Design of Rocking Wall-Frames with a View to Collapse Prevention, Self-Alignment and Repairability. *Struct. Des. Tall Spec. Build.* **2016**, *25*, 93–111. [CrossRef]
37. Casapulla, C.; Giresini, L.; Lourenço, P.B. Rocking and Kinematic Approaches for Rigid Block Analysis of Masonry Walls: State of the Art and Recent Developments. *Buildings* **2017**, *7*, 69. [CrossRef]
38. Aghagholizadeh, M.; Makris, N. Response Analysis of Yielding Structures Coupled to Rocking Walls with Supplemental Damping. *Earthq. Eng. Struct. Dyn.* **2021**, *50*, 2672–2689. [CrossRef]
39. Xiang, M.; Xiong, F.; Lu, Y.; Ge, Q.; Yan, H.; Ran, M. Structural Displacement Ratios for Seismic Evaluation of Structures on Rocking Shallow Foundations. *Buildings* **2022**, *12*, 174. [CrossRef]
40. Giouvanidis, A.I.; Dong, Y. Seismic Loss and Resilience Assessment of Single-Column Rocking Bridges. *Bull. Earthq. Eng.* **2020**, *18*, 4481–4513. [CrossRef]
41. Piras, S.; Palermo, A.; Saiid Saiidi, M. State-of-the-Art of Posttensioned Rocking Bridge Substructure Systems. *J. Bridge Eng.* **2022**, *27*, 03122001. [CrossRef]
42. Pollino, M.; Bruneau, M. Dynamic Seismic Response of Controlled Rocking Bridge Steel-Truss Piers. *Eng. Struct.* **2008**, *30*, 1667–1676. [CrossRef]
43. Agalianos, A.; Psychari, A.; Vassiliou, M.F.; Stojadinovic, B.; Anastasopoulos, I. Comparative Assessment of Two Rocking Isolation Techniques for a Motorway Overpass Bridge. *Front. Built Environ.* **2017**, *3*, 47. [CrossRef]
44. Thiers-Moggia, R.; Málaga-Chuquitaype, C. Seismic Protection of Rocking Structures with Inerters. *Earthq. Eng. Struct. Dyn.* **2019**, *48*, 528–547. [CrossRef]
45. Li, S.; Tsang, H.-H.; Lam, N. Seismic Protection by Rocking with Superelastic Tendon Restraint. *Earthq. Eng. Struct. Dyn.* **2022**, *51*, 1718–1737. [CrossRef]
46. Dimitrakopoulos, I.; Dejong, M. Seismic Overturning of Damped Rocking Structures. 2011. Available online: https://repository.hkust.edu.hk/ir/Record/1783.1-52590 (accessed on 15 January 2023).
47. Pan, X.; Málaga-Chuquitaype, C. Seismic Control of Rocking Structures via External Resonators. *Earthq. Eng. Struct. Dyn.* **2020**, *49*, 1180–1196. [CrossRef]
48. Acikgoz, S.; Ma, Q.; Palermo, A.; DeJong, M.J. Experimental Identification of the Dynamic Characteristics of a Flexible Rocking Structure. *J. Earthq. Eng.* **2016**, *20*, 1199–1221. [CrossRef]
49. Acikgoz, S.; Dejong, M.J. The Interaction of Elasticity and Rocking in Flexible Structures Allowed to Uplift. *Earthq. Eng. Struct. Dyn.* **2012**, *41*, 2177–2194. [CrossRef]
50. Vassiliou, M.F.; Mackie, K.R.; Stojadinović, B. Dynamic Response Analysis of Solitary Flexible Rocking Bodies: Modeling and Behavior under Pulse-like Ground Excitation. *Earthq. Eng. Struct. Dyn.* **2014**, *43*, 1463–1481. [CrossRef]
51. Reggiani Manzo, N.; Vassiliou, M.F. Displacement-Based Analysis and Design of Rocking Structures. *Earthq. Eng. Struct. Dyn.* **2019**, *48*, 1613–1629. [CrossRef]
52. Vassiliou, M.F.; Truniger, R.; Stojadinović, B. An Analytical Model of a Deformable Cantilever Structure Rocking on a Rigid Surface: Development and Verification. *Earthq. Eng. Struct. Dyn.* **2015**, *44*, 2775–2794. [CrossRef]
53. Chopra, A.K. *Dynamics of Structures-Theory and Applications to Earthquake Engineering*, 4th ed.; Prentice Hall: Hoboken, NJ, USA, 2004.
54. Martelli, A.; Forni, M. Seismic Isolation and Other Anti-Seismic Systems Recent Applications in Italy and Worldwide. *J. Anti-Seism. Syst. Int. Soc.* **2010**, *1*, 75–123.
55. D'Amato, M.; Gigliotti, R.; Laguardia, R. Seismic Isolation for Protecting Historical Buildings: A Case Study. *Front. Built Environ.* **2019**, *5*, 87. [CrossRef]
56. Li, S.; Tsang, H.-H.; Lam, N. Seismic Internal Force in Rocking Shear Frame with Superelastic Tendon Restraint. *Earthq. Eng. Struct. Dyn.* **2023**, *52*, 2272–2293. [CrossRef]
57. Morin, D.J. *Introduction to Classical Mechanics: With Problems and Solutions*; Cambridge University Press: Cambridge, UK, 2008; ISBN 9780521876223.

58. Acikgoz, S.; DeJong, M.J. Analytical Modelling of Multi-Mass Flexible Rocking Structures. *Earthq. Eng. Struct. Dyn.* **2016**, *45*, 2103–2122. [CrossRef]
59. *GB 50260-2013*; Code for Seismic Design of Electrical Installations. Ministry of Housing and Urban-Rural Development: Beijing, China, 2013.
60. PEER Pacific Earthquake Engineering Research Center. Available online: http://peer.berkeley.edu/products/strong_ground_motion_db.html (accessed on 28 January 2021).
61. Lachanas, C.G.; Vamvatsikos, D.; Dimitrakopoulos, E.G. Intensity Measures as Interfacing Variables versus Response Proxies: The Case of Rigid Rocking Blocks. *Earthq. Eng. Struct. Dyn.* **2023**, *52*, 1722–1739. [CrossRef]
62. Bantilas, K.E.; Kavvadias, I.E.; Tyrtaiou, M.; Elenas, A. Hilbert–Huang Transform Based Seismic Intensity Measures for Rocking Response Assessment. *Appl. Sci.* **2023**, *13*, 1634. [CrossRef]
63. Bantilas, K.E.; Kavvadias, I.E.; Vasiliadis, L.K.; Elenas, A. Seismic Fragility and Intensity Measure Investigation for Rocking Podium Structures under Synthetic Pulse-like Excitations. *Earthq. Eng. Struct. Dyn.* **2021**, *50*, 3441–3459. [CrossRef]

buildings

MDPI

Article

Study on the Damping Efficiency of a Structure with Additional Viscous Dampers Based on the Shaking Table Test

Xiang Lan [1,2], Longfei Zhang [1,2,*], Baifeng Sun [3,4,*] and Wen Pan [5]

1 College of Architecture and Civil Engineering, Kunming University, Kunming 650214, China; lx523947015@kmu.edu.cn
2 Institute of Engineering Earthquake Resistance and Disaster Reduction, Kunming University, Kunming 650214, China
3 China-Pakistan Belt and Road Joint Laboratory on Smart Disaster Prevention of Major Infrastructures, Ministry of Science and Technology, Nanjing 211189, China
4 Yunnan Design Institute Group Co., Ltd., Kunming 650238, China
5 Faculty of Civil Engineering and Mechanics, Kunming University of Science and Technology, Kunming 650051, China
* Correspondence: zhanglongfei08052003@kmu.edu.cn (L.Z.); sunbaifeng@ydi.cn (B.S.)

Abstract: This study specifically focuses on the damping efficiency of a damped structure with additional viscous dampers. A two-layer steel frame structure with eight sets of viscous dampers is used to conduct a series of seismic simulation shaking table tests, including a non-damped structure without dampers and two damped structures with dampers placed at 1/2 and 1/6 of the beam span, respectively. By conducting these tests, the energy dissipation, force, and displacement of the damper, as well as the parameters of the structure such as floor displacement and acceleration, are obtained. The main damping efficiency indicators of the damped structure are calculated, including the additional damping ratio, inter-story displacement utilization rate, as well as the reduction rate of the vertex displacement and the base shear relative to the non-damped structure. The study shows that the viscous dampers exhibit full hysteresis loops and a strong energy dissipation capacity in the structure. The seismic response of the vertex displacement and base shear in the damped structure is significantly smaller than that in the non-damped structure. Under different seismic levels, including frequent earthquakes, occasional earthquakes, and rare earthquakes, the damping effect of the dampers placed at 1/2 of the beam span is significantly better than that placed at 1/6 of the beam span. For example, the additional damping ratio for the *X*-direction artificial wave REN is 19% and 11%, 20% and 13%, and 13% and 11%, respectively. The patterns for inter-story displacement utilization ratio, reduction rate of the vertex displacement, and reduction rate of the base shear are similar. The research findings strongly indicate that the damped structure with additional viscous dampers exhibits excellent damping efficiency. In future damping design, designers need to fully consider the placement of viscous dampers within the beam span.

Keywords: energy dissipation structure; viscous damper; placement arrangement; damping efficiency; shaking table test

1. Introduction

Energy dissipation structures are designed to reduce the impact of seismic or other external forces on a building by using a series of energy dissipation systems. These systems can absorb and dissipate energy, effectively reducing the vibration response of the structure and preventing or minimizing structural damage [1,2]. There are four main types of energy dissipation control: active, semi-active, passive, and intelligent [3,4]. Passive control devices are further classified into velocity-dependent dampers, displacement-dependent dampers, and hybrid dampers based on their relationship with velocity and displacement. They can also be categorized as rate-dependent or rate-independent dampers based on their dependence on velocity [5,6]. Viscous dampers [7–9] are the most commonly used velocity-based dissipative devices that utilize the damping characteristics of viscous materials to reduce structural vibration response. They generate a damping force when subjected to external forces, which can absorb and dissipate energy. Compared to other types of dampers, viscous dampers have several advantages, including a higher additional damping ratio, stable and reliable damping characteristics within the frequency response range, and relatively simple structures resulting in low installation and maintenance costs. As a result, viscous dampers have been widely applied in practical engineering. There are various installation methods for viscous dampers in damped structures, including supported and cantilever wall types. The supported type includes single diagonal support, V-shaped support, and inverted V-shaped support. However, from the perspective of functional use in building structures, designers and construction units prefer viscous dampers arranged in a cantilever wall type as they do not obstruct doors and windows and they minimize the impact on the building's functionality.

To optimize the implementation of viscous dampers in damped structures, numerous researchers have conducted thorough studies on the damping efficacy of various types of damped structures. Hwang et al. [10] proposed a new calculation formula to accurately predict the additional damping ratio of mid-high-rise buildings. Xie et al. [11] studied the effectiveness and optimization design of non-linear viscous dampers in inelastic structures. Gherbi et al. [12] minimized the design time by estimating the damping coefficient of non-linear fluid viscous dampers. The calculation methods for the additional damping ratio have been a hot topic of research. There are non-dimensional parameter methods based on non-linear viscous dampers [13], simplified calculation methods based on structural characteristics, excitation frequency, and damper parameters [14], as well as non-classical modal damping ratio calculation methods based on the known state-space representation of the dynamic system [15]. The seismic performance of structures with additional viscous dampers has been significantly improved [16]; therefore, viscous dampers have also been applied in steel structures [17], super-tall building structures [18], and even in structural retrofitting [19,20].

As is well known, seismic excitation is essentially a random vibration. Tubaldi et al. [21,22] investigated the probabilistic response assessment and non-stationary stochastic response of structures. Zhang et al. [23] proposed a structural analysis method based on Lyapunov differential equations. Su et al. [24] explored the optimal design of non-linear viscous dampers for large-scale structures under non-stationary seismic excitation, employing a stochastic approach. Rodolfo et al. [25] conducted a probabilistic evaluation of high-rise buildings' dynamic response using transfer functions. Experimental research on the damped structures can be conducted using pseudo-static tests [26]. Alternatively, a more direct approach is to utilize a seismic simulation shaking table for dynamic testing [27,28]. Nevertheless, there have been only a few shaking table tests that have specifically focused on the damped structure with additional viscous dampers and the non-damped structure without any additional viscous dampers.

Scholars have also conducted extensive research on the damping efficiency of optimal control and the most effective placement of dampers in damped structures. Singh et al. [29] suggested a gradient optimization technique, utilizing inter-story displacement, base shear, or floor acceleration as performance indicators. Aguirre et al. [30] evaluated and compared the optimal distribution of spatial dampers for linear and non-linear systems. Parcianello et al. [31] studied the placement of dampers aiming to reduce the displacement and acceleration of frame structures under seismic action. Whittle et al. [32] investigated the impact of five distinct configurations of viscous dampers on the seismic performance of structures. In order to achieve relatively uniform parameters for each damper size in the structure, Lopez [33] proposed a simplified algorithm based on sequential search. Takewaki [34,35], Lin et al. [36], and Aydin et al. [37] conducted research on the optimal placement of dampers considering the minimum transfer function, critical excitation, damping coefficients, and seismic rehabilitation. However, no studies have been found on the damping efficiency of damped structures when dampers are placed at different positions within the same beam span.

In the field of engineering, it is common to reveal the laws of things through the in-depth study of specific phenomena. This process usually begins with observing the phenomena, followed by establishing the corresponding theoretical models through logical reasoning and theoretical analysis. Numerical simulation methods, such as finite element analysis, are then used to verify the related laws, and appropriate experiments are designed for further validation. Finally, the results are modified based on feedback from actual engineering applications to achieve higher applicability accuracy. The main objective of this experiment is to compare and analyze the experimental results of the structure with and without the attached viscous damper. The vibration frequency and velocity under actual working conditions are simulated, and the response of the damper is measured to evaluate its damping effect. This study examines the damping efficiency of structures equipped with viscous dampers and investigates how damper placement affects the damping efficiency across beam spans. The analysis focuses on key damping efficiency indicators, including the additional damping ratio, reduction in the vertex displacement, reduction in the base shear, and inter-story displacement utilization rate. The findings of this research will serve as valuable guidance and reference for design professionals when incorporating viscous dampers into future damped structure designs.

2. Shaking Table Test Design

2.1. Test Equipment

The viscous damper is a type of damper that is dependent on velocity. When used in building construction projects, its frequency typically falls between 1 and 3 Hz, and the velocity range is usually between 100 mm/s and 600 mm/s. Therefore, traditional quasi-static tests or static tests are not suitable for evaluating the performance of the structure. This experiment utilized a dynamic loading system based on the seismic simulation shaking table platform at the Engineering Seismic Research Institute of Kunming University of Science and Technology. The damping effect on structures was evaluated through the loading of earthquake waves onto structural systems. The fundamental parameters of the shaking table are presented in Table 1.

Table 1. Basic parameters of the shaking table.

Projects	Basic Parameters
Tabletop dimensions	4 m × 4 m
Maximum load capacity	30 tons
Vibration direction	Bi-directional three degrees of freedom (*X* and *Y* translational motion, and *Z* rotational motion)
Vibration waveforms	Various types of waves, including regular waves, random waves, and simulated seismic waves.
The maximum acceleration	±1.0 g (Payload of 20 tons), ±0.8 g (Payload of 30 tons)
Maximum velocity	800 mm/s
Maximum displacement	±125 mm
Frequency range	0.1 Hz~100 Hz

2.2. Selection of Materials and Their Material Property Testing

The steel used was conventional Q235 H-section steel commonly found on the market. The H-section steel consisted of different cross-section sizes from the same manufacturer and batch. The three main types were HW100 × 100, HW125 × 125, and HW175 × 175, with actual dimensions shown in Table 2. According to the standard of "Metallic materials-Tensile testing-Part 1: Method of test at room temperature" [38], samples were taken from the web position of the H-section steel as shown in Figure 1a. Then, the specimens were prepared as shown in Figure 1b for tensile testing, and the values of Young's modulus, yield strength, tensile strength, and elongation at fracture of the specimens were measured, as presented in Table 3.

Table 2. Actual dimensions of various types of H-section steel profiles.

Types	Width b/mm	Height h/mm	The Thickness of the Web Plate/mm	The Thickness of the Flange/mm
HW100 × 100	100	100	5.3	7.0
HW125 × 125	125	125	6.0	8.0
HW175 × 175	175	175	6.4	9.3

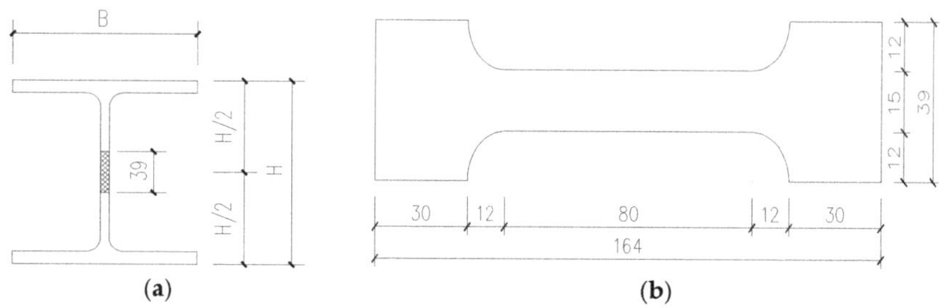

Figure 1. Sample preparation and design diagram for material testing specimens. (**a**) Sampling in the shadowed area. (**b**) Test specimen design diagram.

Table 3. Test results of specimen material properties.

Types	Test Specimen Number	Measured Yield Strength f_y/MPa	Measured Tensile Strength f_u/MPa	$f_u/f_y \geq 1.25$	$f_y/235 \leq 1.30$	Young's Modulus E/GPa	Elongation at Fracture δ/%
HW 100 × 100	No.1	263	376	1.43	1.12	195	31
	No.2	264	378	1.43	1.12	193	27
	No.3	256	375	1.47	1.09	190	31
	Average	261	376	1.44	1.11	193	30
HW 125 × 125	No.1	262	429	1.64	1.11	202	28
	No.2	261	430	1.65	1.11	203	27
	No.3	261	425	1.63	1.11	204	27
	Average	261	428	1.64	1.11	203	27
HW 175 × 175	No.1	260	424	1.63	1.10	204	27
	No.2	258	419	1.62	1.10	212	26
	No.3	262	427	1.63	1.12	209	28
	Average	260	423	1.63	1.11	208	27

The specimens were tested for tensile strength using an electronic universal testing machine with a capacity of 100 kN, as illustrated in Figure 2.

From Table 3, it can be observed that the material test results for the three types of H-section steel specimens in the experiment meet all the specified criteria.

2.3. Experimental Model Fabrication

In order to achieve the purpose of this experiment, a two-story single-span steel frame structure system model was constructed. The height of each floor was 3.6 m, and the span in both the X- and Y-directions was 3.0 m. The cross-section dimensions of the four columns were all HW175 × 175, and the column bases were connected by welding. The cross-section dimensions of the beams in the strong axis direction of the columns (X-direction) were HW100 × 100, and the cross-section dimensions of the beams in the weak axis direction of the columns (*Y*-direction) were HW125 × 125. The beams and columns were connected by "T" type keys with bolt connections. The damper was connected to the beam through bolts, with the upper and lower columns of the intermediate column. The upper and lower columns of the intermediate column were made of steel plates with a width of 400 mm and a thickness of 10 mm. The out-of-plane stiffness was satisfied by structural measures. In order to connect the lower column of the intermediate column, a ground beam layer was specifically designed in the model. The elevation from the apex of the ground beam to the base of the column measured 0.40 m, as depicted in Figure 3. The connection configuration between the beam–column joints and the damper installation method are clearly demonstrated. The shaking table test models are presented in Figure 4; specifically, Figure 4a illustrates the undamped structural model without dampers, while Figure 4b portrays the damping structural model equipped with dampers. The model weight was determined by the combination of the self-weight of components and additional counterweights. According to the requirement of 3.0 kN/m^2, the counterweights were uniformly arranged on the floor slabs of the model. The final total weight of the model was 6.7 tons.

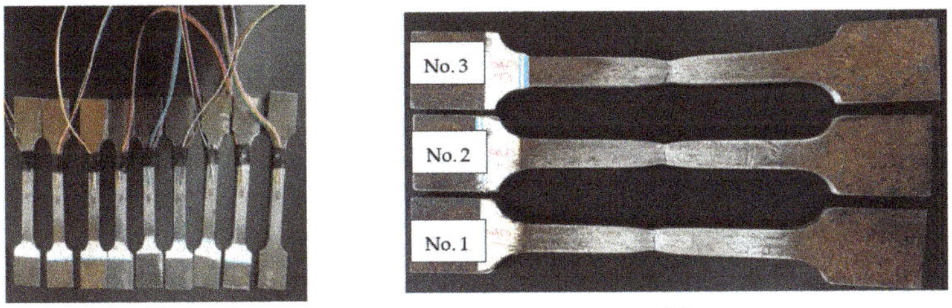

(a) (b)

Figure 2. Comparison of test specimens before and after tensile testing. (**a**) Before the test. (**b**) After the experiment.

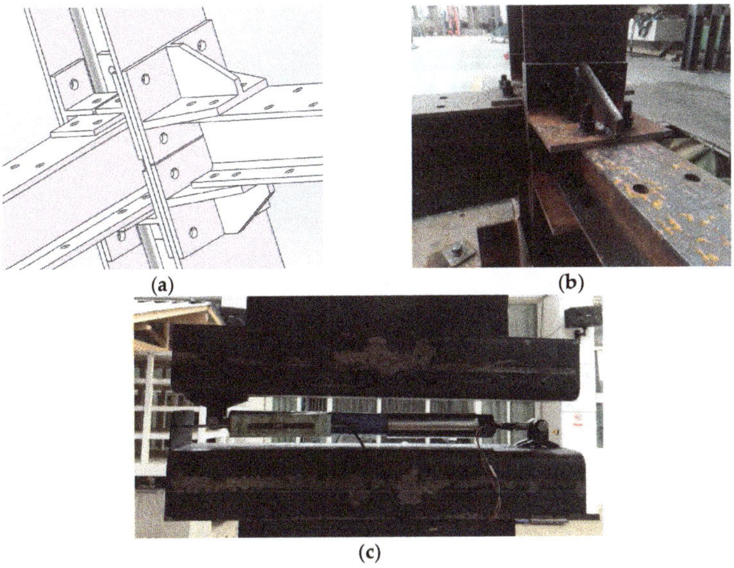

(a) (b) (c)

Figure 3. Diagram of connection of beam–column joints and installation of dampers. (**a**) Assembly opening and positioning diagram of beam–column joints. (**b**) Assembly physical diagram of beam–column joints. (**c**) Physical diagram of damping device installation.

(**a**) (**b**)

Figure 4. Seismic simulation shaking table test models. (**a**) Non-damping structural model. (**b**) Damping structural model. The letters *E*, *S*, *W*, and *N* in the figure, respectively, represent the four cardinal directions: east, south, west, and north.

2.4. Damping Device Design Parameters and Arrangement Positions

The experiment employed eight sets of dampers, with four sets of Type I dampers and four sets of Type II dampers. Five sets of each type of damper were prepared. The design parameters of the dampers mainly included the damping coefficient and damping index. Typically, the total damping coefficient required for a seismic structure is derived based on three parameters: the energy dissipation of the damper, the strain energy of the structure, and the additional damping ratio of the structure. Subsequently, the damping coefficient for each damper is obtained by evenly distributing the total damping coefficient among the dampers. The damping index is usually selected between 0.25 and 0.50 based on practical engineering experience [39]. The final design parameters of the dampers for this experiment are shown in Table 4. The dimensional and physical appearance of the dampers are illustrated in Figure 5.

Table 4. Damping device design parameters.

Types	Damping Coefficient C/kN·(s/mm)$^{\alpha}$	Damping Index α	Design Stroke/mm	Design Load/kN	Quantity/Set	Quantity in Reserve/Set
I	0.4	0.3	±30	2.25	4	1
II	0.3	0.45	±20	3.33	4	1

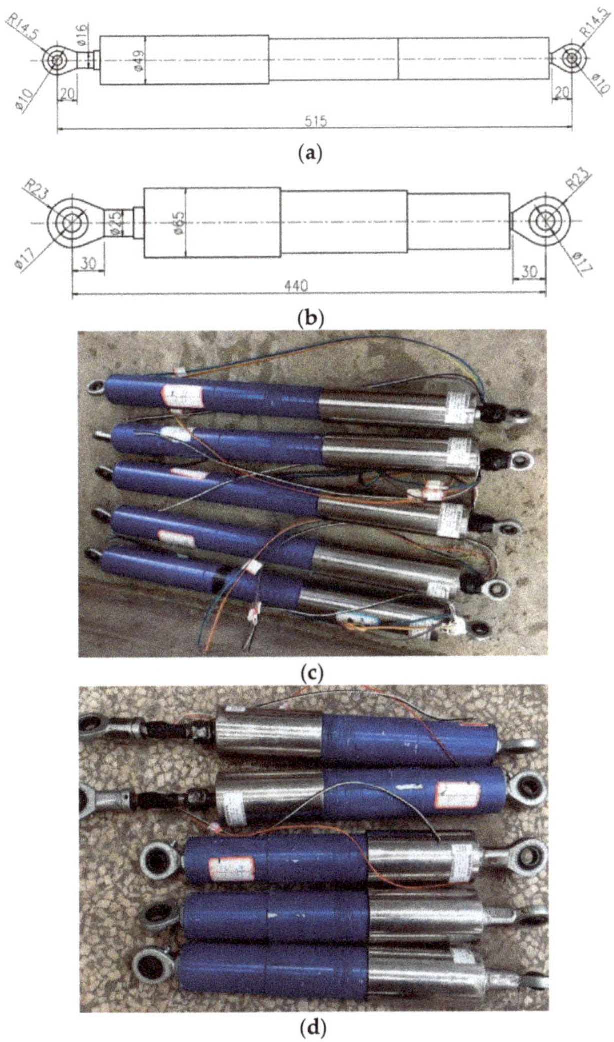

Figure 5. Damping device appearance, structural dimensions, and physical diagram. (**a**) Appearance dimensions of Type I damper. (**b**) Appearance dimensions of Type II damper. (**c**) Physical photo of Type I damper. (**d**) Physical photo of Type II damper.

In accordance with the stipulations of "Technical specification for seismic energy dissipation of buildings" [40], the mechanical performance test of dampers was conducted. The following were the specific steps: (1) Using the sine excitation method, the loading system of the testing machine was controlled by the input displacement $u = u_0\sin(\omega t)$, where u_0 is the designed displacement of the damper and $\omega = 2\pi f_1$ is the circular frequency corresponding to the structural fundamental frequency f_1. (2) Frequency f1 was applied to the damper, with displacement amplitudes of $0.1u_0$, $0.2u_0$, $0.5u_0$, $0.7u_0$, $1.0u_0$, and $1.2u_0$, respectively. Five continuous cycles were completed and the force–displacement hysteresis loops of the damper were plotted. (3) The damping coefficient and damping index corresponding to the third cycle under each operating condition were obtained through regression fitting, and they were taken as the measurement values. Figures 6 and 7 show one representative damper for each of the two types, namely I-4 and II-1. The measured parameters of all dampers are shown in Table 5.

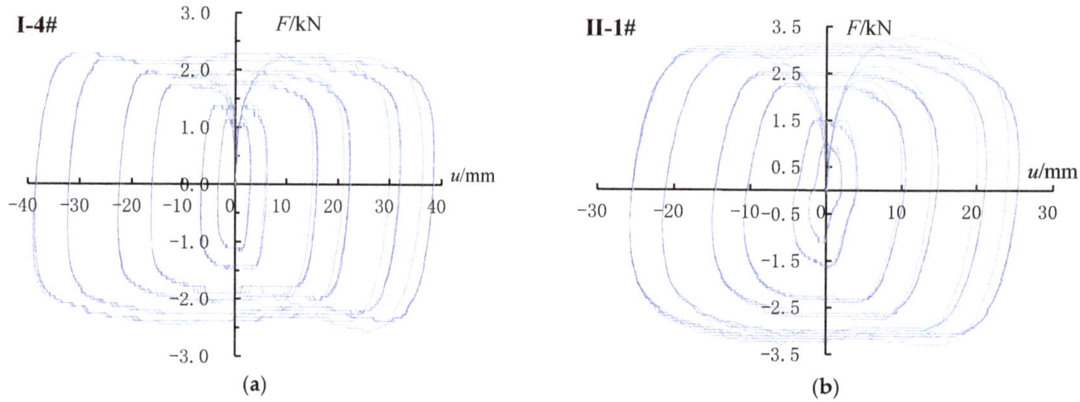

Figure 6. Damping device hysteresis loops measured in practice. (**a**) Hysteresis loops of non-linear viscous damper I-4. (**b**) Hysteresis loops of non-linear viscous damper II-1. Note: "#" in the figure indicates the number, such as I-4# represents No. I-4. The same meaning applies elsewhere in the text.

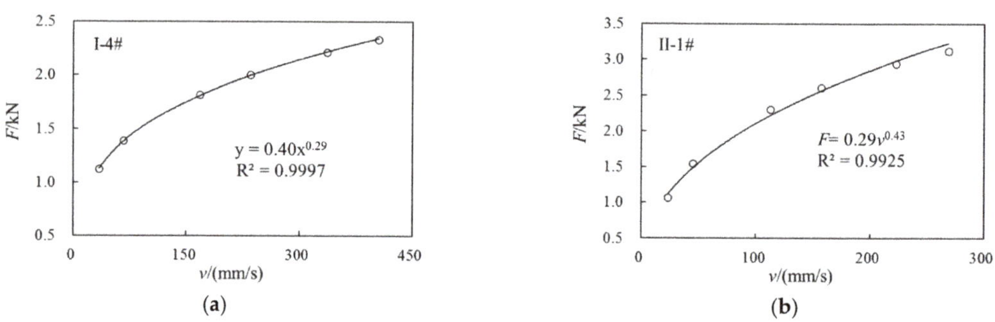

Figure 7. The measured values of the dampers' damping coefficient and index. (**a**) I-4 damper. (**b**) II-1 damper.

Table 5. Damper measured parameters.

Types	Damping Coefficient C	Types	Damping Coefficient C
I-1	0.52 kN/(mm/s)$^{0.28}$	II-1	0.29 kN/(mm/s)$^{0.43}$
I-2	0.20 kN/(mm/s)$^{0.38}$	II-2	0.21 kN/(mm/s)$^{0.49}$
I-3	0.28 kN/(mm/s)$^{0.34}$	II-3	0.37 kN/(mm/s)$^{0.40}$
I-4	0.40 kN/(mm/s)$^{0.29}$	II-4	0.27 kN/(mm/s)$^{0.44}$
I-5	0.36 kN/(mm/s)$^{0.30}$	II-5	0.31 kN/(mm/s)$^{0.42}$

In the experimental model, the dampers in the east–west direction are Type I, with the second floor designated as I-2 and I-4, and the first floor as I-1 and I-5. The dampers in the north–south direction are Type II, with the second floor designated as II-1 and II-3, and the first floor as II-4 and II-2. Figure 8 shows the arrangement of dampers at the 1/2 position within the beam span. We conducted experimental tests on the damped structures corresponding to the damper arrangement at both the 1/2 position and the 1/6 position within the beam span.

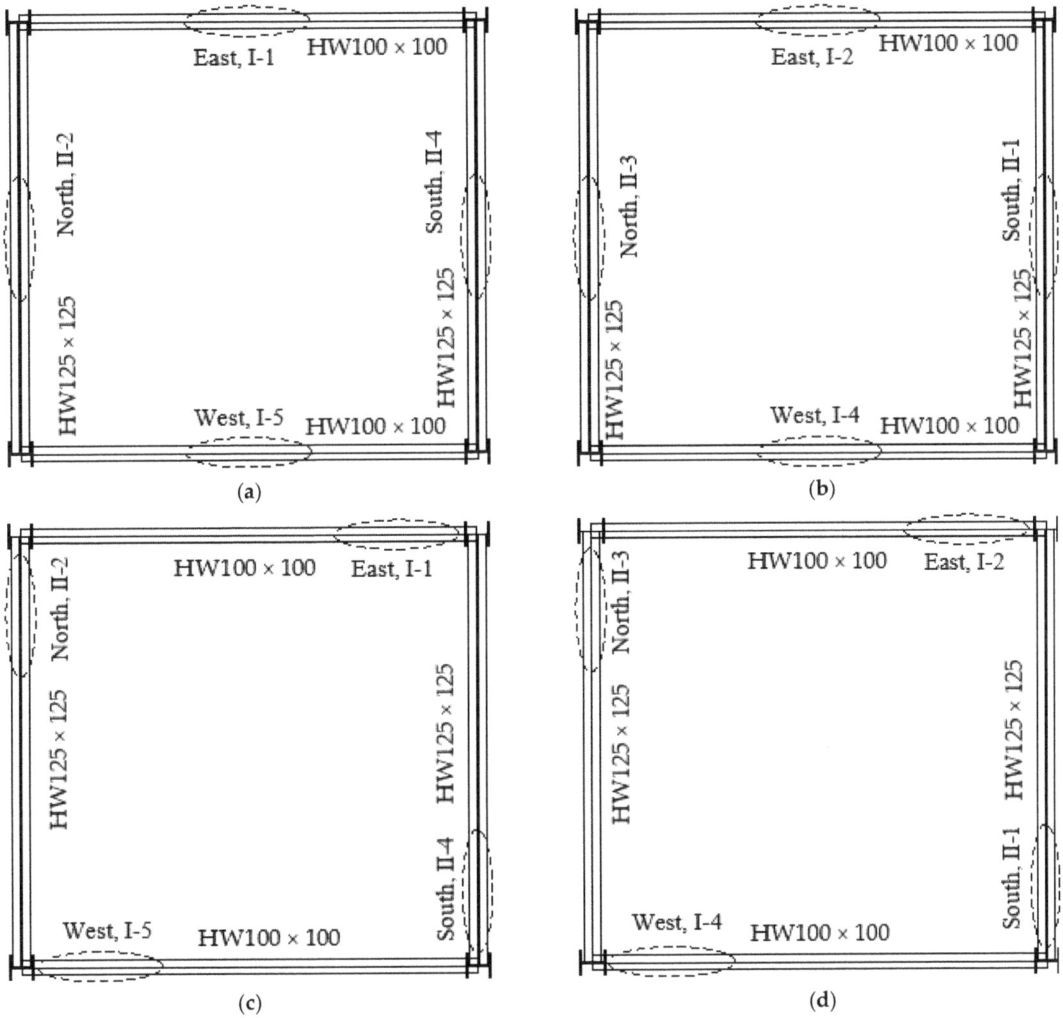

Figure 8. Damping device arrangement position. (**a**) The first layer of dampers is positioned at 1/2 span. (**b**) The second layer of dampers is positioned at 1/2 span. (**c**) The first layer of dampers is positioned at 1/6 span. (**d**) The second layer of dampers is positioned at 1/6 span. Note: The unmarked column sections in the figure are all HW175 × 175. The dotted ellipses denote the placement of viscous damper.

2.5. Sensor Placement

The experiment mainly considered two aspects: the input of seismic excitation and the output of a structural response. In this experiment, control was achieved by inputting iteratively converged displacements into the seismic simulation shaking table, while data collection was performed using acceleration sensors and displacement sensors for the output. The acceleration sensors were used to collect the acceleration values of the table and the floors. On the shaking table surface, one was arranged in the X-direction and another was arranged in the Y-direction. At the position of the ground beam layer, one was arranged in the X-direction and another was arranged in the Y-direction. In the first and second layers of the structure, there were two arrangements in the X-direction and one arrangement in the Y-direction. Therefore, a total of ten acceleration sensors were arranged according to the positions shown in Figure 9a. The primary purpose of arranging displacement sensors was to collect the displacement of the dampers and obtain the displacement time-history curves of the dampers under seismic action, and then plot the hysteresis loops together with the force time-history curves of the dampers. Secondly, the floor displacement of the structure was measured. Two displacement sensors were installed on each floor to measure the inter-story displacements, with one installed in the X-direction and the other in the Y-direction. The top pin of the sensor was placed on the outer surface of the cantilever square steel column. Displacement sensors for measuring damper displacement were installed on both the southern and western floors. Therefore, a total of eight displacement sensors were arranged, as shown in Figure 9b.

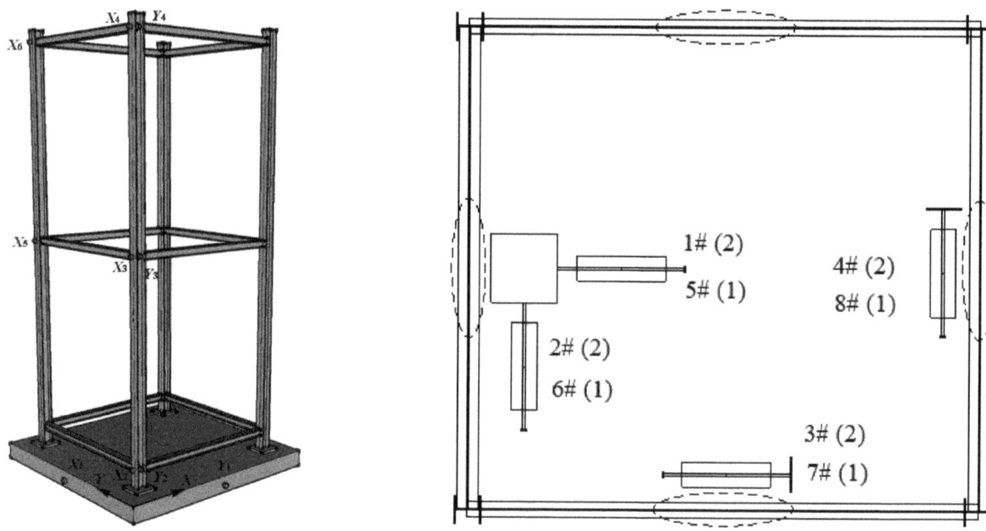

(a) (b)

Figure 9. The placement of displacement and acceleration sensors. (**a**) Positioning of acceleration sensors. (**b**) Positioning of displacement sensors. In (**a**), "X1" represents the first acceleration sensor in the X-direction, "Y1" represents the first acceleration sensor in the Y-direction, and so on. In (**b**), "1#" denotes the position of the first displacement sensor, and the numerical value within parentheses indicates the floor where the displacement sensor is located. The cantilevered square steel column is used to support the top pin of the displacement sensors.

2.6. Experimental Loading Conditions

The experiment considered three seismic levels: frequent earthquakes, occasional earthquakes, and rare earthquakes. The corresponding peak ground accelerations were 0.07 g, 0.20 g, and 0.40 g, respectively. According to the "Code for Seismic Design of Buildings" [41], two natural waves and one artificial wave were selected to input into the seismic simulation shaking table system. The seismic wave records were standardized to peak values, as depicted in Figure 10a. To adjust for various seismic levels, these records were amplified using a scaling factor. However, the time axis of the records remained unaltered. The information derived from the natural wave records is presented in Table 6.

The time-history curves of the seismic waves and their response spectrum curves under the frequent earthquake level are shown in Figure 10b.

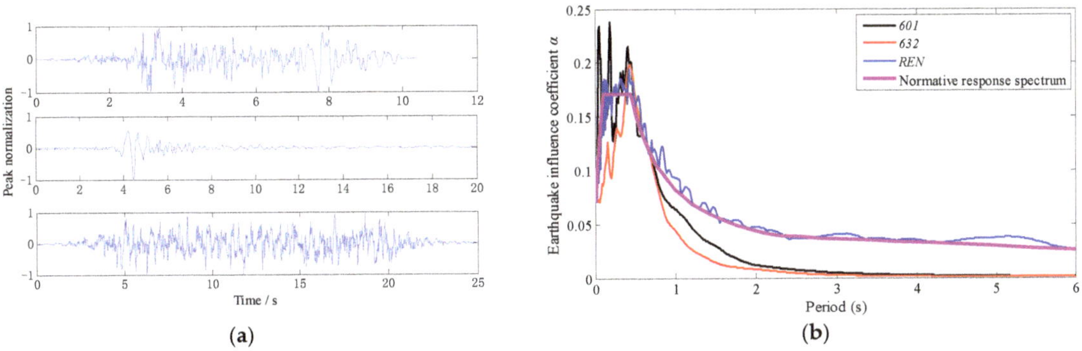

Figure 10. The time-history curves and response spectrum curves of the input seismic waves during the experiment. (**a**) Seismic wave time-history curves. (**b**) Response spectrum curves.

Table 6. Basic information of natural seismic waves.

Numbered	Earthquake Name	Occurrence Time	Recording Station	PGA (m/s^2)
601	Imperial Valley-07	15 October 1979	USGS STATION 412	3.20
632	Northridge-05	17 January 1994	SYLMAR-COUNTY HOSPITAL PARKING LOT	0.90

There were a total of 22 loading conditions in the experimental test, mainly considering three seismic waves, the *X*- and *Y*-axes of the structure, including the levels of frequent earthquakes, occasional earthquakes, and rare earthquakes, as well as white noise conditions, as shown in Table 7. The experiments were divided into three groups: the non-damped structure without additional dampers, the damped structure with additional dampers placed at the 1/2 position within the beam span, and the damped structure with dampers placed at the 1/6 position within the beam span. All three groups of shaking table tests were loaded according to the conditions in Table 7. The three sets of experiments conducted white noise testing under the first working condition and obtained consistent structural frequency results of 2.0 Hz in the X-direction and 1.5 Hz in the *Y*-direction. This indicates that the stiffness along the strong axis (*X*-direction) of the H-shaped steel column section was larger, while the stiffness along the weak axis (*Y*-direction) was smaller. Furthermore, the structural frequency remained unchanged after the fourth white noise test.

Table 7. Experimental loading conditions. W1, W2, W3, and W4 represent the first, second, third, and fourth instances of white noise, respectively. The peak ground accelerations loaded during the tests are considered to be relatively small at 0.07 g.

Operating Condition Numbers	Operating Condition Name	Seismic Levels	Seismic Waves	Direction	PGA/g
1	W1	the first white noise		*XY*	0.07
2	FX-601	Frequent earthquake	601	*X*	0.07
3	FY-601		601	*Y*	0.07
4	FX-632		632	*X*	0.07
5	FY-632		632	*Y*	0.07
6	FX-REN		REN	*X*	0.07
7	FY-REN		REN	*Y*	0.07

Table 7. *Cont.*

Operating Condition Numbers	Operating Condition Name	Seismic Levels	Seismic Waves	Direction	PGA/g
8	W2	the second white noise		*XY*	0.07
9	OX-601	Occasional earthquake	601	*X*	0.20
10	OY-601		601	*Y*	0.20
11	OX-632		632	*X*	0.20
12	OY-632		632	*Y*	0.20
13	OX-REN		REN	*X*	0.20
14	OY-REN		REN	*Y*	0.20
15	W3	the third white noise		*XY*	0.07
16	RX-601	Rare earthquake	601	*X*	0.40
17	RY-601		601	*Y*	0.40
18	RX-632		632	*X*	0.40
19	RY-632		632	*Y*	0.40
20	RX-REN		REN	*X*	0.40
21	RY-REN		REN	*Y*	0.40
22	W4	the fourth white noise		*XY*	0.07

3. Test Results Analysis

3.1. Hysteresis Loops of the Dampers

To facilitate the comparative study of the dampers' energy dissipation effect at various locations, their hysteresis energy dissipation loops were tested and plotted under frequent, occasional, and rare earthquakes. The hysteresis loops of damper No. I-4 were plotted in the X-direction, and the hysteresis loops of damper No. II-1 was plotted in the Y-direction, as shown in Figure 11.

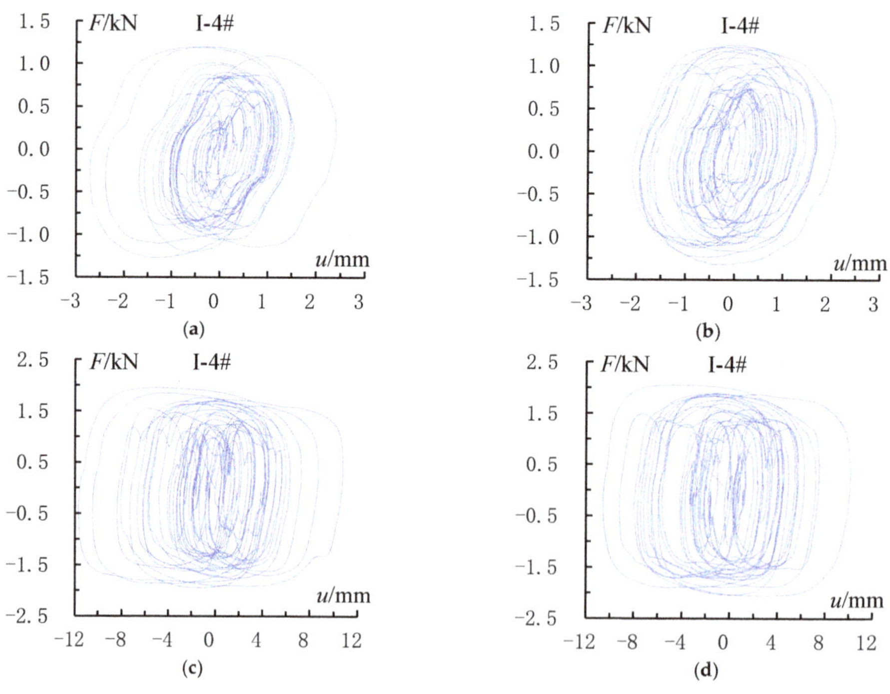

Figure 11. *Cont.*

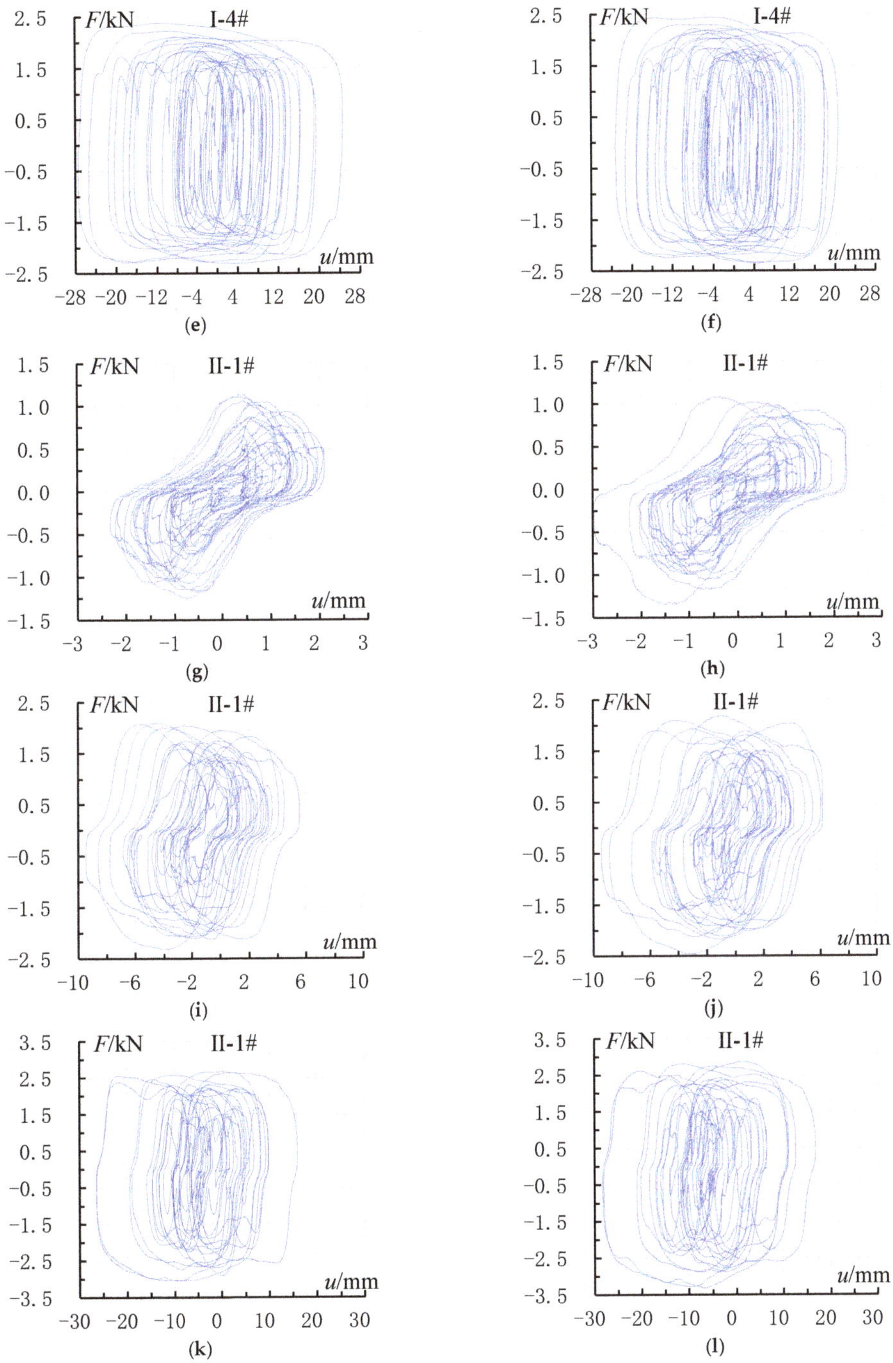

Figure 11. Hysteresis loops of dampers under different arrangement positions and seismic levels. (**a**) Earthquake is frequently encountered at the 1/2 position in the *X*-direction. (**b**) Earthquake is frequently encountered at the 1/6 position in the *X*-direction. (**c**) Earthquake is occasionally encountered

at the 1/2 position in the X-direction. (**d**) Earthquake is occasionally encountered at the 1/6 position in the X-direction. (**e**) Earthquake is rarely encountered at the 1/2 position in tshe X-direction. (**f**) Earthquake is rarely encountered at the 1/6 position in the X-direction. (**g**) Earthquake is frequently encountered at the 1/2 position in the Y-direction. (**h**) Earthquake is frequently encountered at the 1/6 position in the Y-direction. (**i**) Earthquake is occasionally encountered at the 1/2 position in the Y-direction. (**j**) Earthquake is occasionally encountered at the 1/6 position in the Y-direction. (**k**) Earthquake is rarely encountered at the 1/2 position in the Y-direction. (**l**) Earthquake is rarely encountered at the 1/6 position in the Y-direction.

When the damper works under seismic excitation, the degree of fullness of the hysteresis loops and the size of the area reflected in the force and displacement are important indicators to measure the energy dissipation effect of the damper. By studying and comparing the hysteresis loops of the I-4 damper arranged in the X-direction and the II-1 damper arranged in the Y-direction in Figure 11, the following conclusions can be drawn:

(1) The hysteresis loops of the dampers are uniformly full at different seismic levels, indicating strong energy dissipation capacity. With the increase in seismic level, the energy dissipation capacity of the dampers increases rapidly.
(2) The area of the hysteresis loops when the damper is arranged at the 1/2 position within the beam span, i.e., the energy dissipation effect of the damper, is significantly larger than that when the damper is arranged at the 1/6 position within the beam span.
(3) The force and displacement of the I-4 and II-1 dampers under rare earthquakes are in good agreement with the design parameters in Table 4.

3.2. *Additional Damping Ratio*

The additional damping ratio is a crucial parameter for evaluating the damping efficiency of a structural system. Viscous dampers are a prime example of velocity-dependent dampers. In accordance with the prevailing standards in China [40,41], the effective additional damping ratio provided by dampers to the structural system can be calculated using the following formula.

$$\xi_{\mathrm{d}} = \frac{\sum\limits_{j=1}^{n} W_{cj}}{4\pi W_{\mathrm{s}}} \tag{1}$$

where ξ_{d} represents the additional effective damping ratio provided by dampers to the structure; W_{cj} is the energy dissipated by the jth damper in the structure during one cycle of reciprocating motion with an inter-story displacement Δu_j; n is the number of dampers; W_{s} is the total strain energy of the damped structure under the expected displacement.

The structural model's frequency remained constant throughout the initial phase, frequent earthquakes, occasional earthquakes, and rare earthquakes. This suggests that the structure was still in an elastic or near-elastic state. Consequently, the shaking table experiment results can be calculated separately for additional damping ratios in the X- and Y-directions for three seismic waves under frequent, occasional, and rare earthquake conditions using Equation (1), as illustrated in Figure 12.

From Figure 12a, it can be observed that the stiffness of the structure in the X-direction is higher, and its natural frequency is also higher, with a value of 2.0 Hz. The following points can be derived:

(1) The additional damping ratio varies significantly with different locations of damper arrangement.
(2) Regardless of whether the earthquake is frequent, occasional, or rare, the additional damping ratio of the damper arrangement at the 1/2 position within the beam span is significantly greater than that at the 1/6 position within the beam span. For example, for the artificial seismic wave REN, the additional damping ratios are 19% and 11%,

20% and 13%, and 13% and 11% at different seismic levels and different arrangement positions, respectively.

(3) The overall trend is that the additional damping ratio decreases with increasing seismic level, especially when the damper is arranged at the 1/2 position of the beam span. For instance, the additional damping ratios of the natural seismic wave 601 under frequent, occasional, and rare earthquake conditions are 21%, 17%, and 13%, respectively.

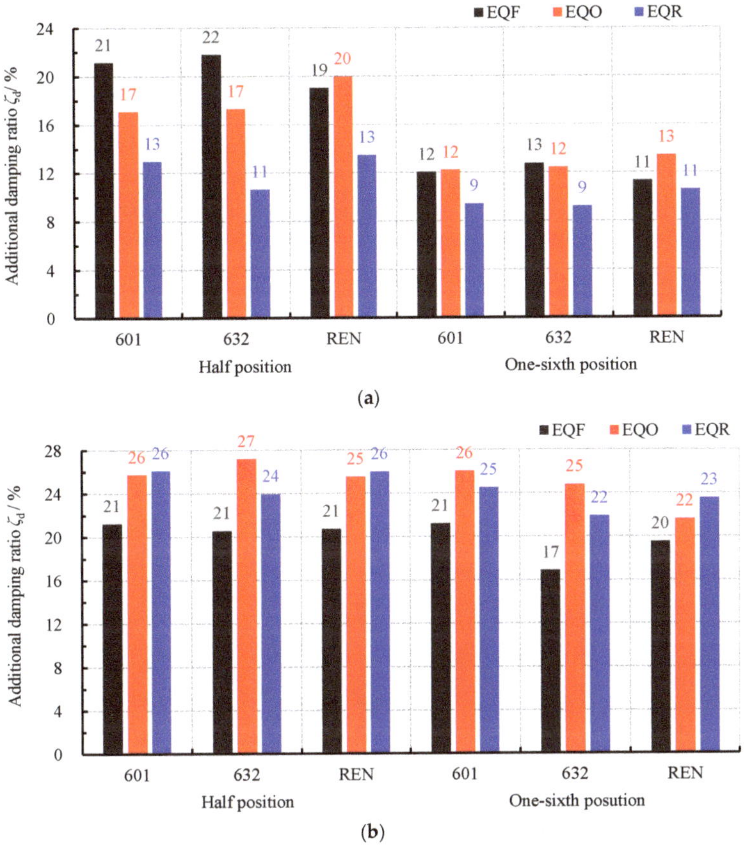

Figure 12. Comparison of the additional damping ratios. (**a**) The additional damping ratio in the X-direction. (**b**) The additional damping ratio in the Y-direction. EQF represents structures subjected to frequent earthquakes, EQO represents structures subjected to occasional earthquakes, and EQR represents structures subjected to rare earthquakes. In the following figures provided in this article, when using these three abbreviations, they all have the same meaning.

From Figure 12b, it can be observed that the stiffness of the structure in the Y-direction is relatively low, and the natural frequency is also low, with a value of 1.5 Hz. The difference in additional damping ratio with different damper arrangement positions is not as significant as in the X-direction for different seismic levels.

In conclusion, it can be inferred that the additional damping ratio provided by the damper to the structure is closely related to the lateral stiffness of the structure under different arrangement positions and seismic levels. The larger the lateral stiffness, the greater the difference and regularity in the additional damping ratio, as in the X-direction.

Conversely, the smaller the lateral stiffness, the smaller the difference and less obvious the regularity in the additional damping ratio, as in the *Y*-direction.

3.3. Inter-Story Displacement Utilization Rate

In damped structures, the displacement of a damper under seismic excitation is an important indicator of their energy dissipation, which can be clearly observed from the hysteresis loops of dampers. By testing the displacement of floors and the corresponding damper displacement under seismic action, the ratio between the two can be calculated under frequent earthquakes, occasional earthquakes, and rare earthquakes. The definition of the inter-story displacement utilization rate is as follows: the ratio between the displacement of a damper and the floor displacement of the floor where the damper is located in a damped structure. The calculation expression is as follows:

$$\eta = \frac{u_{\mathrm{d}}}{u} \times 100\% \tag{2}$$

where η represents the inter-story displacement utilization rate; u represents the displacement of the floor where the damping device is located; u_{d} represents the displacement of the damper.

The experiment investigated the inter-story displacement utilization rate under different arrangements of dampers for frequent, occasional, and rare earthquakes. This newly proposed evaluation index, which can effectively measure the energy dissipation effect of dampers, can be used to assess the damping efficiency of damped structure. The inter-story displacement utilization rate obtained by calculation according to Equation (2) is shown in Figure 13.

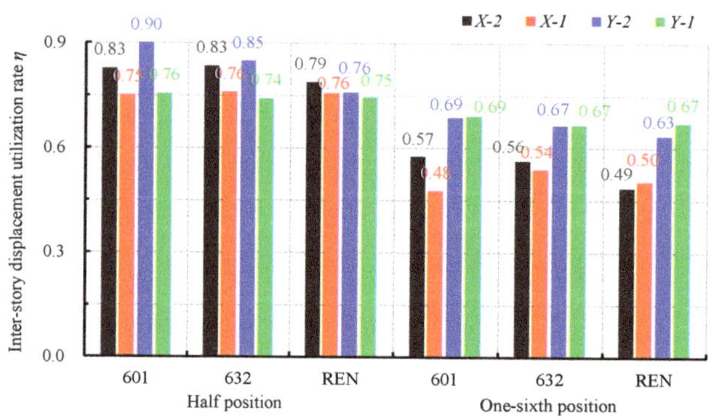

(a)

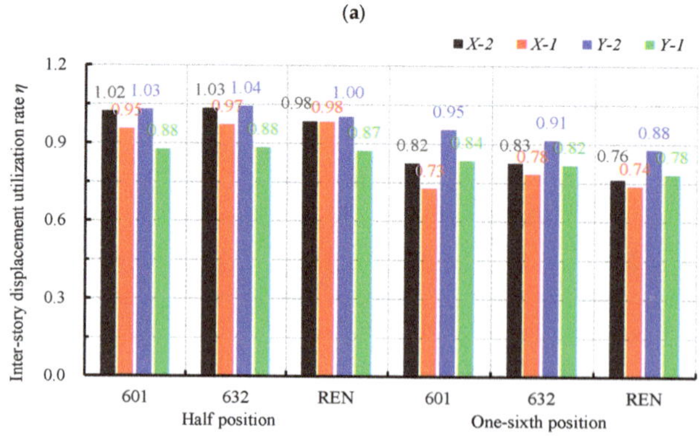

(b)

Figure 13. *Cont.*

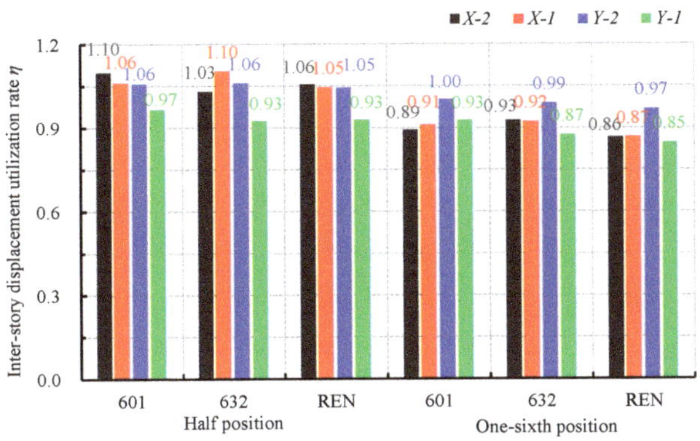

(c)

Figure 13. Comparison of the inter-story displacement utilization rate. (**a**) The inter-story displacement utilization rate under frequent earthquakes. (**b**) The inter-story displacement utilization rate under occasional earthquakes. (**c**) The inter-story displacement utilization rate under rare earthquakes. *X*-2 represents the inter-story displacement utilization rate of the second floor in the *X*-direction. The meanings of other labels on the diagrams are similar.

From Figure 13, the following can be observed:

(1) There is a significant difference in the inter-story displacement utilization rate for different damper placement positions, regardless of whether it is subjected to frequent, occasional, or rare earthquakes. The inter-story displacement utilization rate is significantly higher when the dampers are placed at the 1/2 position within the beam span compared to the 1/6 position within the beam span.

(2) With the increase in seismic levels, the inter-story displacement utilization rate also noticeably increases. For the 1/2 position, the inter-story displacement utilization rate ranges from 0.74 to 0.90 for frequent earthquakes, from 0.87 to 1.04 for occasional earthquakes, and from 0.93 to 1.10 for rare earthquakes.

(3) It can also be observed from the figure that the inter-story displacement utilization rate of the second floor is greater than that of the first floor.

3.4. Reduction Rate of Vertex Displacement

In comparison to non-damped structures, damped structures with additional viscous dampers exhibited a reduced displacement response for each floor. This led to a significant decrease in the vertex displacement of the structure, thereby enhancing its seismic performance. Consequently, the reduction rate of the vertex displacement served as an important indicator to measure the damping efficiency of the structure. This is calculated by comparing the vertex displacement of a damped structure to that of a non-damped structure and is expressed as a ratio. The calculation formula for this ratio is as follows:

$$\mu_{\mathrm{u}} = \frac{u_0 - u}{u_0} \times 100\% \tag{3}$$

where μ_{u} represents the reduction rate of the vertex displacement, u_0 represents the vertex displacement in the non-damped structure, and u represents the vertex displacement in the damped structure.

The reduction rate of the vertex displacement and the averages of three seismic waves in both the X- and Y-directions for structures under frequent, occasional, and rare earthquakes were calculated using Equation (3), as illustrated in Figure 14. Figure 14c presents the average vertex displacements of three seismic waves corresponding to the structure's *X*- and *Y*-directions with dampers arranged at the 1/2 and 1/6 positions within the beam span, respectively. When these values were compared with the average vertex displacement of non-damped structures, the comparison results revealed a distinct contrast.

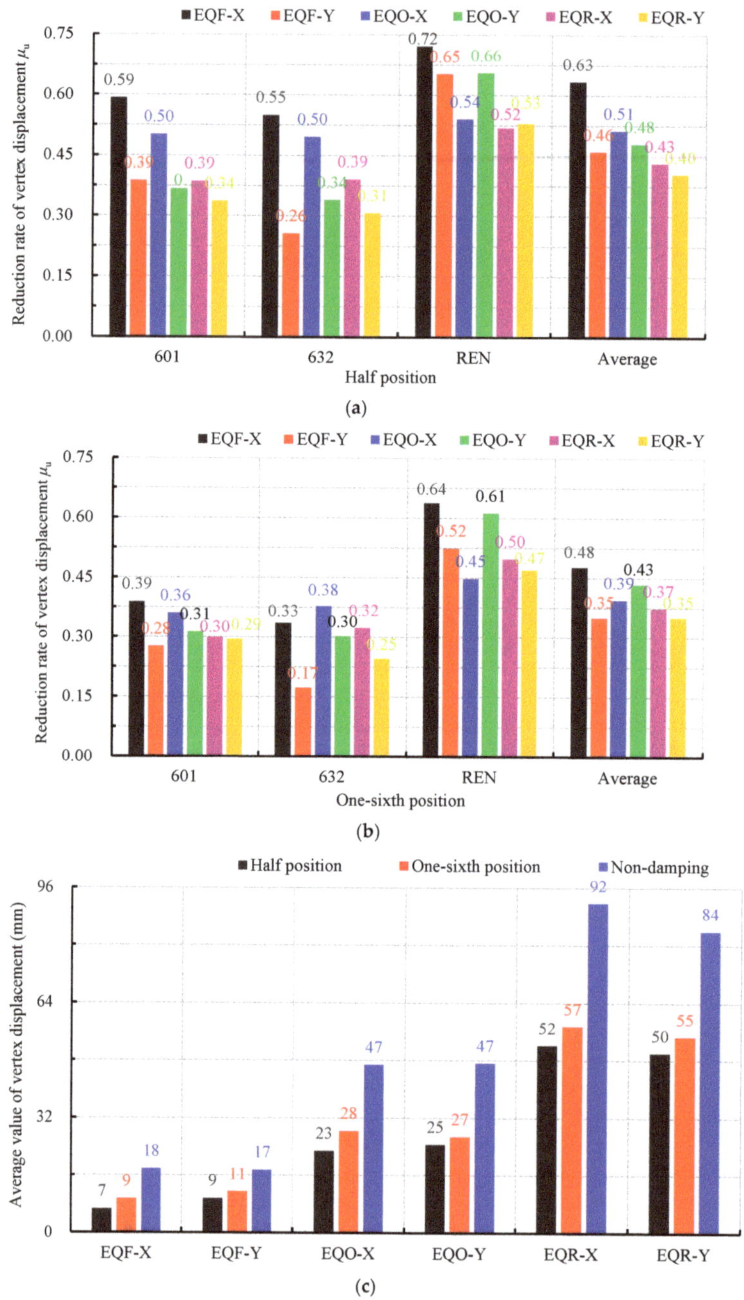

Figure 14. Comparison of vertex displacement reduction rate. (**a**) The reduction rate of vertex displacement at 1/2 position under different seismic levels. (**b**) The reduction rate of vertex displacement at 1/6 position under different seismic levels. (**c**) The average value of vertex displacement at different positions under different seismic levels.

As can be seen from Figure 14, the following is evident:

(1) The reduction rate of the vertex displacement in the damped structure is significant. When the dampers are arranged at the 1/2 position within the beam span, the reduction rates of the vertex displacement for the three waves range from 0.26 to 0.72, while when the dampers are arranged at the 1/6 position within the beam span, the reduction rates of the vertex displacement for the three waves range from 0.17 to 0.64. The average reduction rates of the vertex displacement for the three seismic waves range from 0.40 to 0.63 and 0.35 to 0.48, respectively.
(2) The reduction rate of the vertex displacement under the action of the artificial wave REN is significantly higher than that of the other two natural waves, 601 and 632.
(3) From Figure 14c, it can be seen that the vertex displacement of the damped structure is significantly smaller than that of the non-damped structure. For example, under rare earthquakes, the vertex displacement in the *X*-direction for the damped and non-damped structures is 50-plus mm and 92 mm, respectively, while in the *Y*-direction, it is 50-plus mm and 84 mm, respectively. This indicates that the damped structure with additional viscous dampers has excellent damping efficiency. Moreover, the vertex displacement when the dampers are arranged at the 1/2 position within the beam span is smaller than that when the dampers are arranged at the 1/6 position, indicating that the closer the damper is to the middle of the beam span, the better the damping efficiency of the structure.

3.5. Reduction Rate of Base Shear

The reduction rate of the base shear is a crucial indicator for assessing the damping efficiency of a structure. When a damped structure is equipped with additional viscous dampers, its response to seismic excitation decreases, leading to a significant reduction in the base shear of the structure and improving its seismic performance. The reduction rate of the base shear is calculated by dividing the reduction in the base shear of the damped structure by that of the non-damped structure. The formula for calculating this rate is as follows:

$$\mu_V = \frac{V_0 - V}{V_0} \times 100\% \tag{4}$$

In the equation, μ_V represents the base shear reduction rate, V_0 represents the base shear of the non-damped structure, and V represents the base shear of the damped structure.

The reduction rate of the base shear and the averages of three seismic waves corresponding to the structure's *X*- and *Y*-directions under frequent earthquakes, occasional earthquakes, and rare earthquakes were calculated, according to Equation (4), as shown in Figure 15. Figure 15c presents the average base shear of the three seismic waves corresponding to the structure's *X*- and *Y*-directions when dampers were arranged at the 1/2 and 1/6 positions within the beam span, respectively. When these values were compared with the average base shear of non-damped structures, the comparison results revealed a distinct contrast.

The following conclusions can be drawn from Figure 15:

(1) The reduction rate of the base shear in the damped structure is significant. When dampers are arranged at the 1/2 position within the beam span, the reduction rates of the base shear for the three waves range from 0.16 to 0.46. When dampers are arranged at the 1/6 position within the beam span, the reduction rates of the base shear for the three waves range from 0.12 to 0.41. The average reduction rates of the base shear for the three waves are between 0.28 and 0.34, and between 0.22 and 0.27, respectively.
(2) The reduction rate of the base shear under the artificial wave REN is significantly higher than that of the other two natural waves, 601 and 632.
(3) From Figure 15c, it can be observed that the base shear of the damped structure is significantly smaller than that of the non-damped structure. For example, under rare earthquake conditions, the base shear in the *X*-direction for the damped and

non-damped structures are 30-plus kN and 46 kN, respectively, and in the Y-direction, they are 20-plus kN and 30 kN, respectively. This indicates that the damped structure with additional viscous dampers has excellent damping efficiency. Moreover, the base shear when dampers are arranged at the 1/2 span position is smaller than that when dampers are arranged at the 1/6 span position, indicating that the closer the damper is to the middle of the beam span, the better the damping efficiency of the structure.

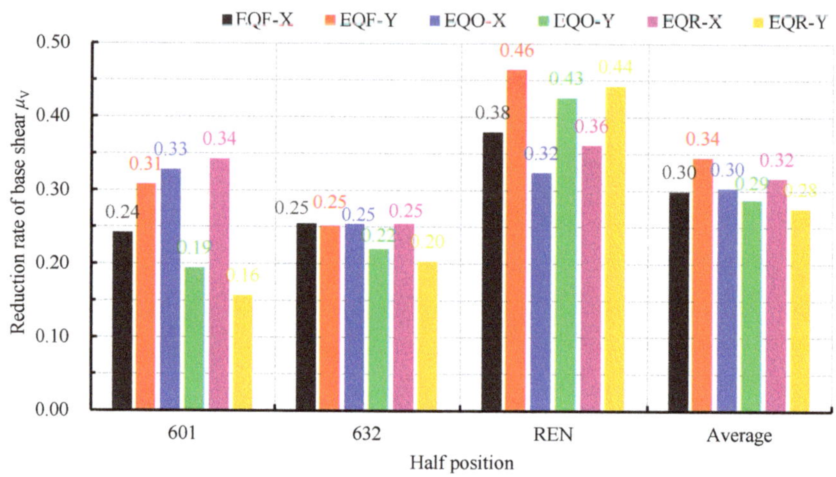

(a)

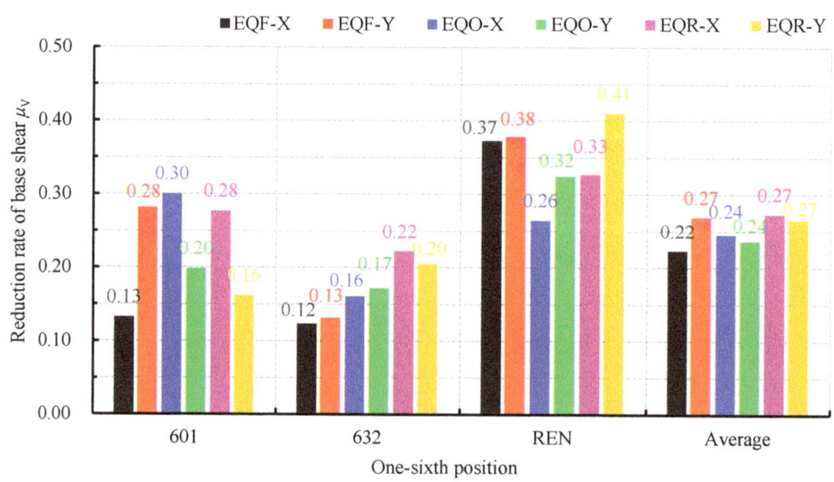

(b)

Figure 15. *Cont.*

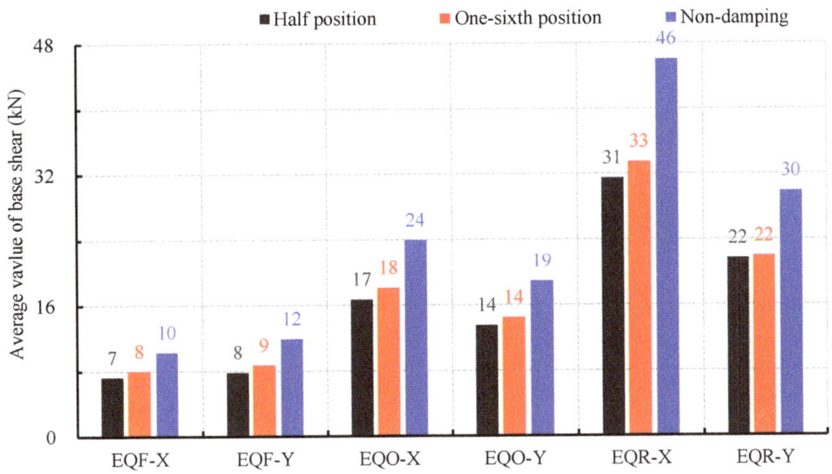

(c)

Figure 15. Comparison of base shear reduction rate. (**a**) The reduction rate of base shear at 1/2 position under different seismic levels. (**b**) The reduction rate of base shear at 1/6 position under different seismic levels. (**c**) The average value of base shear at different positions under different seismic levels.

From Figures 14 and 15, it can be observed that the reduction rate of the base shear in the damped structure is significantly smaller than the reduction rate of the vertex displacement. This is an important principle that designers need to pay attention to when conducting the damping design.

4. Conclusions

This study utilized a two-story steel structure for shaking table testing, examining the response of both non-damped and damped structures to seismic excitation. Additionally, the impact of varying the position of viscous dampers on the damping efficiency of damped structures was explored. The key findings are summarized below.

(1) The response of damped structures and non-damped structures under different seismic levels is significantly different. The vertex displacement was reduced from 17–18 mm to 7–11 mm for frequent earthquakes, from 47 mm to 23–28 mm for occasional earthquakes, and from 84–92 mm to 50–57 mm for rare earthquakes. The base shear was reduced from 10–12 kN to 7–9 kN for frequent earthquakes, from 19–24 kN to 14–18 kN for occasional earthquakes, and from 30–46 kN to 22–31 kN for rare earthquakes. This fully demonstrated that the damped structures with additional viscous dampers had good damping effects.

(2) From the hysteresis loops of the damper, it can be seen that the viscous damper added to the structure effectively dissipated seismic energy. This further significantly reduced the seismic response of the structure. Compared with the non-damped structures, the average reduction rates of the vertex displacement and base shear for the three waves of damped structures were 0.35–0.63 and 0.22–0.34, respectively. This fully demonstrated that the damped structures using viscous dampers exhibited excellent damping efficiency.

(3) The arrangement position of dampers within the beam span significantly affected the damping efficiency of the damped structure. Overall, the damping effect was significantly better when the dampers were placed at the 1/2 position of the beam span compared to the 1/6 position. The hysteresis loops representing the energy dissipation of the dampers illustrated this phenomenon. As for the additional damp-

ing ratio of the structures, taking the artificial wave REN in the X-direction as an example, they were 19% and 11% for frequent earthquakes, 20% and 13% for occasional earthquakes, and 13% and 11% for rare earthquakes under the two placement positions, respectively. The inter-story displacement utilization rate was 0.74–0.90 and 0.48–0.69, 0.87–1.04 and 0.73–0.95, and 0.93–1.10 and 0.85–1.0 for the two positions under frequent, occasional, and rare earthquakes, respectively. The average reduction rate of the vertex displacement for the three waves was 0.40–0.63 and 0.35–0.48 for the two placement positions, respectively. The average reduction rate of the base shear for the three waves was 0.28–0.34 and 0.22–0.27 for the two placement positions, respectively.

(4) Introducing a novel evaluation metric known as the inter-story displacement utilization rate, which is primarily utilized to gauge the displacement utilization of dampers in damped structures based on the floor's displacement where the damper is located. This study revealed that the closer the damper was positioned to the center of the beam span, the higher the inter-story displacement utilization rate. Furthermore, as seismic levels increased, so did the inter-story displacement utilization rate. When a centrally placed damper experienced significant seismic activity, the inter-story displacement utilization rate approached 1.0, indicating that the damper's displacement can reach a level comparable to that of the floor.

Author Contributions: Conceptualization, X.L., L.Z., B.S. and W.P.; methodology, X.L.; software, X.L.; validation, X.L., L.Z. and B.S.; formal analysis, X.L.; investigation, L.Z.; resources, X.L.; data curation, X.L.; writing—original draft preparation, X.L.; writing—review and editing, X.L.; visualization, L.Z.; supervision, B.S.; project administration, X.L.; funding acquisition, X.L. All authors have read and agreed to the published version of the manuscript.

Funding: This research was supported by Yunnan Provincial Basic Research Special Project (By X.L., Grant No. 202101AU070059), Research Project of Introducing Talents in Kunming University (By X.L., Grant No. XJ20210006) and China-Pakistan Belt and Rosad Joint Laboratory on Smart Disaster Prevention of Major Infrastructures (By B.S., Grant No. 2022CPBRJL-10).

Data Availability Statement: All the data utilized in the research has been fully presented in the paper. No additional data has been employed.

Conflicts of Interest: Author Baifeng Sun was employed by the company Yunnan Design Institute Group Co., Ltd. The remaining authors declare that the research was conducted in the absence of any commercial or financial relationships that could be construed as a potential conflict of interest.

References

1. Lin, W.H.; Chopra, A.K. Earthquake response of elastic SDF systems with nonlinear fluid viscous dampers. *Earthq. Eng. Struct. Dyn.* **2002**, *31*, 1623–1642. [CrossRef]
2. Soong, T.T.; Spencer, B.F., Jr. Supplemental energy dissipation: State-of-the-art and state-of-the-practice. *Eng. Struct.* **2002**, *24*, 243–259. [CrossRef]
3. Titirla, M.D. A state-of-the-art review of passive energy dissipation systems in steel braces. *Buildings* **2023**, *13*, 851. [CrossRef]
4. Constantinou, M.C.; Symans, M.D. Experimental study of seismic response of buildings with supplemental fluid dampers. *Struct. Design Tall Build.* **1993**, *2*, 93–132. [CrossRef]
5. Vaiana, N.; Sessa, S.; Marmo, F.; Rosati, L. Nonlinear dynamic analysis of hysteretic mechanical systems by combining a novel rate-independent model and an explicit time integration method. *Nonlinear Dyn.* **2019**, *98*, 2879–2901. [CrossRef]
6. Vaiana, N.; Napolitano, C.; Rosati, L. Some recent advances on the modeling of the hysteretic behavior of rate-independent passive energy dissipation devices. In Proceedings of the 8th International Conference on Computational Methods in Structural Dynamics and Earthquake Engineering, Athens, Greece, 28–30 June 2021; pp. 2435–2445.
7. Zoccolini, L.; Bruschi, E.; Cattaneo, S.; Quaglini, V. Current trends in fluid viscous dampers with semi-active and adaptive behavior. *Appl. Sci.* **2023**, *13*, 10358. [CrossRef]
8. Lu, Z.; Wang, Z.; Zhou, Y.; Lu, X. Nonlinear dissipative devices in structural vibration control: A review. *J. Sound Vib.* **2018**, *423*, 18–49. [CrossRef]
9. Lavan, O. On the efficiency of viscous dampers in reducing various seismic responses of wall structures. *Earthq. Eng. Struct. Dyn.* **2012**, *41*, 1673–1692. [CrossRef]

10. Hwang, J.S.; Huang, Y.N.; Yi, S.L.; Song, Y.H. Design formulations for supplemental viscous dampers to building structures. *J. Struct. Eng.* **2008**, *134*, 22–31. [CrossRef]
11. Xie, Y.; Zhang, J.; Xi, W. Effectiveness evaluation and optimal design of nonlinear viscous dampers for inelastic structures under pulse-type ground motions. *Earthq. Eng. Struct. Dyn.* **2018**, *47*, 2802–2820. [CrossRef]
12. Gherbi, A.; Belgasmia, M. A simplified design strategy of nonlinear fluid viscous dampers for MDOF structures. *Iran. J. Sci. Technol. Trans. Civ. Eng.* **2022**, *46*, 857–864. [CrossRef]
13. Diotallevi, P.P.; Landi, L.; Dellavalle, A. A methodology for the direct assessment of the damping ratio of structures equipped with nonlinear viscous dampers. *J. Earthq. Eng.* **2012**, *16*, 350–373. [CrossRef]
14. Xu, W.; Du, D.; Wang, S.; Li, W. A new method to calculate additional damping ratio considering the effect of excitation frequency. *Adv. Civ. Eng.* **2020**, *2020*, 3172982. [CrossRef]
15. Occhiuzzi, A. Additional viscous dampers for civil structures: Analysis of design methods based on effective evaluation of modal damping ratios. *Eng. Struct.* **2009**, *31*, 1093–1101. [CrossRef]
16. Rama, R.K.; Ansu, M.; Iyer, N.R. A methodology of design for seismic performance enhancement of buildings using viscous fluid dampers. *Struct. Control Health Monit.* **2014**, *21*, 342–355. [CrossRef]
17. Domenico, D.D.; Hajirasouliha, I. Multi-level performance-based design optimisation of steel frames with nonlinear viscous dampers. *Bull. Earthq. Eng.* **2021**, *19*, 5015–5049. [CrossRef]
18. Ding, J.; Wang, S.; Wu, H. Seismic performance analysis of viscous damping outrigger in super high-rise buildings. *Struct. Design Tall Spec. Build.* **2018**, *27*, e1486. [CrossRef]
19. Dargush, G.F.; Sant, R.S. Evolutionary aseismic design and retrofit of structures with passive energy dissipation. *Earthq. Eng. Struct. Dyn.* **2005**, *34*, 1601–1626. [CrossRef]
20. Lavan, O.; Amir, O. Simultaneous topology and sizing optimization of viscous dampers in seismic retrofitting of 3D irregular frame structures. *Earthq. Eng. Struct. Dyn.* **2014**, *43*, 1325–1342. [CrossRef]
21. Tubaldi, E.; Ragni, L.; Dall'Asta, A. Probabilistic seismic response assessment of linear systems equipped with nonlinear viscous dampers. *Earthq. Eng. Struct. Dyn.* **2015**, *44*, 101–120. [CrossRef]
22. Tubaldi, E.; Kougioumtzoglou, I.A. Nonstationary stochastic response of structural systems equipped with nonlinear viscous dampers under seismic excitation. *Earthq. Eng. Struct. Dyn.* **2015**, *44*, 121–138. [CrossRef]
23. Zhang, Y.; Xu, W.; Du, D.; Wang, S. Stochastic optimization of dissipation structures based on Lyapunov differential equations and the full stress design method. *Buildings* **2023**, *13*, 665. [CrossRef]
24. Su, C.; Li, B.; Chen, T.; Dai, X. Stochastic optimal design of nonlinear viscous dampers for large-scale structures subjected to non-stationary seismic excitations based on dimension-reduced explicit method. *Eng. Struct.* **2018**, *175*, 217–230. [CrossRef]
25. Rodolfo, J.T.G.; Yeudy, F.V.A.; Ramon, G.D. Probabilistic estimation of the dynamic response of high-rise buildings via transfer functions. *Eng. Struct.* **2024**, *302*, 117299.
26. Wang, Z.; Zhao, J.; Chen, X.; Liu, S.; Ma, B. Experimental investigation of seismic response of precast concrete panels with castellated keys support pillar connections under in-plane cyclic loading. *Sci. Rep.* **2023**, *13*, 21764. [CrossRef] [PubMed]
27. Chang, K.C.; Lin, Y.Y.; Chen, C.Y. Shaking table study on displacement-based design for seismic retrofit of existing buildings using nonlinear viscous dampers. *J. Struct. Eng.* **2008**, *134*, 671–681. [CrossRef]
28. Bogdanovic, A.; Rakicevic, Z.; Noroozinejad Farsangi, E. Shake table tests and numerical investigation of a resilient damping device for seismic response control of building structures. *Struct. Control Health Monit.* **2019**, *26*, e2443. [CrossRef]
29. Singh, M.P.; Moreschi, L.M. Optimal seismic response control with dampers. *Earthq. Eng. Struct. Dyn.* **2001**, *30*, 553–572. [CrossRef]
30. Aguirre, J.J.; Almazan, J.L.; Paul, C.J. Optimal control of linear and nonlinear asymmetric structures by means of passive energy dampers. *Earthq. Eng. Struct. Dyn.* **2013**, *42*, 377–395. [CrossRef]
31. Parcianello, E.; Chisari, C.; Amadio, C. Optimal design of nonlinear viscous dampers for frame structures. *Soil Dyn. Earthq. Eng.* **2017**, *100*, 257–260. [CrossRef]
32. Whittle, J.K.; Williams, M.S.; Karavasilis, T.L.; Blakeborough, A. A comparison of viscous damper placement methods for improving seismic building design. *J. Earthq. Eng.* **2012**, *16*, 540–560. [CrossRef]
33. Lopez, G.D. A simple method for the design of optimal damper configurations in MDOF structures. *Earthq. Spectra.* **2001**, *17*, 387–398.
34. Takewaki, I. Optimal damper placement for minimum transfer functions. *Earthq. Eng. Struct. Dyn.* **1997**, *26*, 1113–1124. [CrossRef]
35. Takewaki, I. Optimal damper placement for critical excitation. *Probabilist. Eng. Mech.* **2000**, *15*, 317–325. [CrossRef]
36. Lin, T.K.; Hwang, J.S.; Chen, K.H. Optimal distribution of damping coefficients for viscous dampers in buildings. *Int. J. Struct. Stab. Dyn.* **2017**, *17*, 1750054. [CrossRef]
37. Aydin, E.; Boduroglu, M.H.; Guney, D. Optimal damper distribution for seismic rehabilitation of planar building structures. *Eng. Struct.* **2007**, *29*, 176–185. [CrossRef]
38. *GB/T 228.1-2021*; Chinese Standard, Metallic Materials-Tensile Testing–Part 1: Method of Test at Room Temperature. Chinese Architecture and Building Press: Beijing, China, 2021.
39. Lan, X.; Pan, W.; Zhang, L.; Yu, W.; Wu, K. Application and research of viscous fluid damper in frame structure. *Sichuan Build. Sci.* **2022**, *48*, 10–17. (In Chinese)

40. *JGJ 297-2013;* Chinese Standard, Technical Specification for Seismic Energy Dissipation of Buildings. Chinese Architecture and Building Press: Beijing, China, 2013.
41. *GB 50011-2016;* Chinese Standard, Code for Seismic Design of Buildings. Chinese Architecture and Building Press: Beijing, China, 2016.

buildings

MDPI

Review

Structural Vibration Comfort: A Review of Recent Developments

Weiping Xie and Yumeng Hua *

School of Civil Engineering and Architecture, Wuhan University of Technology, Wuhan 430070, China; xiewp@whut.edu.cn
* Correspondence: huaym@whut.edu.cn

Abstract: With continuous improvements in the social economy and living standards of individuals, the vibration comfort of building structures has gradually been emphasized by academic and engineering communities, such as vehicle-induced vibrations in buildings near traffic, human-induced vibrations in large-span structures, wind-induced vibrations in super-high-rise buildings, and machinery-induced structural vibrations. Comfort-based structural analysis is distinct from traditional safety-based structural analysis, and its theoretical systems and unified guidelines have not yet been established. This paper reviews recent research on structural vibration comfort, including major load categories and their impacts, comfort-based structural analysis, evaluation methods, and vibration-mitigation measures. By presenting the shortcomings of the existing research, potential topics for future study are suggested.

Keywords: vibration comfort; serviceability; environmental vibration; vehicle-induced vibration; human-induced vibration; wind-induced vibration; machinery-induced vibration; structural analysis; comfort evaluation; vibration control

1. Introduction

With the upgrading of construction technology, application of lightweight and high-strength materials, realization of new structural systems, and pursuit of architectural aesthetics, modern engineering structures tend to develop in light, flexible, large-span, and towering designs. This change has resulted in an apparent decrease in the structural self-vibration frequency, which is close to the frequency of traffic vehicles, wind, dynamic machinery, and people's daily activities; thus, it is prone to resonance and generates a significant dynamic response. Such structural vibration is generally not sufficient to cause safety problems, but it can cause uncomfortable feelings for the people in the building, which is a comfort problem. People in such vibrating environments can appear nervous, annoyed, or panicky, which may result in mental illness over time. Therefore, in the design and construction of "sensitive buildings" (e.g., buildings near traffic, large-span structures, super-high-rise structures, and buildings with dynamic machinery inside), the structural dynamic response is predicted or measured, and vibration control measures are applied if the response levels exceed comfort standards. Such engineering cases have increased in number in recent years.

Research on vibration comfort began in 1931, when Reiher and Meier conducted pioneering experiments, developed preliminary conclusions on how vibration frequency affects human comfort, and provided a paradigm for comfort studies [1]. Subsequently, scholars in the United States, Europe, and Japan have conducted numerous studies and issued a series of specifications and guidelines [2–7]. Griffin's team conducted a representative study, investigating the effects of frequency, direction, and duration of vibration, human posture, and other factors of the human vibration perception threshold through a number of experiments [5]. The Millennium Bridge vibration event in London, UK, in 2000 brought widespread attention to the comfort of pedestrian bridge structures by scholars and engineers [8,9], subsequently drove the study of structural comfort caused by human-induced loads, and provided a strong impetus to research in this field. Researchers have accumulated numerous results on structural vibration comfort.

High-speed railways and urban rail transits that have developed rapidly in China since the 1980s facilitate resident travel, but also induce environmental vibrations that have caused complaints from nearby residents [10,11]. With improvements in quality of life, people's demand for living environments has also increased. Environmental vibration is defined as an environmental ground motion of small amplitude resulting from natural and/or anthropogenic causes (traffic, human-induced excitation, wind, and machinery excitation) [12]. To address this problem, Chinese scholars have focused attention on research into vibration comfort.

In addition to comfort problems, environmental vibrations cause negative impacts, including secondary noise, failure of precision instruments, damage to non-structural components, safety of ancient buildings, and wildlife preservation. Most of these issues do not involve structural safety and primarily affect architectural functions or structural serviceability. In engineering, these non-safety issues are sometimes all included in comfort analyses, but some of the above analyses do not conform to the definition of comfort. Environmental vibrations are the cause of comfort problems; however, comfort problems are not the only hazards caused by environmental vibrations. In the literature regarding structural vibration comfort, the term "serviceability" is used, but it differs from the connotation of comfort. In the International Organisation for Standardisation (ISO) publication ISO 5805-1997 Mechanical vibration and shock—Human exposure—Vocabulary [13], "comfort" is defined as the subjective state of well-being or absence of mechanical disturbance in relation to the induced environment (mechanical vibration or repetitive shock). This indicates that "comfort" is the subjective feeling of vibration by the user in the structure. "Serviceability" has a broad scope. The ISO publication ISO 10137: 2007 Bases for design of structures—Serviceability of buildings and walkways against vibrations [14] describes the receivers in vibration serviceability problems, which can encompass the building structure (or components such as beams, slabs, walls, and windows, among others), contents of the building (instrument and machinery among others), or human occupants of the building. It indicates that "serviceability" includes not only human comfort but also vibration-induced damage to structures or components (e.g., safety of ancient buildings, non-structural component cracking) and the regular operation of precision instruments.

This paper reviews the progress of research on structural vibration comfort. Section 2 systematically illustrates the comfort problems induced by loads of traffic vehicles, humans, wind, and dynamic machinery. Section 3 discusses comfort-based structural analysis methods and describes the key modeling and calculation details. Section 4 summarizes the comfort evaluation methods. Section 5 illustrates the existing vibration mitigation measures employed in vibration comfort control. Section 6 suggests potential topics and directions for study in light of the shortcomings of current research.

2. Load Categories and Impacts on Vibration Comfort

Statistics in China in 2020 indicate that the sources that induce vibration comfort problems include traffic (26.4%), humans (34.5%), wind (19.1%), dynamic machinery (12.7% commonly and 5.8% for construction), and others (1.5%) [15]. Load categories differ in terms of source characteristics, action position, corresponding sensitive structure type, and dynamic response law. This section surveys the vibration comfort problems induced by each of the four load categories.

2.1. Vehicle-Induced Vibration

Vehicle-induced vibration is the dynamic response of structures caused by the movement of trains, such as high-speed, subway, and maglev trains, or motor vehicles, such as cars, buses, and heavy-duty trucks. In this subsection, research on the vehicle load model, vibration propagation in soil and soil–structure interaction (SSI), and a simplified calculation approach are presented.

2.1.1. Vehicle Load Model

Research on vehicle-load models began with theoretical and experimental studies of vehicles crossing bridges by Willis and Stokes in 1849 [16,17]. The initial classical theoretical methods simplified the vehicle as a moving constant force, moving harmonic force, moving mass, or moving sprung mass and used analytical methods to obtain dynamic responses, as shown in Figure 1a–d. In the moving mass model illustrated in Figure 1c, the force applied on the ground is represented by Equation (1), considering the lateral inertial of the moving object. By disregarding the lateral inertia, Equation (1) simplifies to Equation (2), transforming the model into the moving load shown in Figure 1a [18,19].

$$F(x,t) = M\left(g - \frac{\mathrm{d}^2 w}{\mathrm{d}t^2}\right)\delta(x - x_M) \tag{1}$$

$$F(x,t) = Mg\delta(x - x_M) \tag{2}$$

where g is the gravitational acceleration; $w = w(x,t)$ denotes the lateral deformation of the underlying structure; δ denotes the Dirac delta function; x_M represents the longitudinal coordinate of the moving mass on the ground.

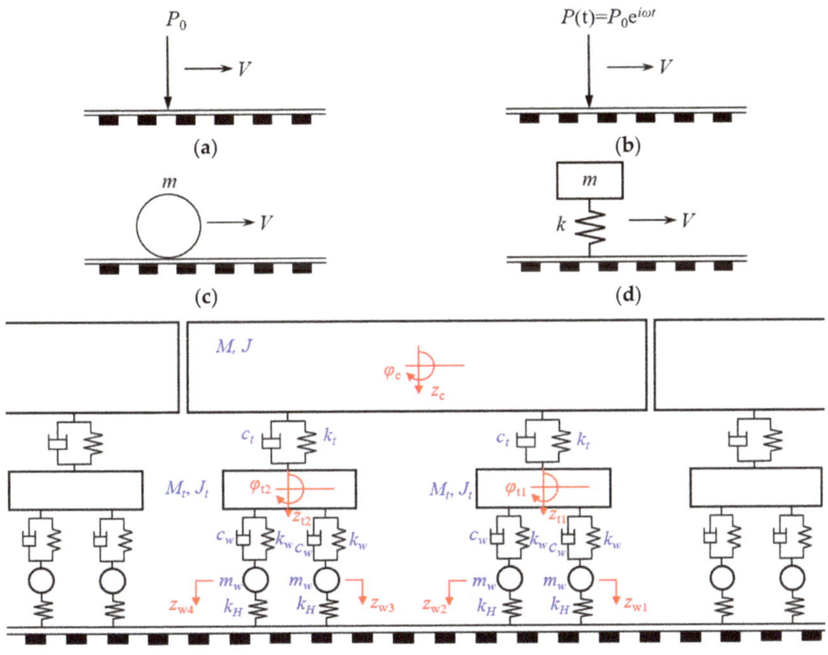

Figure 1. Vehicle load models: (**a**) moving constant force; (**b**) moving harmonic force; (**c**) moving mass; (**d**) moving sprung mass; (**e**) 10-DOF multi-rigid-body model (DOFs are marked in red; mass, stiffness, and damping coefficients are marked in blue).

Since the 1970s, with the development of computer and numerical techniques, vehicle models based on multi-rigid-body dynamics have gradually become the mainstream methods. Chu et al. [20] initially simplified the railway vehicle as a three-degree-of-freedom model system consisting of the car body and wheel–axle sets, after which the degree of freedom (DOF) of developed vehicle models was continuously increased. Figure 1e shows a plane 10-DOF train model popularly used in the prediction of vehicle-induced environmental vibration [10]. This method considers the train–track (or car–road) coupling effect, sets the wheel–rail (or road) contact relationship, uses track irregularity (or road surface roughness) as the main source of excitation, and calculates the equations of motion using iterative or energy methods. Over the years, this vehicle modeling method and its computational theories have tended to be complete and are now applicable to various vehicle types, such as trains and cars, and many related monographs have been published [10,21,22].

Research in this field has mainly been conducted by researchers in the disciplines of vehicle, railway, and bridge engineering. Their primary objective in developing vehicle load models and dynamic analyses was to determine passenger comfort or the safety and durability of train running, railways, or bridges. Table 1 lists the typical train load models in recent years. Therefore, their vehicle load models are sufficiently detailed, with the latest model having 42 DOFs [23]. However, when the purpose of dynamic analysis is turned to the comfort of building structures, theoretically, a vehicle model with fewer DOFs and simplified track and wheel–rail contact relationships can satisfy the accuracy requirement for comfort analyses because of the filtering properties of soil for high-frequency vibrations and the low-frequency characteristics of structures. In the existing literature on comfort analysis, the train models used for calculations include several types of multi-rigid-body models, such as plane models (e.g., 10 DOFs [10,24]) and space models (e.g., 35 DOFs [25,26]). Furthermore, the track irregularity spectrum corresponds to the vertical irregularity of a single track or the vertical, horizontal, alignment, and rail gauge irregularities of double tracks, according to the selected plane or space vehicle model. In the above literature, accurate structural responses have been obtained using different vehicle load models. However, an overly refined vehicle and track model can result in low calculation efficiency. For comfort-based analysis, the selection of suitable vehicle load models to balance the calculation accuracy and efficiency remains to be investigated.

Table 1. Typical train load models and development objectives.

Development Objective	References
running safety	Ma et al., 2022 [27]; Zeng et al., 2022 [28]; Ju et al., 2023 [29]; Zhao et al., 2023 [30]; Tang et al., 2023 [31]; Li et al., 2024 [32]; Jiang et al., 2024 [33]
safety and durability of railways or bridges	Gou et al., 2023 [34]; Chen et al., 2024 [35]; Zhang et al., 2024 [36]
passenger comfort	Gou et al., 2023 [34]; Xin et al., 2023 [37]; Li et al., 2024 [32]
environmental vibration	Qu et al., 2022 [38]; Ren et al., 2023 [26]; Xu, 2023 [39]; Hu et al., 2024 [40]; Malmborg et al., 2024 [41]

2.1.2. Vibration Propagation in Soil and SSI

Buildings sensitive to vehicle-induced vibration can be classified as "vehicle-structure" and "vehicle-soil-structure" types based on the difference in the vibration propagation paths from the source to the structure. The difference between the two is whether the vibrations propagate primarily in the soil. The representative buildings of the first type include integrated station–bridge structures [42–45], buildings with carriageways inside [46,47], and railway overtrack buildings [48–50], among others. Their common characteristic in structural construction is that the track or carriageway is rigidly connected to the structure, and the vibration is primarily propagated within the structure. For instance, railway trains can cause comfort problems in waiting rooms and commercial areas within stations. Arriving and departing trains in stations have low speeds, and thus, integrated station–bridge structures will be more significantly impacted than the stations separated from rail bridges. The second type of building is characterized by vehicle-induced vibrations that must propagate in the soil and then into the structure. Buildings near railways or roads mostly belong to this type [51,52]. This type of building is more common and numerous, and there are still frequent complaints and exceeded standards owing to current imperfections in environmental vibration assessment and structural design methods. For vehicle-induced vibrations in the second type of building, the vibration propagation behavior in the soil and SSI is critical.

The deformation of soil caused by vehicle-induced environmental vibrations is generally in the elastic phase. Waves in the soil are elastic waves that propagate as body and surface waves. This issue began with Lamb's work in 1904, that is, the wave solution for an elastic semi-infinite spatial body acting on its surface or interior using a point or line load [53]. Subsequent researchers considered more complex soil and vibration source properties to extend and deepen Lamb's problem and initially obtained a vehicle-induced vibration propagation attenuation law in soil layers [54–57]. However, the complex dynamic properties of soil, difficulty in detecting inhomogeneous soil layers, and effects of surrounding structures result in difficulties in calculations and predictions. Prediction methods for environmental vibration include empirical formulas [58,59], analytical or semi-analytical methods [60,61], and numerical simulations [62–65], among which the finite element model (FEM) is the more accurate and commonly used prediction method [63–65]. With further research, the modeling of the dynamic properties of natural soil has progressed from elastic bodies [53] to viscoelastic bodies [66], and then to saturated or unsaturated porous media [67]. Although these existing studies have effectively improved accuracy, soil models still have many assumptions compared to natural soils. At present, field measurements remain an effective approach for obtaining vibration attenuation laws for the geological conditions of a region. Therefore, the current environmental impact assessment (EIA) standard in China states that tests are preferentially used to obtain vibration attenuation laws over distance [68].

Vehicle-induced environmental vibrations were input from the structural foundation. Therefore, the SSI becomes a non-negligible issue in the calculations. The SSI issue initially received attention in seismic response analyses [69]. Compared to earthquakes, vehicle-induced environmental vibrations are mainly focused in the vertical direction and have smaller amplitudes and higher frequencies. Under these two excitations, the soil properties and contact relationships with the structural foundation are different. Currently, the impact of SSI in the context of vehicle-induced vibrations has not been adequately addressed. Most existing models assume that the SSI is negligible, with the soil and structure subsystems being uncoupled during calculations. This uncoupled approach is more suited to far-fault ground motions, where the receivers are distant from the source [70]. However, in the case of vehicle-induced vibration, shorter source–receiver distances and higher frequencies result in greater interaction [71,72]. The effects of SSI are mainly reflected in three aspects: changing the original vibration field, elastic boundaries on the structure, and coupled damping [63–75]. Yao et al. [76] calculated that the vibration of building floors could decrease by 10–12 dB when the nonlinear contact interaction between soil and building foundation is considered. The interaction is influenced by dynamic soil characteristics, structural foundation types, structural mass, source–receiver distances, among others [73, 77,78]. Coulier et al. [77] investigated that the interaction can be significant when the source–receiver distance is smaller than the dilatational wavelength in the soil for underground railway tunnels, or smaller than six Rayleigh wavelength for railway track at grade, through a case study of a building on a shallow foundation. Numeric contact elements or simplified spring and damper components can be utilized to model the SSI. Aubry and Clouteau [79] proposed a sub-structuring approach to simulate the SSI, which has been applied to structural vibrations under traffic excitation. Pyl et al. [80] first studied the SSI using a 3D model under road traffic excitation. Colaço et al. [72] used the equivalent stiffness and damping method to consider SSI in calculating railway-induced vibrations.

Vibration propagation in soil and the SSI cause non-uniform excitation of vehicle vibration loads on the structure. Previously, uniform excitation methods have been used to apply vibrations at the ground or structural bottom to a structure calculation model [81,82]. However, Hua et al. [83] indicated that the impact of non-uniform excitation should be considered when calculating the structural dynamic response induced by subways. The non-uniform excitation effect can result in differences in the distribution of structural responses, and its load model and calculation methods need to be studied further.

2.1.3. Simplified Calculation Approach

For comfort-based structural analyses, the primary shortcoming of current vehicle-induced vibration load models is the calculation efficiency rather than accuracy. The accuracy of the load models developed for vehicle comfort and track safety satisfied the requirements of structural comfort analyses. However, the large number of DOFs in soil FEM, the iterative method, and the short calculation time steps result in a large amount of computation and increase the difficulty of modeling, debugging, and calculation of the model, which makes it difficult to extend this calculation method to structural design.

To improve calculation efficiency, researchers have proposed solutions from several perspectives. Zeng et al. [84] proposed a time-integration analysis scheme for the vehicle–bridge-interaction problem to prove the computational efficiency of existing algorithms; Wang et al. [85] developed a multi-point synchronous algorithm to solute the large sparse linear equations of the train–track–soil coupled system, resulting in a five- to tenfold increase in computation efficiency; Touhei [86] proposed a 2.5-dimension (2.5D) FEM to use wave number transformations to track and soil along the line direction. However, this method is challenging to apply to building structures [87]. Currently, simplified load models and efficient calculation methods are pressing requirements.

2.2. Human-Induced Vibration

Human-induced vibrations are a dynamic response of structures to human activity. In this subsection, the human-induced load model, human–structure interaction (HSI), and crowd load model are presented.

2.2.1. Human-Induced Load Model and HSI

Several landmark events, such as the Millennium Bridge vibration event (London, UK, 2000) [8] and the Techno Mart building vibration event (Seoul, Korea, 2011) [88], have attracted extensive attention regarding the comfort problem induced by humans. Human-induced loads generally have a low intensity and thus cause significant dynamic responses by exciting structural resonance. Because the main frequency of human-induced loads is 1–3 Hz, light and flexible engineering structures, such as pedestrian bridges, connective corridors, large-span buildings, and cantilevered buildings, are the primary study subjects [89,90]. In the structural design of such buildings, it is recommended that the fundamental frequency of the structure should exceed 3 Hz, thus avoiding human-induced comfort problems [14,91]. Some studies have proposed that high-frequency floor slabs may also experience human-induced vibration comfort problems [92]; however, this phenomenon requires more engineering cases to be verified.

The human body is an adaptive system, and human behavior is intensely subjective and random, resulting in complex human-induced loads. The establishment of human-induced loads is based on experiments and statistics on human body dynamics parameter calibration and plantar force measurement with different body postures. Furthermore, HSI is a central issue in the study of human-induced vibrations [93]. In particular, when light and strong building materials are applied, the structural frequency decreases, and the ratio of human–structure mass increases, resulting in a more significant effect of HSI. The relevant tests indicate that humans can alter the mass distribution and damping of the structural dynamic system [94–97]. The coupling mechanism and applicable conditions remain to be addressed to develop simplified calculation methods for structural design.

By simplifying the dynamic characteristics of the human body and the HSI, researchers have developed human-induced load models that correspond to various human activities, including walking, running, jumping, and bouncing, among others [89,98,99]. Table 2 lists part of the typical human-induced load models. In particular, the vertical walking load is the most common type of human-induced load and has received the most attention from researchers [89]. The time history of walking force during single-foot landing is double-peaked "M" shapes, consistent with the physical process from heel landing to toe off the ground when walking. In the frequency domain, the walking force has main harmonic and multi-order subharmonic peaks. Types of load models include the Fourier series model, spring-mass-damping model, and bipedal walking model, among others, as listed in Table 2.

Table 2. Typical human-induced load models.

Human Activities	Model Type	Originator, Year
walking (vertical load)	Fourier series model	Zivanovic et al., 2007 [100]; Chen et al., 2014 [101]
	spring-mass-damping model	Silva et al., 2013 [102]; Shahabpoor et al., 2023 [103]
	bipedal walking model	Whittington et al., 2009 [104]; Kim et al., 2011 [105]; Qin et al. 2013 [106]
	multibody model	Máca et al., 2011 [107]
	power spectrum model	Brownjohn et al., 2004 [108]; Piccardo et al., 2012 [109]; Ferrarotti et al., 2016 [110]; Chen et al., 2019 [111]; Wang et al., 2020 [112]
walking (lateral load)	synchronization-based model	Fujino et al., 1993 [113]; Dallard et al., 2001 [114]; Nakamura et al., 2004 [115]; Strogatz et al., 2005 [116]; Blekherman, 2005 [9]; Piccardo et al., 2008 [117]
	self-excitation model	Pizzimenti et al., 2005 [118]; Ingólfsson et al., 2011 [119];
	inverted pendulum model	Macdonald, 2008 [120]; Bocian et al. [121]; Carroll et al., 2011 [122]
running	Fourier series model	Racic et al., 2014 [123]
jumping	Fourier series model	Ellis et al., 2004 [124]
	power spectrum model	Xiong et al., 2018 and 2021 [125,126];
bouncing	Fourier series model	Ji et al., 1994 [127]; Parkhouse et al. 2006 [128]; Duarte et al., 2009 [129];
	SMD model	Dougill et al., 2006 [130]; Wang et al., 2019 [131]

The modeling of human-induced loads is generally based on tests performed on the rigid ground. However, when a human is in a vibrational environment or on a flexible structure, the plantar force can differ from the test results on rigid ground because of the HSI and the spontaneous regulation of the human body system [132]. This influencing factor should be studied to refine the existing human-induced load models.

Human-induced load models applied to residential buildings with common functions have been extensively studied. In recent years, people have expected buildings to have more functionality; therefore, the number of new-functional and comprehensive buildings has increased. A human-induced load model must correspond to a building function. For instance, some special-shaped landscape bridges require staircase walking loads [133]; comprehensive buildings with indoor gymnasiums require sports loads (e.g., basketball, badminton, and volleyball); and comprehensive buildings with performance halls inside require human wave and swing loads. With the promotion and construction of these new buildings, novel human-induced load models required for structural design have become a future research trend.

2.2.2. Crowd Load Model

Because of simplifications in modeling, existing single-human-induced load models cannot completely simulate the dynamic characteristics of the human body yet. However, the mass ratio of one single person to a structure is too small to induce comfort problems. In engineering, the comfort problems of buildings are generally caused by crowd load excitation. Therefore, to improve the accuracy of human-induced vibration calculations, the primary objective is to consider crowd behavior in load modeling instead of further refining the single-human model.

Human dynamic parameters are related to factors such as gender, age, height, and weight, and their distribution in a crowd is strongly random. Moreover, crowd behavior patterns are influenced by building functions, spatial arrangements, and unexpected events [134]. Researchers initially combined single-human models into a crowd model, considered the statistical laws of random factors, and used the Monte Carlo method for a time-domain analysis to calculate the structural response [135]. Methods in the traffic planning field, such as the discrete element theory (DET) [136] and the social force model [137,138], have been used to simulate pathways of crowd activity. However, time-domain load models exhibit natural disadvantages in stochastic analyses. The high computational cost of the Monte Carlo method makes it difficult to use. Presently, the development of time-domain crowd load models has stagnated.

Frequency-domain crowd load models represented by power spectrum models have become a trend because of their advantages in stochastic analyses [108–112,125,126,139,140]. Chen's team contributed outstanding research results using this method [111,112,125,126,140], and the proposed reaction spectrum method was adopted for China's building floor design standard [91]. However, for the current power spectrum model of crowd load, the influence of various random crowd behaviors and the effect of HSI need further consideration.

In addition, with emerging technologies, such as smartphones, big data, motion capture, and neural networks, human-induced load models based on big data have study and application prospects [15,139–142].

2.3. Wind-Induced Vibration

Wind-induced vibration is the dynamic response of a structure to fluctuating wind. Fluctuating wind loads are horizontal dynamic loads with random characteristics that act on building structures. Because of the fluid–structure interaction, the structural dynamic behavior caused by wind loads is complex and includes downwind and crosswind vibrations (e.g., vortex-induced vibration).

Researchers have conducted measurements and formulated different wind speed spectra, such as Karman, Davenport, Harris, Kaimal, and Simiu wind spectra, to analyze downwind vibrations. Among these, the Davenport wind spectrum is based on the measured data of multiple regions and, therefore, has wide applicability [143]. Equation (3) delineates the expression of the power spectrum $S_v(n)$. By using the wind speed spectrum, pulsating wind speeds can be simulated through methods like linear filtering, harmonic superposition, or other techniques. Subsequently, the pulsating wind pressure load is derived from the correlation between wind pressure and speed.

$$S_v(n) = 4K\overline{v}_{10}{}^2 \frac{x^2}{(1+x^2)^{4/3}} \tag{3}$$

where $x = 1200\, n/\overline{v}_{10}$; n is frequency; $\overline{v}_{10}$ is the mean hourly wind speed at the standard reference height of 10 m; K is the drag coefficient for the surface.

The complexity of crosswind vibrations and wind loads in tall buildings surpasses that of downwind vibrations, primarily due to factors like vortex-induced resonance. Developing a mathematical model that accurately describes vortex-induced forces is an effective approach to address this problem, albeit challenging. The complexity of vortex-induced vibrations stems from the shedding of wake vortices at different stages within the lock-in range, and various structural boundary conditions lead to the emergence of different modes. Several simplified mathematical models for vortex-induced forces have been proposed. For instance, the Ruscheweyh model describes vortex-induced forces as harmonic forces [144] in Equation (4). However, the simple model lacks the representation of key properties of vortex-induced vibrations, such as the locking phenomenon and self-limiting amplitudes. Bishop [145], Tamura [146], Vickery [147], Ehsan [148], Larsen [149] et al. proposed more intricate and refined expressions to depict the relationship between frequency, amplitude, and phase of vortex-induced forces and structural parameters. The key fluid parameters in these models must be assumed in advance or identified from measured vortex-induced vibration responses of the structure. Currently, simulation based on computational fluid dynamics is the main technique to calculate vortex-induced forces [150,151].

$$F_v = \frac{1}{2}\rho V^2 D C_L \sin(\omega_s t + \varphi) \quad (4)$$

where F_v is the vortex-induced force; ρ is the density of air; V is the wind speed; D is the structural cross-section dimensions in crosswind direction; C_L is the root mean square of the lift coefficient; ω_s is the circular frequency; t is the time; φ is the phase difference between structural displacement response and vortex-induced force.

Early research on the structural wind resistance design of high-rise buildings focused mainly on structural safety and less on comfort [152]. With the development of urban construction, the number of high-rise and super-high-rise structures is increasing, and wind-induced vibration comfort has often determined the structural design of these buildings. Studies of wind-induced vibration comfort have become increasingly critical.

Traditional methods for safety-based wind-resistant structural analysis are unsuitable for comfort-based analysis, and the understanding of wind-induced comfort calculation and evaluation methods has not been unified. The evaluation index in China uses the downwind and crosswind peak acceleration on top of structures [153,154], with the aim of controlling the vibration at the peak position of the structural first-order mode. However, the tops of high-rise or super-high-rise buildings are often unmanned activity rooftop platforms, masts for lightning protection, or antennae, where comfort problems do not occur. Moreover, wind loads can excite high-order modes of the structure under specific circumstances, which may result in maximum vibration occurring in the middle floors of the building. The ISO and some countries, such as Japan, stipulate that the maximum peak acceleration in the downwind and crosswind directions of all occupied floors must be evaluated [155,156]. In addition, the maximum direction of the wind-induced vibration requires further discussion. When the mass center of the structure does not coincide with the stiffness center, the structure has torsional vibration modes, and the maximum direction of the horizontal vibration can form an angle with the downwind and crosswind directions. Furthermore, under wind load excitation, vertical vibrations of the floor slabs occurred with horizontal vibrations because of the smaller out-of-plane stiffness of the slabs. Therefore, the vibration directions used for the evaluation, such as downwind, crosswind, maximum horizontal, vertical, or whole-body vibrations, require further determination.

The recent SEG Plaza building vibration event (Shenzhen, China, 2021) brought wind-induced comfort problems into public view and attracted the attention of researchers and engineers [157–160]. Experts investigated the event, and the cause was found to be the coupling of the vortex-induced resonance of the mast and the change in the dynamic characteristics of the building and the mast. This event may become a landmark in research on wind-induced vibration comfort. Discussing the vibration mechanism, structural analysis theory, evaluation methods, and other issues can significantly promote the study of wind-induced vibration comfort and update structural design and evaluation systems.

2.4. Machinery-Induced Vibration

Machinery-induced vibration is the dynamic response of structures to dynamic machinery. According to the function and use scenario, the possible machinery includes living equipment (elevators [161], water pumps, power transformers [162], and fans, among others), industrial production equipment (ball mills, bridge cranes [163], and punch machinery, among others), and engineering construction equipment (dynamic compaction machinery [164], rock excavation [165], and pile drivers, among others). Owing to the shortage of urban land and intensive building planning, the distance between this dynamic machinery and the living environment is gradually reduced, even inside buildings. In general, the load of dynamic machinery is not complex. However, different types of dynamic machinery have different characteristics in the time and frequency domains, and their corresponding load models and modeling techniques are also different. Based on the generation mechanism of machinery-induced vibrations and load characteristics, dynamic machinery loads can be divided into three types, and commonly used load modeling techniques are listed in Table 3.

Table 3. Dynamic machinery load types and modeling techniques.

Type No.	Machinery Examples	Generation Mechanism	Load Characteristic	Modeling Techniques
I	elevator traction machinery, power transformer, water pump, fan, ball mill	reciprocating or rotating mechanisms	smooth stochastic process; fixed frequency components; spectrum contains main and multi-order harmonic frequencies	simple harmonic forces; Fourier series model; power spectrum model
II	elevator car, bridge crane	machines move along fixed tracks or routes on the structure	moving load (related to mass, speed, track irregularities, and contact relationships)	moving force, multi-rigid body model
III	punch machinery, dynamic compaction machinery, pile driver	impact behavior during machinery operation	pulse load	impact force, impact process numerical simulation

There are fewer research cases involving dynamic machinery loads and induced structural dynamic response compared to the other categories of loads mentioned above, such as vehicle-induced and human-induced loads. Moreover, because of the large number of machinery categories, most machinery categories have not been developed into mature load models. In engineering, measured vibration data at the vibration source of machinery are used directly as the input. With people paying more attention to comfort problems, it is necessary to establish standardized load models for common dynamic machinery. In addition, structural calculation models, evaluation methods, and isolation measures for machinery-induced vibrations require further investigation.

3. Comfort-Based Structural Analysis Method

With the development of numerical simulation technology and the upgrading of computer performance, the FEM has become the mainstream method for dynamic analysis in structural design. Compared with safety-based structural simplification models for strong vibration analysis, comfort-based structural model simplification methods should be different.

Strong vibration analysis focuses on the bearing capacity or deformation properties of the main load-bearing skeleton of a structure under ultimate loads such as earthquakes. The purpose is to avoid the loss of the seismic or gravity-bearing capacity of the entire structure because of damage to parts of the structure or components. Instead, vibration comfort problems are within the scope of weak vibration analysis. This focuses on the serviceability of structures under micro-amplitude vibrations. Its purpose is to service the people in the structure and consider the comfort of the people in the vibration environment as a criterion. Under micro-amplitude vibrations, the structure has small deformation and is in a state of linear elasticity. Therefore, the comfort-based model should reflect the dynamic characteristics of the objective structure under micro-amplitude vibrations and accurately calculate the acceleration of the floor where people are located. In addition, the median perception threshold of fit persons for the weighted acceleration peak magnitude is approximately 0.015 m/s^2 [166,167], indicating that the accuracy of comfort-based analysis should also achieve this magnitude.

Based on the above principles, He and Xie [168] proposed a sophisticated calculation model for large-span railway station structures based on the evaluation of vibration serviceability. Based on the strong vibration model of the structure, the method models the non-structural components, considers the local construction and boundary conditions, modifies the calculation parameters such as the live load coefficient and damping ratio, and finally obtains a refined elasticity FEM of the structure during the normal service state. Furthermore, by validating additional engineering structures, the modeling method can be suitable for more structural comfort problems [43,161,162]. The following subsection presents the critical differences between the comfort-based structural analysis model and the traditional strong vibration model.

3.1. Structural Floor Slab

The floor slab is the structural component where people are located and the source of the human body receiving vibrations, and therefore, is the location that standards specify for evaluation. The dynamic characteristic of structural floor slabs is an emphasis in modeling.

For a structural floor, a strong vibration model generally adopts the rigid floor assumption and ignores its out-of-plane stiffness. Instead, in comfort-based structural analysis, the main structure, including the structural floor slabs, must be simplified as elastic elements. Researchers have generally accepted this opinion and developed various elasticity calculation models [163,169]. Application of conventional modeling techniques has been shown to be adequate for typical floor slabs such as reinforced concrete slabs. However, novel slab materials or composite floor types have recently been developed with excellent performance and complex construction [170–172]. For each of these applications, it is necessary to study the dynamic behavior by developing suitable modeling techniques.

3.2. Non-Structural Components

Non-structural components are non-load-bearing components in the structural design, including flooring finishes, infill walls, and curtain walls, among others. Strong vibration models generally ignore the effects of non-structural components on the global stiffness and damping of the structure. The model simplified these components to additional mass or load composite values. Instead, when the structure is in a state of normal use, some non-structural components can change the structural mass, stiffness distribution, and damping properties, which can significantly affect the dynamic characteristics of the structure under micro-amplitude vibrations, thus affecting the dynamic response [173]. The influence of various non-structural components on the dynamic characteristics of structures has become a topic of interest [174]. Two types of non-structural components with significant impacts are the non-structural floor layer and wall, which are explained in the following.

- Non-structural layer of floors

The non-structural floor layer is the functional and decorative layer laid on the structural floor slab, such as a thermal insulation layer (foam board), levelling layer (cement mortar), and decorative surface layer (tiles, timber, marble, rubber, and elevated flooring). These non-structural layers may increase floor thickness by more than 50%. Research indicates that an insulation layer can significantly increase the damping ratio of floors [168]. Different materials and types of decorative surfaces have different effects on the dynamic characteristics of floor slabs [174–177]. These effects should be considered in the calculations to modify the dynamic characteristics of the floors.

- Non-structural wall

Non-structural walls, such as infill and curtain walls, have no load-bearing functions in their structural design. In comfort-based structural analysis, lightweight partition walls can be equivalently modeled in terms of mass and stiffness; infill walls can transmit inter-story vibrations, increase the stiffness and damping of floors, alter the modal shapes [168,178], and increase the global lateral stiffness of the structure [179]; curtain walls have weak confinement effects between layers in addition to the effect of mass [180]. At present, the interaction mechanism between non-structural walls and the main structure remains to be investigated. Infill walls are constructed directly on the floor beams. Curtain walls are hung on the building facade using steel studs. The dynamic behavior of the connection between non-structural walls and the main structure is central to such studies. The connection stiffness and the vibration propagation law under weak vibration require further study.

In summary, the importance of non-structural components in comfort-based analyses has gradually been accepted and considered in modeling. However, the refined modeling for non-structural components increases the complexity of the numerical models and decreases the calculation efficiency. Non-structural components are varied and complex, and new component categories are constantly being developed and applied. Research on various non-structural components requires further discussion. Quantifying the effects of various non-structural components and developing simplified methods are practical approaches to consider non-structural components in structural design.

3.3. Local Construction and Boundary

Local constructions include window and door openings, structural joints, and bolted joints, among others. The vibration characteristics of these constructions under micro-amplitude vibrations differ from those under strong vibration conditions. For instance, window and door openings can decrease the mass and stiffness of the infill wall and alter its effect on the structural dynamic characteristics; structural joints under strong vibrations divide the structure completely, whereas those under micro-amplitude vibrations can transmit vibrations, and thus be simplified semi-rigid spring connections; trusses and bolted connections under strong vibration are considered to be axially loaded members and hinges, whereas the bolted nodes of trusses almost never rotate under micro-amplitude vibrations, thus be modeled as beam members and rigid or semi-rigid, respectively.

Modeling of the boundary depends on the selection of the calculation domain. Based on the vibration propagation features under different loads and structures, selecting the calculation domain and reasonably addressing the boundary can improve calculation efficiency while balancing accuracy requirements. For instance, the following can be performed: in vehicle-induced vibration, modeling the building structure and setting up equivalent stiffness and damping at the base of the structure to consider the SSI [78]; in human-induced vibration, modeling partial floors and constraining boundaries for calculations [169]; and in wind-induced vibration, modeling the superstructure, ignoring the foundations, and constraining the model bottom as the fixed end [157]. Current numerical models may have redundant parts that affect the calculation efficiency, and the vibration propagation mechanism requires further investigation.

3.4. Other Dynamic Parameters

This subsection adds other dynamic parameters that are not discussed above but should not be ignored in comfort-based structural analyses.

- Additional mass

Placing large-mass items on the floor, such as large decorations, fixed equipment, and furniture, can change the mass distribution and decrease the fundamental frequency of the floor [181].

- Dynamic elastic modulus

Elastic modulus can impact the natural frequency of structures, which is a key parameter in modeling [182,183]. In comfort-based structural analyses, the elastic modulus of the material at small deformations, that is, the dynamic elastic modulus, should be adopted. Compared to the static elastic modulus obtained by the axial compression experiment of prisms, the dynamic elastic modulus was obtained by the ultrasonic pulse method, representing the relationship between the stress and strain of the material under dynamic loading. For instance, the dynamic elastic modulus of concrete is greater than its static elastic modulus [184]. Bachmann et al. [12] proposed using the dynamic modulus of elasticity to predict the fundamental frequency of the floors in calculating human-induced vibration comfort. Therefore, the technical standard in China stipulates that for reinforced concrete floor or steel-concrete composite floor, the elastic modulus used in the calculation should be 1.2–1.35 times larger than the static elastic modulus to consider the dynamic elastic modulus [91].

- Damping ratio

Damping reduces the structural response. Structural damping mechanisms are complex, and the current theories are not fully comprehensive. In general structures, damping is mainly related to the material properties and other damper mechanisms used in the building. When a structure has great deformation transitioning from elasticity to plasticity, its damping can increase significantly. Consequently, the damping ratio in comfort-based structural analyses should be lower than in seismic analyses. Furthermore, even when structural deformation remains within the elastic range, the dissipation of damping energy during motion is influenced by the extent of deformation, thus exhibiting a relevance of the value of the damping ratio to the vibration amplitude [185,186]. Although a definitive relationship has yet to be established, it is generally observed that more minor excitation results in smaller vibration amplitudes, corresponding to a lower damping ratio. Therefore, in comfort-based structural analyses, the damping ratio under micro-amplitude vibration of the structure should be adopted, which is different from strong vibration analyses. Measured damping ratios of steel and concrete structures are generally about 0.01–0.02 during the normal service stage (micro-amplitude vibration) [168,182,187]. In addition, the wind-induced vibration of high-rise buildings is sensitive to the damping ratio. Hu et al. [157] measured a super-high-rise building and observed a sudden decrease in the damping ratio when wind-induced structural resonance occurred, which may be related to the aerodynamic damping effect [188].

- Construction and measurement error

Construction errors during the building process are common, such as those involving the properties of materials and the dimensions of beams, columns, and slabs. This results in the dynamic characteristic of a completed structure that differs from the design expectations. Construction errors are generally not considered in current structural modeling. In addition, ambient temperature changes can influence the dynamic characteristics of structures such as pedestrian bridges, resulting in fluctuations in the measured self-vibration frequency [189]. Wang et al. [190] monitored a reinforced concrete slab for over a year and observed that possible temperature changes of 0~50 °C can lead to an error of more than 15% in the self-vibration frequency of the tested slab. Errors in identifying structural dynamic characteristics and response measurements can impact the validation of structural models. Therefore, when the dynamic characteristics of the model are not consistent with the measurements, the effect of construction and measurement errors should be considered, in addition to the errors caused by modeling simplification and algorithms.

In summary, the comfort-based structural analysis aims to accurately calculate the vibration of the floors in the user activity area. Its principle is modeling the key members (component, connection, and boundary, among others) in detail and simplifying others based on the effect of each member under micro-amplitude vibration. However, this method has not yet covered all types of vibration sources, structures, or construction details, and a systematic method for structural design has not yet been developed.

4. Vibration Comfort Evaluation Method

Vibration comfort is based on human feelings, which are influenced by the following three categories of factors: (a) external vibration factors, including direction, frequency, amplitude, duration, and load return period; (b) external environmental factors, including structural-borne noise, airborne noise, rattling, movement of furniture, and visual effects (e.g., movement of hanging objects); and (c) receiver factors, including person type (gender, age, height, and health, among others), body posture, experience, expectation, activity motivation (e.g., ease or difficulty of the task performed), economic or living situation, and subjective ambiguity.

For these influencing factors, based on biodynamics, psychology, and fuzzy mathematics methods, researchers have statistically analyzed a large amount of test data, and their conclusions have been applied to comfort evaluations. The relationship between comfort and the amplitude and frequency of vibration acceleration has been studied in depth. The human body in buildings is sensitive to vibration in the 1–80 Hz range [191], and the ISO provides frequency weighting methods for vibrations in different directions and body postures [166,167]. Table 4 lists some of the comfort evaluation standards in various countries with their indexes and partial limits.

Table 4. Comfort evaluation standards in various countries.

Standard; Country, Region, or Organization	Vibration Source	Comfort Index	Evaluation Objects and Limits (Part)
ISO 2631-1: 1997; ISO [166,167]	-	RMS; VDV	uncomfortable: "RMS" 0.8 m/s^2 to 1.6 m/s^2
ISO 10137-2007; ISO [14]	vehicle; human; wind; machinery	RMS; VDV; PA	1. vibration induced by vehicles, humans, or machinery: "RMS, VDV" base curve × multiplying factor; 2. wind-induced vibration: "PA" frequency-dependent curve
ISO 6897-1984; ISO [156]	wind	RMS	frequency-dependent curve
GB 10070-88; China [192]	vehicle; machinery	VAL (VL_z)	residential, cultural and educational building (day): 70 dB
JGJ/T 441-2019; China [91]	human; machinery	PA	building floor with walking loads primarily (residential, hospital, office, etc.): 0.05 m/s^2
JGJ/T 170-2009; China [193]	vehicle	VAL (VL_{max})	residential, cultural and educational buildings (day): 65 dB
JGJ 3-2010; China [153]	wind	PA	concrete structure of tall buildings (residential, apartment): 0.15 m/s^2
JGJ 99-2015; China [154]	wind	PA	steel structure of tall buildings (apartment): 0.20 m/s^2
ATC Design Guide 1; USA [194]	human	RMS; PA	1. vertical vibration: "RMS" base curve × multiplying factor; 2. horizontal vibration (office, residential, etc.): "PA" 0.02 m/s^2
AISC Design Guide 11; USA [195]	human	PA	floor and pedestrian bridge: base curve × multiplying factor
EN 1990: 2002 [196]	human	PA	vertical vibration (pedestrian bridge): 0.7 m/s^2
BS 6472-1: 2008; UK [197]	-	VDV	residential building (16 h day): adverse comment possible 0.4 $m/s^{1.75}$ to 0.8 $m/s^{1.75}$
AIJ ES001-V001; Japan [198]	vehicle; human; wind; machinery	PA	building: frequency-dependent curve
AIJ-GEH-2004; Japan [155]	wind	PA	building: frequency-dependent curve

Note: RMS is weighted acceleration root mean square; VDV is vibration dose value; PA is peak acceleration; VAL is weighted vibration acceleration level, and Z-vibration level VL_z is weighted vertical VAL, frequency maximum magnitude of vibration VL_{max} is maximum weighted vertical VAL at the 1/3-octave center frequency.

From Table 4, the comfort indicators of the existing standards are not unified and use different limits, which lack the basis for conversion. Some indicators are tailored to specific application conditions, yet even for the same comfort issue, different indicators may exist. In engineering applications, using different standards can yield divergent assessment outcomes. These comfort evaluation indices can be classified into two categories. For vibrations with a single primary frequency, such as human-induced vibrations and wind-induced vibrations, peak acceleration is generally adopted [14,91,153–155,194–196,198], whereas broadband vibrations, such as vehicle-induced vibrations, are generally described as weighted acceleration root mean square $a_{\mathrm{w,RMS}}$, vibration dose value $a_{\mathrm{w,VDV}}$ [14] or Z-vibration level VL_z [192], as expressed in Equations (5)–(7). Furthermore, when assessing the vibration comfort problems of floors and pedestrian bridges caused by humans, both natural frequency and peak acceleration are commonly employed [199]. For instance, the Chinese standard JGJ/T 441-2019 specifies that for the floor slabs in residential or office buildings primarily excited by walking loads, the first-order vertical self-vibration frequency should be over 3 Hz, and the limit of vertical peak acceleration is 0.05 m/s^2 [91].

$$a_{\mathrm{w,RMS}} = \left[\frac{1}{T}\int_0^T a_{\mathrm{w}}^2(t)\mathrm{d}t\right]^{1/2} \tag{5}$$

$$a_{\mathrm{w,VDV}} = \left[\frac{1}{T}\int_0^T a_{\mathrm{w}}^4(t)\mathrm{d}t\right]^{1/4} \tag{6}$$

where $a_{\mathrm{w}}(t)$ is the weighted acceleration as a function of time; T is the duration of the measurement.

$$VL_z = 20\lg\frac{a_{\mathrm{w,RMS}}}{a_0} \tag{7}$$

where $a_0 = 10^{-6}\mathrm{m/s^2}$ is the reference acceleration.

Most existing evaluation standards do not consider or simplify the vibration duration and load return period of the influencing factors. In ISO 10137-2007, different multiplying factors are adopted for the "continuous vibration and intermittent vibration" and the "impulsive vibration excitation with several occurrences per day" [14], which demands more stringent limits for the first type of vibration source. The ISO standard published in 1985, ISO 2631-1:1985 Evaluation of human exposure to whole-body vibration-Part 1: General requirements (Withdrawn) [200], specified the evaluation limits for exposure times from 1 min to 24 h and an equation for the accumulation of exposure times for intermittent vibration. As the exposure time increases, a person's tolerance to vibration decreases, and the demand for limits becomes more stringent. However, this part was simplified in the revised version of ISO 2631-1:1997, and the definition of "fatigue-decreased proficiency" was removed [166,167]. Despite this, the influence of the vibration duration and load return period on comfort evaluation should not be ignored, and further refined evaluation methods are required.

In principle, the position of comfort evaluation should be located at the maximum vibration of floors in the user activity area, which is related to the load characteristics, load locations, structure types, and building functions. However, the existing evaluation method requires further improvement because of insufficient conclusions regarding the location of the maximum vibration and the lag of the standards with respect to the latest research. The shortcomings include the following: (a) in vehicle-induced vibration, the spatial distribution of the response is uncertain, and the measured maximum locations can be observed on the lower [201,202], middle [203], or top floors [204–206]. Therefore, the influence of the structural global and floor local modes on the structural response requires discussion. (b) Comprehensive evaluation of multi-directional vibration requires consideration, such as the horizontal response induced by vehicles in high-rise buildings increases with height [207], and horizontal wind loads may cause vertical floor vibration. (c) The comfort evaluation limits for different types of structures should be unified. For example, China's wind vibration evaluation standards specify different peak acceleration limits for concrete and steel structures [153,154]. In addition, the vibration duration and load return period should be added to the comfort evaluation system.

Existing comfort tests have some shortcomings that require further refinements, such as insufficient sample capacity, the difficulty in artificially simulating vibration to accurately reflect the actual vibrations, consideration of the psychological state of the subject, the difficulty in quantifying the influence of external environmental factors, and the current lack of standardized test technology. With the widespread adoption of smartphones and the Internet, human body movements and comfort feelings can by conveniently recorded and then collected by uploading them to the cloud via the Internet. Deep learning and artificial intelligence provide efficient algorithms to train collected big data of the important factors under various influential factors [208–210]. Researchers have begun exploring of these emerging technologies in studying comfort indexes. For instance, Chen et al. [211] proposed a smartphone-based evaluation system for pedestrian-induced footbridge vibration comfort; Cao et al. [15,212] designed a smartphone-based application and an online big data approach to investigate vibration serviceability limits in real environments. In addition, the vibration acceleration of structures is an essential indicator for assessing vibration comfort and obtaining reliable acceleration signal data. Various test techniques developed for structural health monitoring could be used to enhance the comfort-based vibration testing approach [213,214].

5. Mitigation Measures for Vibration Comfort

When the structural response exceeds standard limits and can cause discomfort for building occupants, vibration mitigation measures must be implemented. Vibration control technology has mature theories and has evolved alongside the development of new materials and processes. The vibration sources that cause comfort problems lead to structural responses with small amplitudes and possibly high frequencies, making not all techniques or products for the earthquake-resistant design applicable. Depending on the presence or absence of external energy input, these measures can be categorized as active or passive control technologies. In micro-amplitude vibration control for comfort, passive control technology is primarily utilized. This section focuses on the commonly employed measures to address comfort issues, including vibration isolation in railway tracks, vibration isolation barriers in soil, vibration isolation bearings for building structures, and tuned mass dampers (TMD).

5.1. Vibration Isolation in Railway Tracks

Vibrations induced by vehicles and machinery are artificial vibration sources, making isolating these sources an effective control strategy. In railway tracks, many systematic vibration isolation applications have been implemented [215–217]. These isolators are typically installed at the rails, rail fastenings, rail slabs, or ballast beds to mitigate the propagation of vehicle-induced vibrations to the surrounding environment. In recent years, phononic crystal theory has been employed in the design of vibration control techniques. For instance, Hu et al. [218] developed a periodic layered slab track structure. The currently applied vibration isolators are typically effective in attenuating high-frequency vibrations but exhibit limited efficacy in addressing low-frequency components, sometimes even amplifying such vibrations. Low-frequency vibrations demonstrate less attenuation through soil over distances compared to high-frequency vibrations additionally, increasing the risk of resonance with building structures possessing low-frequency natural frequencies. Consequently, enhancing the capacity of track vibration isolators to mitigate low-frequency vibrations has emerged as a primary imperative. Sung et al. [219] developed an additional anti-vibration sleeper track to reduce low-frequency vibrations. Qu et al. [220] introduced a railway track concept inspired by chiral phononic crystals, offering significantly improved low-frequency vibration isolation through an orthogonal polarization coupling mechanism within the chiral subunit cell. Although track vibration isolation measures have been commonly implemented in urban rail transit systems, their adoption in high-speed rail networks remains limited due to safety and durability concerns.

5.2. Vibration Isolation Barrier in Soil

Vibration isolation barriers are constructed within the soil to isolate propagation paths. This strategy is particularly relevant in scenarios involving vibrations generated by vehicles or machinery. Common types of vibration isolation barriers include in-filled trenches [221–223], arrays of piles [224], and wave-impeding blocks (WIB) [225–227], among others. Vibration waves exhibit phenomena such as reflection and diffraction upon encountering these barriers. Consequently, the dimensions of the barrier, including width, length, and depth, must be determined based on the wavelength and the dimensions of the vibration source and the sensitive structure. In recent years, the application of phononic crystal theory for analyzing wave behaviors in periodic structures has gained significant traction. For instance, Pu et al. [228] revisited the surface wave manipulation in periodic trench barriers from the perspective of complex band structures. A limitation of vibration isolation barriers is the large engineering quantity for construction, especially when targeting long wavelengths for isolation, which necessitates extensive lengths and depths. For example, fillers in in-filled trenches may encounter contamination or consolidation, diminishing the vibration mitigation effect.

5.3. Base Vibration Isolation Bearing

With the focus on comfort issues induced by railways and the construction of sensitive building structures like over-track buildings, there is a growing need for vibration isolation bearings to mitigate vehicle-induced vibrations. Traditional isolation bearings, such as laminated rubber bearings, lead rubber bearings, and friction pendulum bearings, have proven effective in mitigating horizontal ground motion caused by earthquakes. However, vehicle-induced vibrations primarily occur in the vertical direction. The mechanism of isolation bearings involves shifting the natural frequency of the superstructure to deviate from the primary frequency of ground motion [229]. Vehicle-induced excitation typically spans a broad frequency range, posing the challenge of frequency shifting. Initially, researchers investigated the impact of traditional seismic isolation bearings on vertical railway-induced vibration isolation [49,230,231]. Subsequently, the concept of integrated control of engineering and seismic vibrations in China has led to a surge in research interest in three-dimensional vibration isolation bearings (3D-VIB) that can effectively address both seismic and vehicle-induced vibrations [232]. Building upon traditional seismic bearings, researchers have proposed various innovative designs and configurations for these bearings. For instance, Cao and Pan et al. [233,234] developed a 3D-VIB using a combination of disc springs and single friction pendulum bearings and another design incorporating a thick laminated rubber bearing and a friction pendulum. Sheng et al. [235] developed a new 3D-VIB, which is a laminated rubber bearing with vertical through-holes filled by a mixture of sand and rubber particles. Liang et al. [236] proposed a 3D-VIB composed of rubber isolation bearings and disc springs. He et al. [237] proposed a hybrid vibration bearing consisting of lead rubber bearing and thick rubber bearing with decoupled horizontal and vertical behaviors. While the aforementioned innovative vibration isolation bearings have demonstrated outstanding performance in both experimental and simulation settings, their practical application effects in engineering remain to be validated. In addition, isolation bearings intended for vehicle-induced vibrations must permit relative motion between the superstructure and substructure under daily micro-amplitude vibrations. It inevitably decreases horizontal stiffness and potentially raises risks of destabilization.

5.4. TMD in Building Structures

TMDs currently serve as a primary technology for mitigating wind-induced vibrations in super tall buildings, human-induced vibrations in large-span floors and pedestrian bridges [238,239], and occasionally, vibrations caused by machinery within structures [240]. The fundamental configuration of TMDs consists of a mass, springs, and dampers and is designed to absorb vibrations at specific frequencies. The target frequency is generally the horizontal first-order frequency for tall buildings or the vertical first-order frequency for floors and bridges. TMDs are installed at the location of the maximum amplitude of the mode shapes. The ratio between the TMD mass and the generalized mass of the to-be-controlled mode defines the maximum vibration reduction effect. Therefore, conventional TMDs often exhibit excessive mass, posing safety risks. Distributed tuned mass dampers

(MTMD) have been proposed to address this issue, and additionally implemented for the control of multi-order structural modes [241]. With the development of new materials and processes, novel TMDs have been proposed, including: (a) TMDs with more complex DOFs, nonlinear stiffness components or novel dampers [242–245]; (b) semi-active tuned mass dampers (SATMDs) [246,247] and active rotary inertia driver (ARID) systems [248]; and (c) tuned mass damper inerters (TMDIs) [249]. Furthermore, the fixed-point law in classical Den Hartog theory has been widely adopted for engineering designs. Using the minimum acceleration of the original structure as the optimization criterion, Ikeda et al. [250] derived the optimal ratios of frequencies and damping ratios are

$$\gamma_{\text{opt}} = \frac{1}{1+\mu} + \left(0.096 + 0.88\mu - 1.8\mu^2\right)\xi + \left(1.34 - 2.9\mu + 3\mu^2\right)\xi^2 \tag{8}$$

$$\xi_{\text{opt}} = \sqrt{\frac{3\mu(1+0.49\mu+3\mu^2)}{8(1+\mu)}} + \left(0.13 + 0.72\mu + 0.2\mu^2\right)\xi + \left(0.19 + 1.6\mu - 4\mu^2\right)\xi^2 \tag{9}$$

where γ_{opt} and ξ_{opt} are the optimal ratios of frequencies and damping ratios, respectively; μ is the mass ratio; ξ is the damping ratio of the structure.

Nevertheless, researchers have continued to investigate the location and parameters of the novel TMDs to optimize the adsorption effect and robustness, such as H_2 norm optimization [251]. In addition, despite the widespread adoption of TMD technology, its efficacy in vibration reduction occasionally falls short of engineering expectations due to various factors: low sensitivity of TMDs, which cannot be activated in case of micro-amplitude vibrations; malfunctions in critical components such as damper oil leaks; inaccurate prediction of structural frequency resulting in the deviation of the parameters of the pre-customized TMD from the optimal values.

6. Conclusions

Compared to traditional seismic and wind resistance analyses, vibration comfort remains a niche research direction. Nevertheless, it covers a broad scope and often involves collaboration across multiple disciplines and fields. In recent years, research on vibration comfort has progressed significantly, and calculation theories and evaluation methods have been established. However, there are still some issues to be addressed. This paper discusses recent developments in this field and presents the existing shortcomings in research and applications. The following are suggested potential topics for future studies regarding the load, structural analysis, evaluation methods, and vibration mitigation measures.

1. Standardized stochastic load models

The main load categories that can cause vibration comfort problems are vehicle-induced, human-induced, fluctuating wind, and dynamic machinery loads. Current load models for structural design are not sufficiently developed for adoption. For fluctuating wind loads, mature power spectrum models have been established; for human-induced loads, relevant studies are ongoing; however, for the various types of dynamic machinery loads and the complex vibration source mechanism of vehicle-induced loads, widely accepted standardized stochastic load models are still lacking.

2. Simplified comfort-based modeling method for structural design

With increasing concern for vibration comfort problems, more building structures require specialised structural design and optimisation based on comfort. However, the current comfort-based modeling methods developed by researchers have not yet achieved the balance of convenience, accuracy, and efficiency required for structural design. Therefore, simplified comfort-based modeling methods for structural design have essential engineering applications.

3. Comfort evaluation method considering duration and load return period

Existing vibration comfort evaluations only consider the amplitude, frequency, and direction of vibration, and ignore other features, such as the vibration duration and load return period. As the types and forms of vibration sources increase, evaluation methodologies require further refinement.

4. Application of novel vibration mitigation measures

With the development of new materials (e.g., polymer materials, magnetorheological materials, and shape memory alloys), innovations in vibration isolator construction, and proposed corresponding vibration isolation design theories, vibration reduction and isolation measures have been upgraded and optimized. Researchers initially use theories, simulations, or experiments to innovate novel technologies and finally promote their application. However, many novel technologies have been devised without considering their implications on the safety and durability of the original structure, which should be emphasized.

Structural vibration comfort is a complex aspect of building design that requires careful consideration to ensure a comfortable environment for occupants. One possible approach is the installation of an alarm system that can monitor and notify managers and occupants of excessive structural vibrations. This system can help detect vibration sources within the building and effectively prevent secondary accidents, such as stampedes caused by excessive panic in crowds. The latest vibration-based structural health monitoring techniques could be feasibly utilized in the design of the system [252,253]. In addition, big data and machine-learning techniques also provide new ideas for research on structural vibration comfort and provide directions for future study. They can be useful in studying the stochastic behavior of loads and structures and the subjective ambiguity in comfort evaluation.

Author Contributions: Writing—original draft preparation, Y.H.; writing—review and editing, W.X. and Y.H.; supervision, W.X.; funding acquisition, W.X. All authors have read and agreed to the published version of the manuscript.

Funding: This research was funded by Natural Science Foundation of China, grant number 52278458.

Data Availability Statement: Data sharing is not applicable.

Conflicts of Interest: The authors declare no conflicts of interest.

References

1. Reiher, H.; Meister, F.J. Human sensitivity to vibrations. *Eng. Res.* **1931**, *2*, 381–386.
2. Miwa, T. Evaluation methods for vibration effect part 1: Measurements of threshold and equal sensation contours of whole body for vertical and horizontal vibrations. *Ind. Health* **1967**, *5*, 183–205. [CrossRef]
3. Jones, A.J.; Saunders, D.J. Equal comfort contours for whole body vertical, pulsed sinusoidal vibration. *J. Sound Vib.* **1972**, *23*, 1–14. [CrossRef]
4. Irwin, A.W. Perception, comfort and performance criteria for human beings exposed to whole body pure yaw vibration and vibration containing yaw and translational. *J. Sound Vib.* **1981**, *76*, 481–497. [CrossRef]
5. Gierke, M.J. *Handbook of Human Vibration*; Academic Press: London, UK, 1990.
6. Kanda, J.; Tamura, Y.; Fujii, K.; Ohtsuki, T.; Shioya, K.; Nakata, S. Probabilistic evaluation of human perception threshold of horizontal vibration of buildings (0.125 Hz to 6.0 Hz). In Proceedings of the Structures Congress XII, Atlanta, GA, USA, 24–28 April 1994.
7. Shioya, K.; Fujii, K.; Tamura, Y.; Kanda, J. Human perception thresholds of two dimensional sinusoidal motion in horizontal plane. *J. Struct. Constr. Eng.* **1994**, *461*, 29–36. [CrossRef] [PubMed]
8. Dallard, P.; Fitzpatrick, T.; Flint, A.; Low, A.; Roche, M. London Millennium bridge: Pedestrian-induced lateral vibration. *J. Bridge Eng.* **2001**, *6*, 412–417. [CrossRef]
9. Blekherman, A.N. Swaying of pedestrian bridges. *J. Bridge Eng.* **2005**, *10*, 142–150. [CrossRef]
10. Xia, H. *Traffic Induced Environment Vibrations and Controls*; Science Press: Beijing, China, 2010.
11. He, Y.; Zhang, Y.; Yao, Y.; He, Y.; Sheng, X. Review on the prediction and control of structural vibration and noise in buildings caused by rail transit. *Buildings* **2023**, *13*, 2310. [CrossRef]

12. Bachmann, H.; Ammann, W.J.; Deischl, F.; Eisenmann, J.; Floegl, I.; Hirsch, G.H.; Klein, G.K.; Lande, G.J.; Mahrenholtz, O.; Natke, H.G.; et al. *Vibration Problems in Structures: Practical Guidelines*; Birkhäuser: Basel, Switzerland, 1995. [CrossRef]
13. *ISO 5805-1997*; Mechanical Vibration and Shock—Human Exposure—Vocabulary. International Organization for Standardization: Geneva, Switzerland, 1997.
14. *ISO 10137-2007*; Bases for Design of Structures—Serviceability of Buildings and Walkways against Vibrations. International Organization for Standardization: Geneva, Switzerland, 2007.
15. Cao, L.; Chen, J. Big data investigation for vibration serviceability using smart phones. *J. Vib. Eng.* **2020**, *33*, 961–970. [CrossRef]
16. Willis, R. Preliminary essay to the appendix B: Experiments for determining the effects produced by causing weights to travel over bars with different velocities. In *Report of the Commissioners Appointed to Inquire into the Application of Iron to Railway Structures*; Grey, G., Ed.; W. Clowes and Sons: London, UK, 1849.
17. Stokes, G.G. Discussion of a differential equation relating to the breaking of railway bridges. *Trans. Camb. Philos. Soc.* **1849**, *8*, 707–735.
18. Yu, G.; Kiani, K.; Roshan, M. Dynamic analysis of multiple-nanobeam-systems acted upon by multiple moving nanoparticles accounting for nonlocality, lag, and lateral inertia. *Appl. Math. Model.* **2022**, *108*, 326–354. [CrossRef]
19. Ma, X.; Roshan, M.; Kiani, K.; Nikkhoo, A. Dynamic response of an elastic tube-like nanostructure embedded in a vibrating medium and under the action of moving nano-objects. *Symmetry* **2023**, *15*, 1827. [CrossRef]
20. Chu, K.H.; Garg, V.K.; Dhar, C.L. Railway-bridge impact: Simplified train and bridge model. *J. Struct. Div.* **1979**, *105*, 1823–1844. [CrossRef]
21. Zhai, W.M. *Vehicle-Track Coupled Dynamics*; Science Press: Beijing, China, 2015; Volume 1.
22. Lei, X.Y. *High Speed Railway Track Dynamics: Models, Algorithms and Application*; Science Press: Beijing, China, 2022.
23. Luo, J.; Zhu, S.; Zhai, W. An advanced train-slab track spatially coupled dynamics model: Theoretical methodologies and numerical applications. *J. Sound Vib.* **2021**, *501*, 116059. [CrossRef]
24. Ma, M.; Liu, W.; Qian, C.; Deng, G.; Li, Y. Study of the train-induced vibration impact on a historic bell tower above two spatially overlapping metro lines. *Soil Dyn. Earthq. Eng.* **2016**, *81*, 58–74. [CrossRef]
25. Qu, S.; Yang, J.; Zhu, S.; Zhai, W.; Kouroussis, G. A hybrid methodology for predicting train-induced vibration on sensitive equipment in far-field buildings. *Transp. Geotech.* **2021**, *31*, 100682. [CrossRef]
26. Ren, Y.; Qu, S.; Yang, J.; Li, Q.; Zhu, B.; Zhai, W.; Zhu, S. An efficient three-dimensional dynamic stiffness-based model for predicting subway train-induced building vibrations. *J. Build. Eng.* **2023**, *76*, 107239. [CrossRef]
27. Ma, C.; Choi, D.H. Stochastic dynamic analysis of the train-track-bridge system under tridirectional spatially correlated ground motions. *Soil Dyn. Earthq. Eng.* **2022**, *160*, 107324. [CrossRef]
28. Zeng, Z.P.; Liu, F.S.; Wang, W.D. Three-dimensional train-track-bridge coupled dynamics model based on the explicit finite element method. *Soil Dyn. Earthq. Eng.* **2022**, *153*, 0267–7261. [CrossRef]
29. Ju, S.H. Derailment of high-speed trains moving on curved and cant rails under seismic loads. *Soil Dyn. Earthq. Eng.* **2023**, *166*, 107757. [CrossRef]
30. Zhao, H.; Wei, B.; Shao, Z.; Xie, X.; Jiang, L.; Xiang, P. Assessment of train running safety on railway bridges based on velocity-related indices under random near-fault ground motions. *Structures* **2023**, *57*, 105244. [CrossRef]
31. Tang, Y.; Zhu, Z.; Ba, Z.; Lee, V.W.; Gong, W. Running safety assessment of trains considering post-earthquake damage state of bridge-track system. *Eng. Struct.* **2023**, *287*, 116187. [CrossRef]
32. Li, Z.; Xu, L.; Wang, W.; Zhang, J.; Wang, J.; Peng, B.; Zeng, Z. On use of train-track-subgrade dynamic model for investigating the train-induced cumulative deformation of subgrade and its dynamic effects. *Appl. Math. Model.* **2024**, *127*, 71–95. [CrossRef]
33. Jiang, Y.; Chi, M.; Yang, J.; Dai, L.; Xie, Y.; Guo, Z. Investigation on the mechanism and measures of derailment of empty freight train passing a turnout in the diverging route. *Eng. Fail. Anal.* **2024**, *156*, 107822. [CrossRef]
34. Gou, H.; Gao, H.; Ban, X.; Meng, X.; Bao, Y. Vibration energy transmission in high-speed train-track-bridge coupled systems. *Eng. Struct.* **2023**, *297*, 117019. [CrossRef]
35. Chen, L.; Wang, Y.; He, Z.; Zhai, Z.; Bai, Y. Research on dynamic characteristics of railway side-cracked slab for train-track coupled system. *Eng. Fail. Anal.* **2024**, *160*, 108241. [CrossRef]
36. Zhang, Q.; Dong, J.; Leng, W.; Zhang, C.; Wen, C.; Zhou, Z. Dynamic stress response in a novel prestressed subgrade under heavy-haul train loading: A numerical analysis. *Constr. Build. Mater.* **2024**, *412*, 134749. [CrossRef]
37. Xin, T.; Wang, J.; Fang, Q.; Mai, H.; Guo, P.; Wang, G. Passenger ride comfort in subway due to new subway excavation below. *Tunn. Undergr. Space Technol.* **2023**, *132*, 104904. [CrossRef]
38. Qu, S.; Yang, J.; Feng, Y.; Peng, Y.; Zhao, C.; Zhu, S.; Zhai, W. Ground vibration induced by maglev trains running inside tunnel: Numerical modelling and experimental validation. *Soil Dyn. Earthq. Eng.* **2022**, *157*, 107278. [CrossRef]
39. Xu, L. An isoparametric element permutation method for railway tunnel-soil interaction modeling in train-track-tunnel-soil dynamic analysis. *Tunn. Undergr. Space Technol.* **2023**, *140*, 105320. [CrossRef]
40. Hu, J.; Zou, C.; Liu, Q.; Li, X.; Tao, Z. Floor vibration predictions based on train-track-building coupling model. *J. Build. Eng.* **2024**, *89*, 109340. [CrossRef]
41. Malmborg, J.; Flodén, O.; Persson, P.; Persson, K. Numerical study on train-induced vibrations: A comparison of timber and concrete buildings. *Structures* **2024**, *62*, 106215. [CrossRef]

42. Zhang, X.; Ruan, L.; Zhao, Y.; Zhou, X.; Li, X. A frequency domain model for analysing vibrations in large-scale integrated building-bridge structures induced by running trains. *Proc. ImechE Part F J. Rail Rapid Transit* **2019**, *234*, 226–241. [CrossRef]
43. Xie, W.; Fang, S.; He, W.; Chen, B. Vibration serviceability evaluation of large-span station structures. In Proceedings of the 5th International Symposium on Environmental Vibration (ISEV2011), Chengdu, China, 20–22 October 2011.
44. Guo, X.; Wang, S. Research on the dynamic response of the multi-line elevated station with "integral station-bridge system". *Buildings* **2024**, *14*, 758. [CrossRef]
45. Wang, T.; Jiang, B.; Sun, X. Train-induced vibration prediction and control of a metro depot and over-track buildings. *Buildings* **2023**, *13*, 1995. [CrossRef]
46. Pridham, B.; Walters, N.; Nelson, L.; Roeder, B. Addressing Parking Garage Vibrations for the Design of Research and Healthcare Facilities. In *Dynamics of Civil Structures*; Caicedo, J., Pakzad, S., Eds.; The Society for Experimental Mechanics, Inc.: Danbury, CT, USA, 2017; pp. 147–157.
47. Pan, T.C.; Mita, A.; Li, J. Vehicle-induced floor vibrations in a multistory factory building. *J. Perform. Constr. Fac.* **2001**, *15*, 54–61. [CrossRef]
48. Tao, Z.; Moore, J.A.; Sanayei, M.; Wang, Y.; Zou, C. Train-induced floor vibration and structure-borne noise predictions in a low-rise over-track building. *Eng. Struct.* **2022**, *255*, 113914. [CrossRef]
49. Zhou, Y.; Ma, K.; Chen, P.; Lu, D.; Wu, H. Investigations on train-induced vibration and vibration control of an over-track building using thick-layer rubber bearings. *Struct. Des. Tall Spec. Build.* **2022**, *31*, e1898. [CrossRef]
50. Zou, C.; Wang, Y.; Moore, J.A.; Sanayei, M. Train-induced field vibration measurements of ground and over-track buildings. *Sci. Total Environ.* **2016**, *575*, 1339–1351. [CrossRef]
51. He, L.; Tao, Z. Building vibration measurement and prediction during train operations. *Buildings* **2024**, *14*, 142. [CrossRef]
52. Haladin, I.; Bogut, M.; Lakušić, S. Analysis of tram traffic-induced vibration influence on earthquake damaged buildings. *Buildings* **2021**, *11*, 590. [CrossRef]
53. Lamb, H. On the propagation of tremors over the surface of an elastic solid. *Philo. Trans. R. Soc. Lond. Ser. A—Math. Phys. Eng. Sci.* **1903**, *72*, 128–130. [CrossRef]
54. Gao, G.Y.; Song, J.; Yang, J. Identifying the boundary between near field and far field in ground vibration caused by surface loading. *J. Cent. South Univ.* **2014**, *21*, 3284–3294. [CrossRef]
55. Kim, D.S.; Lee, J.S. Propagation and attenuation characteristics of various ground vibrations. *Soil Dyn. Earthq. Eng.* **2000**, *19*, 115–126. [CrossRef]
56. Niu, D.; Deng, Y.; Mu, H.; Chang, J.; Xuan, Y.; Cao, G. Attenuation and propagation characteristics of railway load-induced vibration in a loess area. *Transp. Geotech.* **2022**, *37*, 100858. [CrossRef]
57. Yang, Y.B.; Hung, H.H. Soil vibrations caused by underground moving trains. *J. Geotech. Geoenviron. Eng.* **2008**, *134*, 1633–1644. [CrossRef]
58. Sadeghi, J.; Esmaeili, M.H.; Akbari, M. Reliability of FTA general vibration assessment model in prediction of subway induced ground borne vibrations. *Soil Dyn. Earthq. Eng.* **2019**, *117*, 300–311. [CrossRef]
59. With, C.; Bahrekazemi, M.; Bodare, A. Validation of an empirical model for prediction of train-induced ground vibrations. *Soil Dyn. Earthq. Eng.* **2006**, *26*, 983–990. [CrossRef]
60. Cao, Z.G.; Cai, Y.Q.; Sun, H.L.; Xu, C.J. Dynamic responses of a poroelastic half-space from moving trains caused by vertical track irregularities. *Int. J. Numer. Anal. Meth. Geomech.* **2011**, *35*, 761–786. [CrossRef]
61. Sheng, X.; Jones, C.J.C.; Thompson, D.J. A theoretical model for ground vibration from trains generated by vertical track irregularities. *J. Sound Vib.* **2004**, *272*, 937–965. [CrossRef]
62. Hussein, M.F.M.; Hunt, H.E.M. A numerical model for calculating vibration from a railway tunnel embedded in a full-space. *J. Sound Vib.* **2007**, *305*, 401–431. [CrossRef]
63. Degrande, G.; Clouteau, D.; Othman, R.; Arnst, M.; Chebli, H.; Klein, R.; Chatterjee, P.; Janssens, B. A numerical model for ground-borne vibrations from underground railway traffic based on a periodic finite element-boundary element formulation. *J. Sound Vib.* **2006**, *293*, 645–666. [CrossRef]
64. Xu, L.; Zhai, W. Vehicle-track-tunnel dynamic interaction: A finite/infinite element modelling method. *Railw. Eng. Sci.* **2021**, *29*, 109–126. [CrossRef]
65. Ma, M.; Xu, L.; Du, L.; Wu, Z.; Tan, X. Prediction of building vibration induced by metro trains running in a curved tunnel. *J. Vib. Control* **2020**, *27*, 515–528. [CrossRef]
66. Yang, Y.B.; Hung, H.H. A 2.5D finite/infinite element approach for modelling visco-elastic bodies subjected to moving loads. *Int. J. Numer. Methods Eng.* **2001**, *51*, 1317–1336. [CrossRef]
67. Biot, M.A. Generalized theory of acoustic propagation in porous dissipative media. *J. Acoust. Soc. Am.* **1962**, *34*, 1254–1264. [CrossRef]
68. *HJ 453-2018*; Technical Guidelines for Environmental Impact Assessment—Urban Rail Transit. Ministry of Ecological Environment: Beijing, China, 2019.
69. Anand, V.; Kumar, S.R.S. Seismic soil-structure interaction: A state-of-the-art review. *Structures* **2018**, *16*, 317–326. [CrossRef]
70. Bielak, J.; Loukakis, K.; Hisada, Y.; Yoshimura, C. Domain reduction method for three-dimensional earthquake modeling in localized regions, part I: Theory. *Bull. Seism. Soc. Am.* **2003**, *93*, 817–824. [CrossRef]

71. Lopes, P.; Costa, P.A.; Ferraz, M.; Calçada, R.; Cardoso, A.S. Numerical modeling of vibrations induced by railway traffic in tunnels: From the source to the nearby buildings. *Soil Dyn. Earthq. Eng.* **2014**, *61*, 269–285. [CrossRef]
72. Colaço, A.; Barbosa, D.; Costa, P.A. Hybrid soil-structure interaction approach for the assessment of vibrations in buildings due to railway traffic. *Transp. Geotech.* **2022**, *32*, 100691. [CrossRef]
73. Kuo, K.A.; Papadopoulos, M.; Lombaert, G.; Degrande, G. The coupling loss of a building subject to railway induced vibrations: Numerical modelling and experimental measurements. *J. Sound Vib.* **2019**, *442*, 459–481. [CrossRef]
74. Li, X.; Chen, Y.; Zou, C.; Wu, J.; Shen, Z.; Chen, Y. Building coupling loss measurement and prediction due to train-induced vertical vibrations. *Soil Dyn. Earthq. Eng.* **2023**, *164*, 107644. [CrossRef]
75. Edirisinghe, T.L.; Talbot, J.P. The significance of source-receiver interaction in the response of piled foundations to ground-borne vibration from underground railways. *J. Sound Vib.* **2021**, *506*, 116178. [CrossRef]
76. Yao, J.B.; Xia, H.; Zhan, N. Study on the train-induced environmental vibrations considering soil-structure interaction. *Procedia Eng.* **2017**, *199*, 2747–2752. [CrossRef]
77. Coulier, P.; Lombaert, G.; Degrande, G. The influence of source-receiver interaction on the numerical prediction of railway induced vibrations. *J. Sound Vib.* **2014**, *333*, 2520–2538. [CrossRef]
78. Colaço, A.; Costa, P.A.; Castanheira-Pinto, A.; Amado-Mendes, P.; Calçada, R. Experimental validation of a simplified soil-structure interaction approach for the prediction of vibrations in buildings due to railway traffic. *Soil Dyn. Earthq. Eng.* **2020**, *141*, 106499. [CrossRef]
79. Aubry, D.; Clouteau, D. A Subdomain Approach to Dynamic Soil-structure Interaction. In *Recent Advances in Earthquake Engineering and Structural Dynamics*; Davidovici, D., Clough, R.W., Eds.; Ouest Editions/AFPS: Nantes, France, 1992; pp. 251–272.
80. Pyl, L.; Degrande, G.; Clouteau, D. Validation of a source-receiver model for road traffic induced vibrations in buildings. II: Receiver model. *J. Eng. Mech.* **2004**, *130*, 1394–1406. [CrossRef]
81. Ling, Y.; Zhang, Y.; Luo, Q.; He, A. Field measurement and simplified numerical model for vibration response of subway superstructure. *Structures* **2023**, *47*, 313–323. [CrossRef]
82. Erkal, A.; Kocagöz, M.S. Interaction of vibrations of road and rail traffic with buildings and surrounding environment. *J. Perform. Constr. Facil.* **2020**, *34*, 04020038. [CrossRef]
83. Hua, Y.; Xie, W.; Xie, J. Non-uniform excitation method for predicting railway-induced vibrations of buildings near operational subways. *J. Build. Eng.* **2024**, *84*, 108669. [CrossRef]
84. Zeng, Q.; Stoura, C.D.; Dimitrakopoulos, E.G. A localised lagrange multipliers approach for the problem of vehicle-bridge-interaction. *Eng. Struct.* **2018**, *168*, 82–92. [CrossRef]
85. Wang, L.; Zhu, Z.; Bai, Y.; Li, Q.; Costa, P.A.; Yu, Z. A fast random method for three-dimensional analysis of train-track-soil dynamic interaction. *Soil Dyn. Earthq. Eng.* **2018**, *115*, 252–262. [CrossRef]
86. Touhei, T. Impulsive response of an elastic layered medium in the anti-plane wave field based on a thin-layered element and discrete wave number method. *Doboku Gakkai Ronbunshu* **1993**, *465*, 137–144. [CrossRef] [PubMed]
87. Yang, Y.B.; Ge, P.; Li, Q.; Liang, X.; Wu, Y. 2.5D vibration of railway-side buildings mitigated by open or infilled trenches considering rail irregularity. *Soil Dyn. Earthq. Eng.* **2018**, *106*, 204–214. [CrossRef]
88. Lee, S.H.; Lee, K.K.; Woo, S.S.; Cho, S.H. Global vertical mode vibrations due to human group rhythmic movement in a 39 story building structure. *Eng. Struct.* **2013**, *57*, 296–305. [CrossRef]
89. Chen, J. *Human-induced Load and Structural Vibration*; Science Press: Beijing, China, 2016.
90. Wang, H.; Ge, Q.; Zeng, D.; Zhang, Z.; Chen, J. Human-induced vibration serviceability: From dynamic load measurement towards the performance-based structural design. *Buildings* **2023**, *13*, 1977. [CrossRef]
91. *JGJ/T 441-2019*; Technical Standard for Human Comfort of the Floor Vibration. Ministry of Housing and Urban-Rural Development: Beijing, China, 2019.
92. Mohammed, A.S.; Pavic, A.; Racic, V. Improved model for human induced vibrations of high-frequency floors. *Eng. Struct.* **2018**, *168*, 950–966. [CrossRef]
93. Shahabpoor, E.; Pavic, A.; Racic, V. Interaction between walking humans and structures in vertical direction: A literature review. *Shock Vib.* **2016**, *2016*, 3430285. [CrossRef]
94. Salyards, K.A.; Noss, N.C. Experimental evaluation of the influence of human-structure interaction for vibration serviceability. *J. Perform. Constr. Facil.* **2014**, *28*, 458–465. [CrossRef]
95. Shahabpoor, E.; Pavic, A.; Racic, V.; Zivanovic, S. Effect of group walking traffic on dynamic properties of pedestrian structures. *J. Sound Vib.* **2017**, *387*, 207–225. [CrossRef]
96. Zhang, S.; Xu, L.; Qin, J. Vibration of lightweight steel floor systems with occupants: Modelling, formulation and dynamic properties. *Eng. Struct.* **2017**, *147*, 652–665. [CrossRef]
97. Zhou, D.; Ji, T. Free vibration of rectangular plates with continuously distributed spring-mass. *Int. J. Solids Struct.* **2006**, *43*, 6502–6520. [CrossRef]
98. Racic, V.; Pavic, A.; Brownjohn, J.M.W. Experimental identification and analytical modelling of human walking forces: Literature review. *J. Sound Vib.* **2009**, *326*, 1–49. [CrossRef]
99. Nikooyan, A.A.; Zadpoor, A.A. Mass-spring-damper modelling of the human body to study running and hopping—An overview. *Proc. IMechE Part H J. Eng. Med.* **2011**, *225*, 1121–1135. [CrossRef]

100. Živanović, S.; Pavić, A. Probability-based prediction of multi-mode vibration response to walking excitation. *Eng. Struct.* **2007**, *29*, 942–954. [CrossRef]
101. Chen, J.; Wang, H.Q.; Peng, Y.X. Experimental investigation on fourier-series model walking load and its coefficients. *J. Vib. Shock.* **2014**, *33*, 11–15. [CrossRef]
102. Silva, F.T.D.; Brito, H.M.B.F.; Pimentel, R.L. Modeling of crowd load in vertical directiong using biodynamic model for pedestrians crossing footbridges. *Can. J. Civ. Eng.* **2013**, *40*, 1196–1204. [CrossRef]
103. Shahabpoor, E.; Pavic, A. Human-structure dynamic interactions: Identification of two-degrees-of-freedom walking human model. *J. Sound Vib.* **2023**, *569*, 117974. [CrossRef]
104. Whittington, B.R.; Thelen, D.G. A simple mass-spring model with roller feet can induce the ground reactions observed in human walking. *J. Biomech. Eng.* **2009**, *131*, 011013. [CrossRef]
105. Kim, S.; Park, S. Leg stiffness increases with speed to modulate gait frequency and propulsion energy. *J. Biomech.* **2011**, *44*, 1253–1258. [CrossRef]
106. Qin, J.W.; Law, S.S.; Yang, Q.S.; Yang, N. Pedestrian-bridge dynamic interaction, including human participation. *J. Sound Vib.* **2013**, *332*, 1107–1124. [CrossRef]
107. Máca, J.; Valášek, M. Interaction of human gait and footbridges. In Proceedings of the 8th International Conference on Structural Dynamics (EURODYN 2011), Leuven, Belgium, 4–6 July 2011.
108. Brownjohn, J.M.W.; Pavic, A.; Omenzetter, P. A spectral density approach for modelling continuous vertical forces on pedestrian structures due to walking. *Can. J. Civ. Eng.* **2004**, *31*, 65–77. [CrossRef]
109. Piccardo, G.; Tubino, F. Equivalent spectral model and maximum dynamic response for the serviceability analysis of footbridges. *Eng. Struct.* **2012**, *40*, 445–456. [CrossRef]
110. Ferrarotti, A.; Tubino, F. Generalized equivalent spectral model for serviceability analysis of footbridges. *J. Bridge Eng.* **2016**, *21*, 04016091. [CrossRef]
111. Chen, J.; Wang, J.; James, M.W.B. Power spectral-density model for pedestrian walking load. *J. Struct. Eng.* **2019**, *145*, 04018239. [CrossRef]
112. Wang, J.; Chen, J.; Yokoyama, Y.; Xiong, J. Spectral model for crowd walking load. *J. Struct. Eng.* **2020**, *146*, 04019220. [CrossRef]
113. Fujino, Y.; Pacheco, B.M.; Nakamura, S.; Warnitchai, P. Synchronization of human walking observed during lateral vibration of a congested pedestrian bridge. *Earthq. Eng. Struct. Dyn.* **1993**, *22*, 741–758. [CrossRef]
114. Dallard, P.; Fitzpatrick, A.J.; Flint, A.; Bourva, S.L.; Low, A.; Smith, R.M.R.; Willford, M. The London Millennium footbridge. *Struct. Eng.* **2001**, *79*, 17–33.
115. Nakamura, S. Model for lateral excitation of footbridges by synchronous walking. *J. Struct. Eng.* **2004**, *130*, 32–37. [CrossRef]
116. Strogatz, S.H.; Abrams, D.M.; McRobie, A.; Eckhardt, B.; Ott, E. Crowd synchrony on the Millennium bridge. *Nature* **2005**, *438*, 43–44. [CrossRef]
117. Piccardo, G.; Tubino, F. Parametric resonance of flexible footbridges under crowd-induced lateral excitation. *J. Sound Vib.* **2008**, *311*, 353–371. [CrossRef]
118. Pizzimenti, A.D.; Ricciardelli, F. Experimental evaluation of the dynamic lateral loading of footbridges by walking pedestrians. In Proceedings of the 6th International Conference on Structural Dynamics (EURODYN 2005), Paris, France, 4–7 September 2005.
119. Ingólfsson, E.T.; Georgakis, C.T.; Ricciardelli, F.; Jönsson, J. Experimental identification of pedestrian-induced lateral forces on footbridges. *J. Sound Vib.* **2011**, *330*, 1265–1284. [CrossRef]
120. Macdonald, J.H.G. Lateral excitation of bridges by balancing pedestrians. *Proc. R. Soc. A—Math. Phys. Eng. Sci.* **2008**, *465*, 1055–1073. [CrossRef]
121. Bocian, M.; MacDonald, J.H.G.; Burn, J.F. Biomechanically inspired modelling of pedestrian-induced forces on laterally oscillating structures. *J. Sound Vib.* **2012**, *331*, 3914–3929. [CrossRef]
122. Carroll, S.P.; Owen, J.S.; Hussein, M.F.M. Crowd-bridge interaction by combining biomechanical and discrete element models. In Proceedings of the 8th International Conference on Structural Dynamics (EURODYN 2011), Leuven, Belgium, 4–6 July 2011.
123. Racic, V.; Morin, J.B. Data-driven modelling of vertical dynamic excitation of bridges induced by people running. *Mech. Syst. Signal Proc.* **2014**, *43*, 153–170. [CrossRef]
124. Ellis, B.R.; Ji, T. Loads generated by jumping crowds: Numerical modelling. *Struct. Eng.* **2004**, *82*, 35–40. [CrossRef]
125. Xiong, J.; Chen, J.; Caprani, C. Spectral analysis of human-structure interaction during crowd jumping. *Appl. Math. Model.* **2021**, *89*, 610–626. [CrossRef]
126. Xiong, J.; Chen, J. Power spectral density function for individual jumping load. *Int. J. Struct. Stab. Dyn.* **2018**, *18*, 1850023. [CrossRef]
127. Ji, T.; Ellis, B.R. Floor vibration induced by dance-type loads: Theory. *Struct. Eng.* **1994**, *72*, 37–44.
128. Parkhouse, J.G.; Ewins, D.J. Crowd-induced rhythmic loading. *Proc. Inst. Civ. Eng.-Struct. Build.* **2006**, *159*, 247–259. [CrossRef]
129. Duarte, E.; Ji, T. Action of individual bouncing on structures. *J. Struct. Eng.* **2009**, *135*, 818–827. [CrossRef]
130. Dougill, J.W.; Wright, J.R.; Parkhouse, J.G.; Harrison, R.E. Human structure interaction during rhythmic bobbing. *Struct. Eng.* **2006**, *84*, 32–39.
131. Wang, H.Q.; Chen, J.; Nagayama, T. Parameter identification of spring-mass-damper model for bouncing people. *J. Sound Vib.* **2019**, *456*, 13–29. [CrossRef]
132. Ahmadi, E.; Caprani, C.; Živanović, S.; Heidarpour, A. Vertical ground reaction forces on rigid and vibrating surfaces for vibration serviceability assessment of structures. *Eng. Struct.* **2018**, *172*, 723–738. [CrossRef]

133. Kerr, S.C.; Bishop, N.W.M. Human induced loading on flexible staircases. *Eng. Struct.* **2001**, *23*, 37–45. [CrossRef]
134. Helbing, D.; Buzna, L.; Johansson, A.; Werner, T. Self-organized pedestrian crowd dynamics: Experiments, simulations, and design solutions. *Transp. Sci.* **2005**, *39*, 1–24. [CrossRef]
135. He, W.; Xie, W.; Chen, B.; Sun, L. Assessment on vibration serviceability of large-span railway station under crowd load. In Proceedings of the 5th International Symposium on Environmental Vibration (ISEV2011), Chengdu, China, 20–22 October 2011.
136. Carroll, S.P.; Owen, J.S.; Hussein, M.F.M. Modelling crowd-bridge dynamic interaction with a discretely defined crowd. *J. Sound Vib.* **2012**, *331*, 2685–2709. [CrossRef]
137. Helbing, D.; Molnár, P. Social force model for pedestrian dynamics. *Phys. Rev. E* **1995**, *51*, 4282–4286. [CrossRef]
138. Pu, X.; He, T.; Zhu, Q. Considering the effect of obstacles and semi-rigid boundary conditions on the dynamic response of the floor under random crowd-structure interaction. *Acta Mech.* **2023**, *234*, 3821–3841. [CrossRef]
139. Xiong, J.; Chen, J. A generative adversarial network model for simulating various types of human-induced loads. *Int. J. Struct. Stab. Dyn.* **2019**, *19*, 1950092. [CrossRef]
140. Dong, J.K.; Ye, M. Design for random response of structures subject to rhythmic crowd loading. *Buildings* **2023**, *13*, 1085. [CrossRef]
141. Martinelli, L.; Racic, V.; Lago, B.A.D.; Foti, F. Testing walking-induced vibration of floors using smartphones recordings. *Robotics* **2020**, *9*, 37. [CrossRef]
142. Xiong, J.; Cao, Z.; Duan, S.; Cao, B.; Qian, H.; Li, C. A fourier series-based multi-point excitation model for crowd jumping loads. *Buildings* **2023**, *13*, 1782. [CrossRef]
143. Davenport, A.G. The application of statistical concepts to the wind loading of structures. *Proc. Inst. Civ. Eng.* **1961**, *19*, 449–472. [CrossRef]
144. Ruscheweyh, H.; Sedlacek, G. Crosswind vibrations of steel stacks. -critical comparison between some recently proposed codes. *J. Wind Eng. Ind. Aerodyn.* **1988**, *30*, 173–183. [CrossRef]
145. Bishop, R.E.D.; Hassan, A.Y. The lift and drag forces on a circular cylinder oscillating in a flowing fluid. *Proc. R. Soc. Lond. Ser. A—Math. Phys. Sci.* **1964**, *277*, 51–75. [CrossRef]
146. Tamura, Y.; Matsui, G. Wake-oscillator model of vortex-induced oscillation of circular cylinder. In Proceedings of the 5th International Conference, Fort Collins, CO, USA, 8–14 July 1979.
147. Vickery, B.J.; Basu, R.I. Across-wind vibrations of structures of circular cross-section, part i: Development of a mathematical model fort wo-dimensional conditions. *J. Wind Eng. Ind. Aerodyn.* **1983**, *12*, 49–73. [CrossRef]
148. Ehsan, F.; Scanlan, R.H. Vortex-induced vibrations of flexible bridges. *J. Eng. Mech.* **1990**, *116*, 1392–1411. [CrossRef]
149. Larsen, A. A generalized model for assessment of vortex-induced vibration of flexible structures. *J. Wind Eng. Ind. Aerodyn.* **1995**, *57*, 281–294. [CrossRef]
150. Wijesooriya, K.; Mohotti, D.; Amin, A.; Chauhan, K. An uncoupled fluid structure interaction method in the assessment of structural responses of tall buildings. *Structures* **2020**, *25*, 448–462. [CrossRef]
151. Li, Y.; Zhu, Y.; Chen, F.; Li, Q.S. Aerodynamic loads of tapered tall buildings: Insights from wind tunnel test and CFD. *Structures* **2023**, *56*, 104975. [CrossRef]
152. Simiu, E.; Yeo, D. *Wind Effects on Structures: Modern Structural Design for Wind*; John Wiley & Sons, Inc.: Hoboken, NJ, USA, 2019.
153. *JGJ 3-2010*; Technical Specification for Concrete Structures of Tall Building. Ministry of Housing and Urban-Rural Development: Beijing, China, 2010.
154. *JGJ 99-2015*; Technical Specification for Steel Structure of Tall Building. Ministry of Housing and Urban-Rural Development: Beijing, China, 2015.
155. *AIJ-GEH-2004*; Guidelines for the Evaluation of Habitability to Building Vibration. Architectural Institute of Japan: Tokyo, Japan, 2007.
156. *ISO 6897-1984*; Guidelines for the Evaluation of the Response of Occupants of Fixed Structures, Especially Buildings and Off-shore Structures, to Low-frequency Horizontal Motion (0.063 to 1Hz). International Standard Organization: Geneva, Switzerland, 1984.
157. Hu, W.; Teng, D.; Li, J.; Xu, Z.; Wang, Y.; Lu, W.; Li, Z.; Teng, J. Structural Dynamic Parameter Identification of Saige Building Based on Distributed Synchronous Acquisition Method. *J. Build. Struct.* **2022**, *43*, 76–84. [CrossRef]
158. Xu, W.; Li, R.; Qiu, J.; Li, Q.; Yu, Z. Study on wind-induced human comfort of the SEG plaza under local excitation based on wind tunnel test. *Sustainability* **2023**, *15*, 3067. [CrossRef]
159. Zhou, H.; Shao, X.; Zhang, J.; Yao, H.; Liu, Y.; Tan, P.; Chen, Y.; Xu, L.; Zhang, Y.; Gong, W. Real-time hybrid model test to replicate high-rise building resonant vibration under wind loads. *Thin-Walled Struct.* **2024**, *197*, 111559. [CrossRef]
160. Yang, Y.; Ma, F.; Han, Q. Reconstruction of boundary layer wind field at the seg plaza based on dual-lidar measurement and numerical simulation. *J. Wind Eng. Ind. Aerodyn.* **2023**, *223*, 105298. [CrossRef]
161. Xie, W.; Zhu, M.; Yin, Y.; Tang, Z.; Peng, Q.; Cheng, Y. Research on vibration and secondary noise induced by elevator car-guiderail coupling. *Noise Vib. Control* **2023**, *43*, 74–81.
162. Zhang, X.; Wang, Y.; Xie, W. Research on structure vibration and secondary noise caused by 220 kV transformers. *Noise Vib. Control* **2020**, *40*, 188–193.
163. Li, H.; Yang, W.; Liu, P.; Wang, M. Resonance measurement and vibration reduction analysis of an office building induced by nearby crane workshop vibration. *J. Build. Eng.* **2022**, *58*, 105018. [CrossRef]
164. Hwang, J.H.; Tu, T.Y. Ground vibration due to dynamic compaction. *Soil Dyn. Earthq. Eng.* **2006**, *26*, 337–346. [CrossRef]

165. Valeria, L.; Annamaria, L.; Gaetano, E.; Domenico, R.; Giuseppina, U. Vibrations induced by mechanical rock excavation on R.C. buildings in an urban area. *Buildings* **2021**, *11*, 15. [CrossRef]
166. *ISO 2631-1:1997*; Mechanical Vibration and Shock—Evaluation of Human Exposure to Whole Body Vibration Part 1: General Requirements. International Organization for Standardization: Geneva, Switzerland, 1997.
167. *ISO 2631-1:1997/Amd 1:2010*; Mechanical Vibration and Shock—Evaluation of Human Exposure to Whole-body Vibration—Part 1: General Requirements Amendment 1. International Organization for Standardization: Geneva, Switzerland, 2010.
168. He, W.; Xie, W. Study on sophisticated calculation model of large-span railway station structures based on vibration serviceability evaluation. *China Civ. Eng. J.* **2014**, *47*, 13–23. [CrossRef]
169. Zhu, Q.; Liu, K.; Liu, L.; Du, Y.; Zivanovic, S. Experimental and numerical analysis on serviceability of cantilevered floor based on human-structure interaction. *J. Constr. Steel. Res.* **2020**, *173*, 106184. [CrossRef]
170. Xie, Z.; Hu, X.; Du, H.; Zhang, X. Vibration behavior of timber-concrete composite floors under human-induced excitation. *J. Build. Eng.* **2020**, *32*, 101744. [CrossRef]
171. Xue, S.; Zhang, Z.; Zhang, Z.; Zhou, H.; Shen, Y. Effects of strongbacks and strappings on vibrations of timber truss joist floors. *Shock Vib.* **2021**, *2021*, 6630719. [CrossRef]
172. An, Q.; Ren, Q.; Liu, H.; Yan, X.; Chen, Z. Dynamic performance characteristics of an innovative cable supported beam structure—Concrete slab composite floor system under human-induced loads. *Eng. Struct.* **2016**, *117*, 40–57. [CrossRef]
173. Astroza, R.; Ebrahimian, H.; Conte, J.P.; Restrepo, J.I.; Hutchinson, T.C. Influence of the construction process and nonstructural components on the modal properties of a five-story building. *Earthq. Eng. Struct. Dyn.* **2016**, *45*, 1063–1084. [CrossRef]
174. Devin, A.; Fanning, P.J. Non-structural elements and the dynamic response of buildings: A review. *Eng. Struct.* **2019**, *187*, 242–250. [CrossRef]
175. Liang, H.; Xie, W.; Wei, P.; Zhou, Y.; Zhang, Z. The effect of the decorative surface layer on the dynamic properties of a symmetric concrete slab. *Symmetry* **2021**, *13*, 1174. [CrossRef]
176. Reynolds, P.; Pavic, A. Effects of false floors on vibration serviceability of building floors. I: Modal properties. *J. Perform. Constr. Facil.* **2003**, *17*, 75–86. [CrossRef]
177. Jiménez-Alonso, J.F.; Pérez-Aracil, J.; Díaz, H.A.M.; Sáez, A. Effect of vinyl flooring on the modal properties of a steel footbridge. *Appl. Sci.* **2019**, *9*, 1374. [CrossRef]
178. Miskovic, Z.; Pavic, A.; Reynolds, P. Effects of full-height nonstructural partitions on modal properties of two nominally identical building floors. *Can. J. Civ. Eng.* **2009**, *36*, 1121–1132. [CrossRef]
179. Li, B.; Hutchinson, G.L.; Duffield, C.F. The influence of non-structural components on tall building stiffness. *Struct. Des. Tall Spec. Build.* **2011**, *20*, 853–870. [CrossRef]
180. Wang, Z.; Song, L.; Cheng, Z.; Yang, H.; Wen, J.; Qi, M. Finite element model for vibration serviceability evaluation of a suspended floor with and without tuned mass dampers. *Buildings* **2023**, *13*, 309. [CrossRef]
181. Pedersen, L.H.; Frier, C.; Andersen, L. Flooring-systems and Their Interaction with Usage of the Floor. In *Dynamics of Civil Structures*; Caicedo, J., Pakzad, S., Eds.; The Society for Experimental Mechanics, Inc.: Danbury, CT, USA, 2017; pp. 205–211.
182. Setareh, M. Vibration serviceability of a building floor structure. I: Dynamic testing and computer modeling. *J. Perform. Constr. Facil.* **2010**, *24*, 497–507. [CrossRef]
183. Huang, M.; Ling, Z.; Sun, C.; Lei, Y.; Xiang, C.; Wan, Z.; Gu, J. Two-stage damage identification for bridge bearings based on sailfish optimization and element relative modal strain energy. *Struct. Eng. Mech.* **2023**, *86*, 715–730. [CrossRef]
184. Gidrão, G.d.M.S.; Carrazedo, R.; Bosse, R.M.; Silvestro, L.; Ribeiro, R.; Souza, C.F.P.d. Numerical modeling of the dynamic elastic modulus of concrete. *Materials* **2023**, *16*, 3955. [CrossRef]
185. Amabili, M. Nonlinear damping in large-amplitude vibrations: Modelling and experiments. *Nonlinear Dyn.* **2018**, *93*, 5–18. [CrossRef]
186. He, Y.; Liu, Y.; Wu, M.; Fu, J.; He, Y. Amplitude dependence of natural frequency and damping ratio for 5 supertall buildings with moderate-to-strong typhoon-induced vibrations. *J. Build. Struct.* **2023**, *78*, 107589. [CrossRef]
187. Han, Z.; Brownjohn, J.M.W.; Chen, J. Structural modal testing using a human actuator. *Eng. Struct.* **2020**, *221*, 111113. [CrossRef]
188. Onundi, L.O.; Elinwa, A.U.; Matawal, D.S. An experimental determination of damping ratio of a multi-storey building subjected to aerodynamic loadings. *J. Civ. Eng. Constr. Technol.* **2012**, *3*, 127–139. [CrossRef]
189. Luo, J.; Huang, M.; Lei, Y. Temperature effect on vibration properties and vibration-based damage identification of bridge structures: A literature review. *Buildings* **2022**, *12*, 1209. [CrossRef]
190. Wang, Z.; Huang, M.; Gu, J. Temperature effects on vibration-based damage detection of a reinforced concrete slab. *Appl. Sci.* **2020**, *10*, 2869. [CrossRef]
191. *ISO 2631-2:2003*; Mechanical Vibration and Shock—Evaluation of Human Exposure to Wholebody Vibration—Part 2: Vibration in Buildings (1 Hz to 80 Hz). International Organization for Standardization: Geneva, Switzerland, 2003.
192. *GB 10070-88*; Standard of Environmental Vibration in Urban Area. Ministry of Ecology and Environment: Beijing, China, 1988.
193. *JGJ/T 170-2009*; Standard for Limit and Measuring Method of Building Vibration and Secondary Noise Caused by Urban Rail Transit. Ministry of Housing and Urban-Rural Development: Beijing, China, 2009.
194. Applied Technology Council. *ATC Design Guide 1, Minimizing Floor Vibration*; Applied Technology Council: Redwood City, CA, USA, 1999.
195. AISC. *AISC Design Guide 11, Floor Vibrations Due to Human Activity*; American Institute of Steel Construction: Chicago, IL, USA, 2001.

196. *EN 1990: 2002;* Eurocode—Basis of Structural Design. British Standards Institution: London, UK, 2002.
197. *BS 6472-1: 2008;* Guide to Evaluation of Human Exposure to Vibration in Buildings. British Standards Institution: London, UK, 2008.
198. *AIJ ES-V001-2018;* Guidelines for the Evaluation of Habitability to Building Vibration. Architectural Institute of Japan: Tokyo, Japan, 2018.
199. Xiong, J.; Liu, Z.; Duan, S.; Qian, H. A review of evaluation methods of standards for structural vibration serviceability under crowd walking. *Buildings* **2024**, *14*, 675. [CrossRef]
200. *ISO 2631/1-1985;* Evaluation of Human Exposure to Whole-body Vibration—Part 1: General Requirements. International Organization for Standardization: Geneva, Switzerland, 1985.
201. Tao, Z.; Wang, Y.; Sanayei, M.; Moore, J.A.; Zou, C. Experimental study of train-induced vibration in over-track buildings in a metro depot. *Eng. Sturct.* **2019**, *198*, 109473. [CrossRef]
202. Farahani, M.V.; Sadeghi, J.; Jahromi, S.G.; Sahebi, M.M. Modal based method to predict subway train-induced vibration in buildings. *Structures* **2023**, *47*, 557–572. [CrossRef]
203. Hua, Y.; Xie, W.; Chen, B. Research on influence of metro vibration on vertical floor response of buildings. *J. Build. Struct.* **2023**, *44*, 122–129. [CrossRef]
204. Xia, H.; Zhang, N.; Cao, Y. Experimental study of train-induced vibrations of environments and buildings. *J. Sound Vib.* **2005**, *280*, 1017–1029. [CrossRef]
205. Xia, H.; Chen, J.; Wei, P.; Xia, C.; Roeck, G.D.; Degrande, G. Experimental investigation of railway train-induced vibrations of surrounding ground and a nearby multi-story building. *Earthq. Eng. Eng. Vib.* **2009**, *8*, 137–148. [CrossRef]
206. Xia, Q.; Qu, W. Experimental and numerical studies of metro train-induced vibrations on adjacent masonry buildings. *Int. J. Struct. Stab. Dyn.* **2016**, *16*, 1550067. [CrossRef]
207. Yu, Y.; Xie, W.; Song, B. The vibration measurement and evaluation due to the traffic loads. In *Proceedings of the International Seminar on Environmental Vibration: Prediction, Monitoring and Evaluation;* China Communications Press: Hangzhou, China, 2003.
208. Du, H.; Du, S.; Li, W. Probabilistic time series forecasting with deep non-linear state space models. *CAAI T. Intell. Technol.* **2023**, *8*, 3–13. [CrossRef]
209. Wan, S.; Guan, S.; Tang, Y. Advancing bridge structural health monitoring: Insights into knowledge-driven and data-driven approaches. *J. Data Sci. Intell. Syst.* **2023**, *00*, 1–12. [CrossRef]
210. Wang, W.; Sun, Y.; Li, K.; Wang, J.; He, C.; Sun, D. Fully bayesian analysis of the relevance vector machine classification for imbalanced data problem. *CAAI T. Intell. Technol.* **2023**, *8*, 192–205. [CrossRef]
211. Chen, D.; Wu, J.; Yan, Q. A novel smartphone-based evaluation system of pedestrian-induced footbridge vibration comfort. *Adv. Struct. Eng.* **2019**, *22*, 1685–1697. [CrossRef]
212. Cao, L.; Chen, J. Online investigation of vibration serviceability limitations using smartphones. *Measurement* **2020**, *162*, 107850. [CrossRef]
213. Deng, Z.; Huang, M.; Wan, N.; Zhang, J. The current development of structural health monitoring for bridges: A review. *Buildings* **2023**, *13*, 1360. [CrossRef]
214. Zhang, J.; Huang, M.; Wan, N.; Deng, Z.; He, Z.; Luo, J. Missing measurement data recovery methods in structural health monitoring: The state, challenges and case study. *Measurement* **2024**, *231*, 114528. [CrossRef]
215. Sol-Sánchez, M.; Moreno-Navarro, F.; Rubio-Gámez, M.C. The use of elastic elements in railway tracks: A state of the art review. *Constr. Build. Mater.* **2015**, *75*, 293–305. [CrossRef]
216. Ouakka, S.; Verlinden, O.; Kouroussis, G. Railway ground vibration and mitigation measures: Benchmarking of best practice. *Railw. Eng. Sci.* **2022**, *30*, 1–22. [CrossRef]
217. Wang, Y.; He, Z.; Wang, K.; Bai, Y.; Li, P. Comparing dynamic performance between new sleeper-damping and floating-slab track system. *Constr. Build. Mater.* **2023**, *400*, 132588. [CrossRef]
218. Hu, Y.; Cheng, Z.; Shi, Z. Vibration reduction performance of a periodic layered slab track. *Adv. Environ. Vib. Transp. Geeodyn.* **2020**, *66*, 779–791. [CrossRef]
219. Sung, D.; Chang, S.; Kim, S. Effect of additional anti-vibration sleeper track considering sleeper spacing and track support stiffness on reducing low-frequency vibrations. *Constr. Build. Mater.* **2020**, *263*, 120140. [CrossRef]
220. Qu, S.; Ding, W.; Dong, L.; Zhu, J.; Zhu, S.; Yang, Y.; Zhai, W. Chiral phononic crystal-inspired railway track for low-frequency vibration suppression. *Int. J. Mech. Sci.* **2024**, *274*, 109275. [CrossRef]
221. Mahdavisefat, E.; Heshmati, A.; Salehzadeh, H.; Bahmani, H.; Sabermahani, M. Vibration screening by trench barriers, a review. *Arab. J. Geosci.* **2017**, *10*, 513. [CrossRef]
222. Meng, L.; Shi, Z.; Hao, S.; Cheng, Z. Filtering property of periodic in-filled trench barrier for underground moving loads. *Constr. Build. Mater.* **2023**, *400*, 132655. [CrossRef]
223. Yao, J.; Zhao, R.; Zhang, N.; Yang, D. Vibration isolation effect study of in-filled trench barriers to train-induced environmental vibrations. *Soil Dyn. Earthq. Eng.* **2019**, *125*, 105741. [CrossRef]
224. Ai, Z.Y.; Cao, Z. Vibration isolation of row of piles embedded in transverse isotropic multi-layered soils. *Comput. Geotech.* **2018**, *99*, 115–129. [CrossRef]
225. Takemiya, H. Field vibration mitigation by honeycomb WIB for pile foundations of a high-speed train viaduct. *Soil Dyn. Earthq. Eng.* **2003**, *24*, 69–87. [CrossRef]

226. Cao, X.; Zhou, F.; Liu, J.; Ma, Q. Experimental study and numerical analysis for vibration isolation performance on open trench and wave impeding block combined vibration isolation barrier. *Soil Dyn. Earthq. Eng.* **2024**, *177*, 108418. [CrossRef]
227. Chen, J.; Geng, J.; Gao, G.; Luo, W.; Liu, Y.; Li, K. Mitigation of subway-induced low-frequency vibrations using a wave impeding block. *Transp. Geotech.* **2022**, *37*, 100862. [CrossRef]
228. Pu, X.; Shi, Z. Broadband surface wave attenuation in periodic trench barriers. *J. Sound Vib.* **2020**, *468*, 115130. [CrossRef]
229. James, P. Base-isolated buildings: Towards performance based design. *Proc. Inst. Civ. Eng.-Struct. Build.* **2016**, *169*, 574–582. [CrossRef]
230. Pan, P.; Shen, S.; Shen, Z.; Gong, R. Experimental investigation on the effectiveness of laminated rubber bearings to isolate metro generated vibration. *Measurement* **2018**, *122*, 554–562. [CrossRef]
231. Liang, Q.; Zhou, Y.; Wang, D.; Luo, W.; Li, J.; He, Z. Shaking table test of vertical isolation performances of super high-rise structure under metro train-induced vibration. *J. Build. Eng.* **2024**, *82*, 108323. [CrossRef]
232. *T/CECS 1234-2023*; Technical Standard for Integrated Control of Engineering Vibration and Seismic Vibration of Building Engineering. China Association for Engineering Construction Standardization: Beijing, China, 2023.
233. Cao, Y.R.; Pan, P.; Sun, J.B.; Wang, H.S. Mechanical properties and isolation effect of disc spring-single friction pendulum 3D vibration isolation device. *J. Build. Struct.* **2022**, *43*, 44–53. [CrossRef]
234. Cao, Y.; Pan, P.; Wang, H.; Sun, J.; Xiao, G.; Zuo, Z. Development of an innovative three-dimensional vibration isolation bearing. *Eng. Struct.* **2023**, *295*, 116890. [CrossRef]
235. Sheng, T.; Liu, G.; Bian, X.; Shi, W.; Chen, Y. Development of a three-directional vibration isolator for buildings subject to metro- and earthquake-induced vibrations. *Eng. Struct.* **2022**, *252*, 113576. [CrossRef]
236. Liang, Q.; Luo, W.; Zhou, Y.; Lu, Z.; Li, J.; He, Z. Vibration filtering effect of a novel three-dimensional isolation bearing on metro vibration isolation. *Eng. Struct.* **2024**, *301*, 117304. [CrossRef]
237. He, W.; Luo, H.; Chang, W.; Xu, H.; Liu, W.; Zhang, Q. Experiment investigation and in situ test of hybrid vibration bearing system applied to overtrack historical buildings. *Struct. Control Health Monit.* **2022**, *29*, e2921. [CrossRef]
238. Rahimi, F.; Aghayari, R.; Samali, B. Application of tuned mass dampers for structural vibration control: A state-of-the-art review. *Civ. Eng. J.* **2020**, *6*, 1622–1651. [CrossRef]
239. Elias, S.; Matsagar, V. Research developments in vibration control of structures using passive tuned mass dampers. *Annu. Rev. Control* **2017**, *44*, 129–156. [CrossRef]
240. Wang, Z.; Chen, Y.; Hu, M.; Chen, A. Vertical vibration and TMD mitigation of an industrial building floor subjected to machine excitation. *J. Vib. Eng.* **2019**, *32*, 986–995. [CrossRef]
241. Elias, S.; Matsagar, V. Distributed multiple tuned mass dampers for wind vibration response control of high-rise building. *J. Eng.* **2014**, *2014*, 198719. [CrossRef]
242. Li, C.; Pan, H.; Cao, L. Pendulum-type tuned tandem mass dampers-inerters for crosswind response control of super-tall buildings. *J. Wind Eng. Ind. Aerodyn.* **2024**, *247*, 105706. [CrossRef]
243. Cao, H.Q. Combined tuned mass dampers for structural vibration control. *Int. J. Non-Linear Mech.* **2023**, *157*, 104550. [CrossRef]
244. Roozbahan, M.; Turan, G. An improved passive tuned mass damper assisted by dual stiffness. *Structures* **2023**, *50*, 1598–1607. [CrossRef]
245. Li, Y.; Tan, P.; Li, S.; He, H. A novel tuned inerter eddy current damper: Modeling, optimization, and evaluation. *Eng. Struct.* **2023**, *285*, 116026. [CrossRef]
246. Wang, L.; Nagarajaiah, S.; Shi, W.; Zhou, Y. Semi-active control of walking-induced vibrations in bridges using adaptive tuned mass damper considering human-structure-interaction. *Eng. Struct.* **2021**, *244*, 112743. [CrossRef]
247. Wang, L.; Zhou, Y.; Shi, W. Random crowd-induced vibration in footbridge and adaptive control using semi-active TMD including crowd-structure interaction. *Eng. Struct.* **2024**, *306*, 117839. [CrossRef]
248. Zhang, C.W. The active rotary inertia driver system for flutter vibration control of bridges and various promising applications. *Sci China Tech. Sci.* **2023**, *66*, 390–405. [CrossRef]
249. Kang, X.; Huang, Q.; Wu, Z.; Tang, J.; Jiang, X.; Lei, S. A review of the tuned mass damper inerter (TMDI) in energy harvesting and vibration control: Designs, analysis and applications. *Comp. Model. Eng. Sci.* **2024**, *139*, 2361–2398. [CrossRef]
250. Ikeda, K.; Ioi, T. On the dynamic vibration damped absorber of the vibration system. *Trans. Jpn. Soc. Mech. Eng.* **1977**, *43*, 1707–1715. [CrossRef]
251. Liu, Y.; Lin, C.C.; Parker, J.; Zuo, L. Exact H_2 optimal tuning and experimental verification of energy-harvesting series electromagnetic tuned-mass dampers. *J. Vib. Acoust.* **2016**, *138*, 061003. [CrossRef]
252. Kamariotis, A.; Chatzi, E.; Straub, D. A framework for quantifying the value of vibration-based structural health monitoring. *Mech. Syst. Signal Proc.* **2023**, *184*, 109708. [CrossRef]
253. Alarcon, M.; Soto, P.; Hernandez, F.; Guindos, P. Structural health monitoring of South America's first 6-story experimental light-frame timber-building by using a low-cost RaspberryShake seismic instrumentation. *Eng. Struct.* **2023**, *275*, 115278. [CrossRef]

buildings

MDPI

Article

Analysis of Amplification Effect and Optimal Control of the Toggle-Style Negative Stiffness Viscous Damper

Qiang Zhou [1,2,*], Wen Pan [1,2,*] and Xiang Lan [3]

1 Faculty of Civil Engineering and Mechanics, Kunming University of Science and Technology, Kunming 650500, China
2 Yunnan Seismic Engineering Technology Research Center, Kunming 650500, China
3 College of Architecture and Civil Engineering, Kunming University, Kunming 650214, China; lx523947015@kmu.edu.cn
* Correspondence: qzhou19920120@163.com (Q.Z.); 13312039@kust.edu.cn (W.P.)

Abstract: This paper proposes a new toggle-style negative stiffness viscous damper (TNVD), and evaluates the performance of the TNVD with the displacement amplification factor (f_d) and the energy dissipation factor (f_E). Firstly, the composition and characteristics of the TNVD are introduced. Subsequently, the displacement amplification factor is introduced to evaluate the displacement amplification ability of the TNVD, and it is decomposed into a geometric amplification factor and an effective displacement coefficient. Then, based on the geometric amplification factor and effective displacement coefficient, the correlation between the TNVD's displacement amplification ability and inter-story deformation is studied, and an improved TNVD is proposed. By the comparison of the finite element calculation results, it is found that the improved TNVD can utilize the assumption of small structural deformation. After that, the impacts of plentiful aspects, such as the length of the lower connecting rod, the horizontal inclination angle of the lower connecting rod, the inter-story deformation limit, the cross-sectional area of the connecting rod, the damping coefficient, and the negative stiffness on the f_d and f_E of the improved TNVD, are expounded. The research results show that when the length of the TNVD's lower connecting rod remains unchanged, the f_d and f_E present a trend of increasing first and then decreasing with the increase in the horizontal inclination angle of the lower connecting rod. When the inter-story deformation is fixed, there exists an optimal lower connecting rod's length that satisfies a specific relationship to achieve the optimal geometric amplification factor of the TNVD. By adjusting the damping parameters of the TNVD, we can obtain a better effective displacement coefficient greater than 0.95 in the proposed target region. Meanwhile, the f_d and f_E increase with the decrease in the negative stiffness. An optimization strategy for the improved TNVD has been proposed to ensure that the TNVD has the characteristics of operational safety, ideal displacement amplification capability, and energy dissipation capability. Furthermore, a multi-objective control design method with an additional improved TNVD structure is proposed. The vibration reduction effect of the structure with the improved TNVD and the effectiveness of the optimization strategy are verified through examples.

Keywords: energy dissipation; toggle-style negative stiffness damper; displacement amplification factor; energy consumption coefficient; optimal control

1. Introduction

One effective way to control structural vibration is to increase energy dissipation capacity by installing dampers. In contrast to active and semi-active vibration control technology [1,2], this passive vibration control technology is widely applied due to its good vibration reduction effect, low overall cost, and simple maintenance in the later stage. In engineering, the commonly used energy dissipation dampers include viscous dampers, viscoelastic dampers, friction dampers, and metal yield dampers [3–6]. In addition, other scholars studied the application of a new-type high-damping fluid damper with shear thickening fluid [7]. This type of energy dissipation damper is usually connected to the structure using supports or intermediate columns, and the upper limit of its deformation or velocity is the inter-layer deformation or velocity of the structure. Therefore, it is necessary to adopt displacement amplification technology to significantly improve the deformation of dampers, so as to more effectively control the vibration response of the structure.

In order to improve the deformation of dampers, different forms of amplification devices have been proposed successively. Yang proposed the displacement-amplified mild steel bar joint damper using the principle of lever amplification, whose energy dissipation capacity is almost three times that of an ordinary damper [8]. Constantinou et al. [9,10] first proposed a toggle brace damper, and derived a displacement amplification factor (the ratio of the damper's displacement to the structural horizontal displacement) based on the assumption of small structural deformation. Meanwhile, in order to avoid minor changes in the geometric shape of the toggle having a significant impact on the displacement amplification factor, it was recommended to take a range from 2 to 5 for the displacement amplification factor. Taylor [11] connected the toggle to the beam–column node, and proposed an improved toggle brace amplification device. Sigaher [12] further designed a compact scissor-shaped device, and showed by experiments that this device can provide significant damping and effectively control the seismic response of the structure. Subsequently, numerous researchers studied different forms of toggle-style devices [13,14]. The above-mentioned studies on the toggle-style displacement amplification factor were mostly based on the assumption of small structural deformation. The study of Zhang [15] indicated that the displacement amplification factor of the toggle mechanism was related to the inter-story displacement and its direction. Polat [16] suggested controlling the displacement amplification factor based on the assumption of small deformation within a certain range to avoid significant changes in the displacement amplification factor due to large inter-story displacement. Lan [17] proposed a local toggle-brace damper that saves building space, and further adopted a displacement increment amplification factor based on large deformation to compare the displacement amplification ability of the toggle devices with different geometric shapes. For various amplification devices, the amplification ability should be constant and simple for engineering applications, and have high reliability. Regarding structural reliability analysis, She et al. [18] proposed an active learning Kriging method that can achieve high-precision fault probability. The displacement amplification ability of the above-mentioned toggle-brace damper is affected by inter-story deformation, which further increases the complexity of its engineering application. Taking some measures to make sure that the displacement amplification ability of the toggle-brace damper has nothing to do with the amplitude and direction of the inter-story deformation will be beneficial for engineering designers to understand and apply it more conveniently.

Various forms of toggle dampers have significant displacement amplification abilities, while the prerequisite for achieving this amplification function is that the deformation of the toggle itself is relatively small. A lower toggle stiffness will significantly reduce the displacement of the damper. The studies of Wang et al. [19,20] indicated that when the brace connected to the damper is flexible rather than rigid, the negative stiffness device will amplify the displacement of the damper. Therefore, the combination of negative stiffness devices and toggle-style devices can further increase the displacement of dampers and reduce the stiffness requirements of the toggle. Currently, numerous researchers have proposed different forms of negative stiffness devices and applied them to the vibration response control of structures. Nagarajaiah [21,22] proposed a device that achieves negative stiffness through preloading springs. Pasala et al. [23–25] added gap springs on the basis of the negative stiffness device proposed by Nagarajaiah, forming a bilinear elastic negative stiffness system. Meanwhile, Chen et al. [26,27] proposed a negative stiffness device combining pre-compressed springs and inclined planes. Liu [28] proposed a negative stiffness device with self-centering, adopting memory-shaped metals. Zhu et al. [29–31] designed a new negative stiffness device utilizing permanent magnets and turbine damping, and applied it to the vibration response control of cable-stayed cables. The experimental test results indicated that the proposed negative stiffness device could effectively reduce the vibration response of the cable and provide high damping. Tan studied the effect of low temperature on the mechanical properties of lead core high-damping rubber bearings for bridge seismic bearings [32]. Attary et al. [33,34] employed vibration table tests and numerical simulations to study the role of negative stiffness devices as foundation isolators in bridge structures. Sun [35] designed a double triangular damper with equivalent negative stiffness characteristics, and applied it to foundation isolation systems. Numerical simulations displayed that for the displacement of the isolator and roof acceleration of the structure, the double triangular damping device possessed a higher isolation effect than a traditional lead–rubber bearing. Applying negative stiffness devices to energy dissipation structures requires a combination of negative stiffness devices and viscous dampers to simultaneously control the acceleration and displacement of the structure, due to the fact that the negative stiffness device weakens the stiffness of the structure and increases its inter-story displacement. Negative stiffness devices and viscous dampers need to be connected to the structure through different forms of support. However, the impact of the stiffness of the support on traditional viscous damping systems and negative stiffness viscous damping systems is different. Chalarca [36] discussed the effect of the total stiffness of viscous dampers (the series stiffness of support stiffness and viscous damper stiffness) on floor acceleration under far-field earthquakes. The results revealed that the seismic response increases with the decrease in total stiffness value. Chen et al. [37–40] pointed out that lower support stiffness would lead to a reduced energy dissipation in viscous damping systems. However, the studies of Wang et al. [19,20,41] revealed that flexible supports were advantageous in increasing the energy dissipation of the negative stiffness viscous damper systems. Therefore, a negative stiffness damping system and an amplification device can be combined to achieve a dual increase in a damper's energy consumption. This can not only provide greater damping to the structure, but also reduce the stiffness requirements of the amplification device. Presently, there are relatively scarce studies on the combination of these two parts.

In order to further improve the displacement amplification ability of the toggle-style device, this paper proposes a symmetrical toggle-style negative stiffness viscous damper (TNVD) by combining traditional toggle support damping systems and negative stiffness devices. Specifically, this paper proposes to utilize displacement amplification factor f_d to describe the TNVD's displacement amplification capability, and apply energy dissipation factor f_E to describe the TNVD's energy dissipation capability. f_d can be represented by the product of the geometric amplification factor and effective displacement coefficient. The geometric amplification factor is utilized to describe the influence of geometric parameters on the TNVD amplification ability, while the effective displacement coefficient is applied to describe the weakening of the TNVD's amplification ability by the effective stiffness of the toggle's connecting rod. In this paper, the correlation between the TNVD's displacement amplification ability and the inter-story deformation is studied, which is reflected in the influence of the inter-story deformation on both the geometric amplification factor and the effective displacement coefficient. Moreover, an improved TNVD that can apply the small structural deformation assumption is proposed. The impacts of the lower connecting rod's length, the lower connecting rod's horizontal inclination angle, the inter-story deformation limit, the damping coefficient, and negative stiffness on the TNVD's f_d and f_E are expounded. In addition, an optimization strategy of the improved TNVD is put forward to ensure that the improved TNVD can not only be operated safely and effectively, but also have ideal displacement amplification capability and energy dissipation capability. On this basis, furthermore, a multi-objective control design method is proposed. Consequently, the control effects of three TNVD vibration reduction schemes on the seismic response of a nine-story steel frame are compared.

2. Composition and Displacement Amplification Factor of the TNVD

2.1. Composition and Characteristics of the TNVD

Figure 1 reveals the TNVD's installation form in the structure, the composition of the negative stiffness device, and the hysteresis curve after the negative stiffness device is connected in parallel with the viscous damper. The symmetrical toggle-style negative stiffness viscous damper (TNVD) consists of a toggle-style damping system and a negative stiffness device (NSD), and is symmetrically installed as shown in Figure 1a. The composition of the NSD is illustrated in Figure 1b. The upper connecting rod of the TNVD is hinged to the beam–column node, and the lower connecting rod is hinged to the middle of the beam. A viscous damper (VD) and a negative stiffness device (NSD) are connected in parallel to form an NVD, which is hinged at the connection between the upper and lower toggles and the beam–column nodes. The inter-story deformation of the structure causes the upper and lower toggles to move, resulting in deformation of the viscous damper (VD) and NSD. When the piston rod in the NSD moves, the negative stiffness force generated by the preloading spring further amplifies the deformation of the VD. Therefore, the amplification property of the TNVD originates from the deformation amplification of the toggle's connecting rod and the negative stiffness generated by the pre-compressed spring acting on the convex surface.

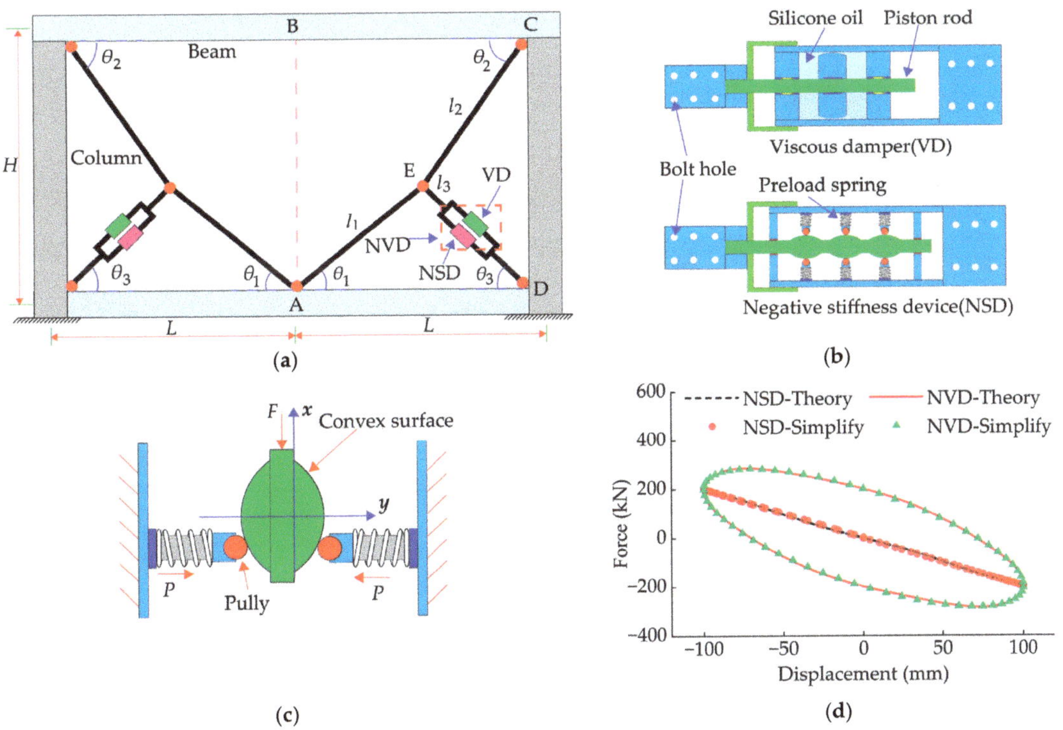

Figure 1. Symmetric toggle-style negative stiffness viscous damper (TNVD): (**a**) TNVD; (**b**) viscous damper (VD) and negative stiffness device (NSD); (**c**) NSD's negative stiffness force; (**d**) hysteresis curves of NSD and NVD.

In Figure 1c, the spring stiffness is k_{SP}; the preload displacement is Δ_{SP}; the convex shape is $y = f(x)$; the number of springs is N; and the negative stiffness force of the NSD is $F(x)$. $F(x)$ is calculated from Equation (1). Figure 1d indicates the hysteresis curves of the NVD and the NSD when they undergo sinusoidal deformation with a circular frequency of 1 rad/s and an amplitude of 100 mm. In Figure 1d, the theoretical curve reveals that the number of springs in the NSD is 12, the spring stiffness k_{SP} is 1.3 kN/mm, the convex surface shape is $f(x) = 50\cos(2\pi/600x) - 7\cos(2\pi/280x)$, the preload displacement Δ_{SP} is 55 mm, and the VD parameter is $c = 2\text{kN}/(\text{mm/s})$. In Figure 1d, the negative stiffness of the simplified curve is −2 kN/mm, and the damping coefficient is $2\text{kN}/(\text{mm/s})$. As is illustrated in Figure 1d, when the deformation of the NVD is controlled within a certain range, the simplified curve is very close to the analytical curve model. Therefore, by means of constructing preload springs and convex shapes reasonably, the NVD can be represented in parallel by a linear negative stiffness element and a viscous element.

$$F(x) = N\frac{f'(x)}{[1+f'(x)]^2}k_{SP}[\Delta_{SP} + f(x) - f(0)] \tag{1}$$

2.2. The TNVD's Displacement Amplification Factor

The basic geometric information of the TNVD is shown in Figure 1a. $2L$ is the span of the frame and H is the height of the frame. l_1 and l_2 are the lengths of the lower and upper connecting rods, respectively. l_3 is the sum of the NVD and support lengths. θ_1, θ_2, and θ_3 are the horizontal inclination angles of the lower connecting rod, upper connecting rod, and NVD, respectively. The elastic modulus of the connecting rod material is E, and the cross-sectional area of the lower and upper connecting rods is equal and is A_1. The constitutive relationship of the VD is $F_{VD} = c\dot{u}_d$, where c is the damping coefficient of the VD, and u_d and $\dot{u}_d$ are the deformation and velocity of the VD, respectively. The constitutive relationship of the NSD is $F_{NSD} = k_N u_d$, where k_N is the negative stiffness of the NSD. We assume that the maximum inter-story deformation of the structure is U_o, and the maximum displacement of the VD on the right side of Figure 1a is u_{dR}. When U_o remains unchanged, the damping coefficient is $c = 0$, and the negative stiffness is $k_N = 0$, the maximum deformation of the right VD is u_{aR}. It is defined as follows:

$$\gamma_R = \frac{u_{aR}}{U_o},\ \eta_R = \frac{u_{dR}}{u_{aR}} \tag{2}$$

This paper calls γ_R the geometric amplification factor, which represents the geometric amplification attribute of the right TNVD. In concept, this amplification attribute is consistent with the damper's displacement amplification factor, $\cos\theta$ (θ is the angle between the slant brace and the horizontal direction) in the classic slant brace connection. η_R is the effective displacement coefficient of the right VD, which represents the impact of the stiffness of the upper and lower connecting rods in the TNVD on the displacement of the VD. In concept, this impact is consistent with the impact of the slant brace stiffness on the displacement of the damper in the classical slant brace connection.

The displacement amplification factors f_{dL} and f_{dR} are defined as the ratios of the displacement of the left and right VDs to the inter-story displacement, respectively. f_{dR} is shown in Equation (3). The total displacement amplification factor, f_d, of the symmetrically installed TNVD is defined as the ratio of the sum of the displacement of the left and right VDs to the inter-story displacement, as shown in Equation (4).

$$f_{dR} = \frac{u_{dR}}{U_o} = \gamma_R\eta_R \tag{3}$$

$$f_d = f_{dR} + f_{dL} \tag{4}$$

2.2.1. The TNVD's Geometric Amplification Factor

The geometric amplification factor γ_R represents the amplification of the VD's displacement by the geometric construction of the TNVD. When the damping coefficient is $c = 0$, and the negative stiffness is $k_N = 0$, the deformation of the right TNVD is revealed, shown in Figure 2.

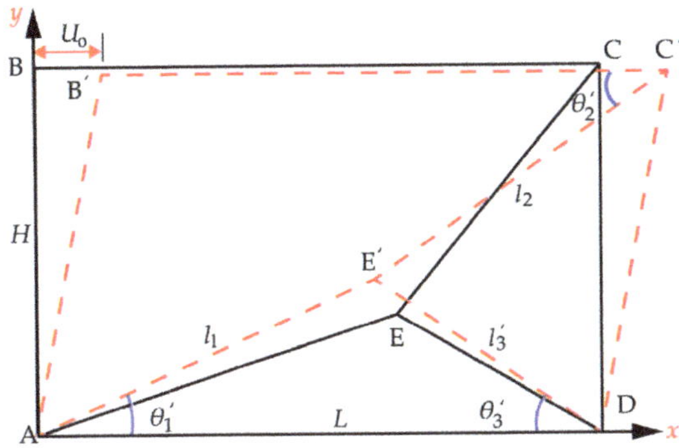

Figure 2. The TNVD's deformation diagram.

When the structural deformation, U_o, is small, the geometric amplification factor derived on the basis of the small structural deformation can be found, described in Reference [14] as follows:

$$\gamma^* = \frac{\sin(\theta_1 + \theta_3)}{\sin(\theta_2 - \theta_1)}\cos(\theta_2) \tag{5}$$

Due to the close correlation between the amplification effect of the toggle-style amplification mechanism and inter-story deformation, it is difficult to reflect this phenomenon by adopting the small deformation analysis results of Equation (5). For this, it is necessary to derive an analytical solution for the geometric amplification factor γ_R. We establish an orthogonal coordinate system, as displayed in Figure 2. When the TNVD deforms in the positive direction of x, U_o is defined as positive. It is assumed that during inter-story deformation, the lengths of beam BC, beam AD, column CD, and the upper and lower connecting rods remain unchanged. When the inter-story deformation is U_o, the node B moves to the node B′, the node C moves to the node C′, and the node E moves to the node E′. The horizontal inclination angle of each component after the TNVD's movement is as follows:

$$\begin{cases} \theta_1' = \arctan\left(\frac{H}{L+U_o}\right) - \arccos\left[\frac{(L+U_o)^2+H^2+l_1^2-l_2^2}{2l_1\sqrt{(L+U_o)^2+H^2}}\right] \\ \theta_2' = \arctan\left(\frac{H}{L+U_o}\right) + \arccos\left[\frac{(L+U_o)^2+H^2+l_2^2-l_1^2}{2l_2\sqrt{(L+U_o)^2+H^2}}\right] \\ \theta_3' = \arcsin\left[\frac{l_1\sin(\theta_1')}{\sqrt{L^2+l_1^2-2Ll_1\cos(\theta_1')}}\right] \end{cases} \tag{6}$$

In the combination of Equations (2) and (6), the geometric amplification factor γ_R of the right TNVD is as follows:

$$\gamma_R = \left|\left[\sqrt{L^2 + l_1^2 - 2Ll_1\cos(\theta_1')} - l_3\right]/U_o\right| \tag{7}$$

In Figure 1a, the left TNVD and right TNVD are symmetric about the AB node. Therefore, when the inter-story deformation is U_o, the geometric amplification factor γ_L of the left TNVD is equal to that of the right TNVD when the inter-story deformation is $-U_o$. U_o in Equation (6) is replaced by $-U_o$ to calculate the horizontal inclination angle θ_1' of the the left TNVD's lower connecting rod, and then it is substituted into Equation (7) to calculate the geometric amplification factor γ_L when the inter-story deformation is U_o.

We define the average value of θ_1' and θ_1 as θ_1^{av}, the average value of θ_2' and θ_2 as θ_2^{av}, and the average value of θ_3' and θ_3 as θ_3^{av}. The approximate value of γ_R is defined as γ_R^S, which is shown in Equation (8). Meanwhile, the calculation approach of γ_L^S is the same as γ_R^S.

$$\gamma_R^S = \frac{\sin(\theta_1^{av} + \theta_3^{av})}{\sin(\theta_2^{av} - \theta_1^{av})}\cos(\theta_2^{av}) \tag{8}$$

In Figure 2, we can see that the distance between node A and node C′ increases when the TNVD deforms in the positive direction of x. When the broken line AE′C′ degenerates into a straight line AC′, the upper and lower connecting rods are tightened, and the TNVD has the risk of failure. In order to ensure the normal and effective operation of the TNVD, there is a relationship that must be defined between the length l_1 of the lower connecting rod, the length l_2 of the upper connecting rod, and the structure's inter-story deformation limit $U_{o\ \mathrm{Max}}$, as follows:

$$l_1 + l_2 \geq \sqrt{(L + U_{o\ \mathrm{Max}})^2 + H^2} \tag{9}$$

In order to facilitate the study of the variation law of γ_R with U_o, it is necessary to limit $U_{o\ \mathrm{Max}}$ within a reasonable range. According to "Technical guideline for maintaining normal functionality of buildings in earthquakes" [42], for steel structures belonging to Class I buildings, the inter-story displacement angle under rare earthquakes must be less than 1/100. Therefore, $U_{o\ \mathrm{Max}} \leq H/100$. In order to intuitively present the variation of γ^*, γ_R, and γ_R^S with U_o, the TNVD is analyzed when the length parameters are fixed and $U_{o\ \mathrm{Max}} = H/100$. The geometric parameters of the TNVD are displayed in Table 1, and the geometric amplification factors are shown in Figure 3.

Table 1. Geometric parameters of the TNVD.

Amplification Device	H(mm)	L(mm)	l_1(mm)	l_2(mm)
TNVD	4000	4000	2800	2888.7

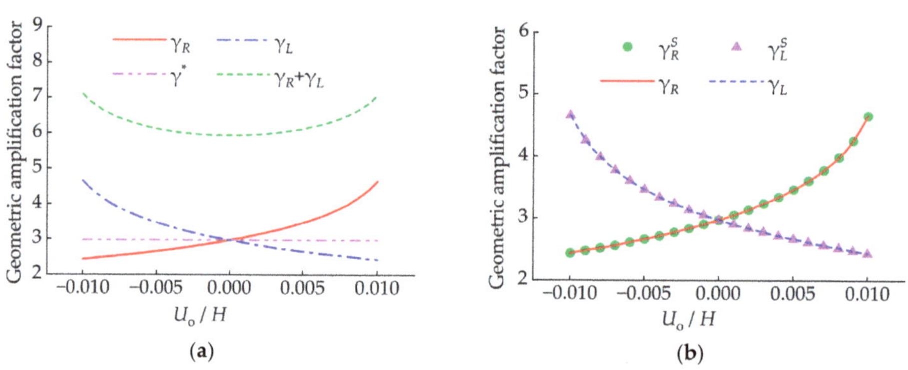

Figure 3. Variations of γ^*, γ_R, γ_L, γ_R^S, and γ_L^S with U_o/H: (**a**) γ^*, γ_R, γ_L, and $\gamma_R + \gamma_L$; (**b**) γ_R^S and γ_L^S.

From Figure 3, it can be seen that the amplitude and direction of the inter-story deformation have a significant impact on γ_R and γ_L; the geometric amplification factor based on the small structural deformation assumption is only equal to the analytical solution calculated by Equation (6) when $U_o = 0$. The sum of γ_R and γ_L of the TNVD installed symmetrically obtains the minimum value when $U_o = 0$. The difference between the approximate value γ_R^S (or γ_L^S) and the analytical solution γ_R (or γ_L) is relatively small. Therefore, the TNVD's γ_R (or γ_L) can be approximated as the small deformation solution γ_R^S (or γ_L^S) corresponding to the average of the TNVD's initial inclination angle and the inclination angle at the U_o position.

2.2.2. The TNVD's Effective Displacement Coefficient

The effective displacement coefficient η_R represents the impact of the stiffness of the upper and lower connecting rods in the TNVD on the displacement of the VD. From Figure 4, it can be seen that the force on each member of the TNVD satisfies the following relationship: $F_1 = F_3 \sin(\theta_2 + \theta_3)/\sin(\theta_2 - \theta_1)$, $F_2 = F_3 \sin(\theta_1 + \theta_3)/\sin(\theta_2 - \theta_1)$. The elastic modulus of the connecting rod steel is E, and the cross-sectional area of the lower connecting rod and the upper connecting rod is equal and is A_1. By means of the unit load method, the deformation of the upper and lower connecting rods in the θ_3 direction is calculated by Equation (10) as follows:

$$\Delta = \int_0^{l_1} \frac{F_3}{EA_1}[\frac{\sin(\theta_2+\theta_3)}{\sin(\theta_2-\theta_1)}]^2 dx + \int_0^{l_2} \frac{F_3}{EA_1}[\frac{\sin(\theta_1+\theta_3)}{\sin(\theta_2-\theta_1)}]^2 dx \tag{10}$$

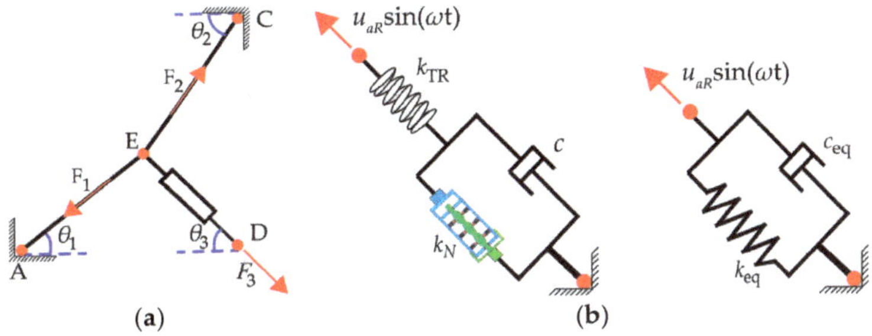

Figure 4. Force analysis of connecting rods and the θ_3 directional mechanical model: (**a**) force analysis of upper and lower connecting rods; (**b**) mechanics model in the θ_3 direction and equivalent model.

The equivalent stiffness k_{TR} of the upper and lower connecting rods of the right TNVD can be found in the following Equation (11):

$$k_{TR} = \frac{F_3}{\Delta} = \frac{EA_1^2}{\alpha l_1 A_1 + \beta l_2 A_1} \tag{11}$$

where $\alpha = [\sin(\theta_2+\theta_3)/\sin(\theta_2-\theta_1)]^2$, $\beta = [\sin(\theta_1+\theta_3)/\sin(\theta_2-\theta_1)]^2$.

The mechanical model of the θ_3 direction is actually a negative stiffness element and a damping element in parallel, which then connect in series with a spring element. This is consistent in form with the mechanical model of viscoelastic damping elements considering support flexibility. According to the definition in Section 2.2, the maximum deformation of the right VD is u_{aR} when U_o remains unchanged, the damping coefficient is $c = 0$, and negative stiffness is $k_N = 0$. In order to reveal the displacement effectiveness of the VD in the θ_3 direction mechanical model, it is assumed that this system is subjected to stable sinusoidal excitation. Without involving dynamic analysis, parameter studies can be conducted on the system in the steady state. We assume that the sinusoidal deformation $u_{aR}\sin(\omega t)$ is generated by the θ_3 directional mechanical model and equivalent model; then, the VD's $u_d(t)$, $\dot{u}_d(t)$, and $F(t)$ are as follows:

$$\begin{cases} u_{dR}(t) = u_{aR}\sin(\omega t) - [k_{eq}u_{aR}\sin(\omega t) + c_{eq}\omega u_{aR}\cos(\omega t)]/k_{TR} \\ \dot{u}_{dR}(t) = \omega u_{aR}\cos(\omega t) - [k_{eq}\omega u_{aR}\cos(\omega t) - c_{eq}\omega^2 u_{aR}\sin(\omega t)]/k_{TR} \\ F(t) = c\dot{u}_{dR}(t) + k_N u_{dR}(t) = k_{eq}u_{aR}\sin(\omega t) + c_{eq}\omega u_{aR}\cos(\omega t) \end{cases} \tag{12}$$

Equation (12) is utilized to solve for k_{eq} and c_{eq} in the equivalent model as follows:

$$\begin{cases} k_{eq} = k_{TR}\frac{(k_N/k_{TR})^2 + k_N/k_{TR} + (c\omega/k_{TR})^2}{(1+k_N/k_{TR})^2 + (c\omega/k_{TR})^2} \\ c_{eq} = \frac{c}{(1+k_N/k_{TR})^2 + (c\omega/k_{TR})^2} \end{cases} \tag{13}$$

According to the fact that the energy consumption in the θ_3 directional mechanics model and in the equivalent model are equal, we obtain as follows:

$$\eta_R = \frac{u_{dR}}{u_{aR}} = \frac{1}{\left[(1 + k_N/k_{TR})^2 + (c\omega/k_{TR})^2\right]^{0.5}} \tag{14}$$

According to Equation (14), η_R is closely related to the equivalent stiffness k_{TR} of the upper and lower connecting rods and the negative stiffness k_N of the NSD. Meanwhile, the presence of negative stiffness k_N increases the η_R. According to Equation (11), inter-story deformation U_o will significantly change the horizontal inclination angles θ_1', θ_2', and θ_3' of each component after the TNVD motion, thereby causing the equivalent stiffness k_{TR} to change with U_o accordingly. Therefore, just like the geometric amplification factor γ_R, the effective displacement coefficient η_R is also related to the inter-story displacement. The equivalent stiffness calculated based on the assumption of small structural deformation is called k_T^*, and the effective displacement coefficient calculated using k_T^* is called η^*. The calculation method of the equivalent stiffness k_{TL} and effective displacement coefficient η_L of the left TNVD can refer to Equations (11) and (14).

In order to visually present the variation law of η_R with U_o, the TNVD of the geometric shape shown in Table 1 is analyzed, and the following parameters are added (see Table 2).

According to Figures 3 and 5, it can be seen that the variation law of the equivalent stiffness k_{TR} or k_{TL} with the inter-story deformation U_o is opposite to that of the geometric amplification factor γ_R or γ_L. When the inter-story deformation U_o increases in the positive direction, γ_R increases, k_{TR} and η_R decrease, γ_L decreases, and k_{TL} and η_L increase. For the TNVD, the effective displacement coefficient varies with the variation in the inter-story deformation, namely, the effect of the equivalent stiffness of the upper and lower connecting rods of the TNVD on the VD's displacement is related to the amplitude and direction of the inter-story deformation. As is illustrated in Figure 5b, the sum f_d of the displacement amplification factor of the TNVD on both sides decreases with the increase in inter-story deformation amplitude. This means that when the inter-story deformation is large, the displacement amplification factor calculated through utilizing the small deformation assumption may overestimate the displacement amplification ability of the symmetric TNVD. The risk of such overestimation also exists for traditional toggle amplification mechanisms, which can be verified by setting the negative stiffness f_d to 0. For various amplification devices, the amplification capacity should be constant and simple for engineering applications. The characteristic that the displacement amplification ability of the symmetric TNVD is affected by inter-story deformation further increases the complexity of its engineering applications.

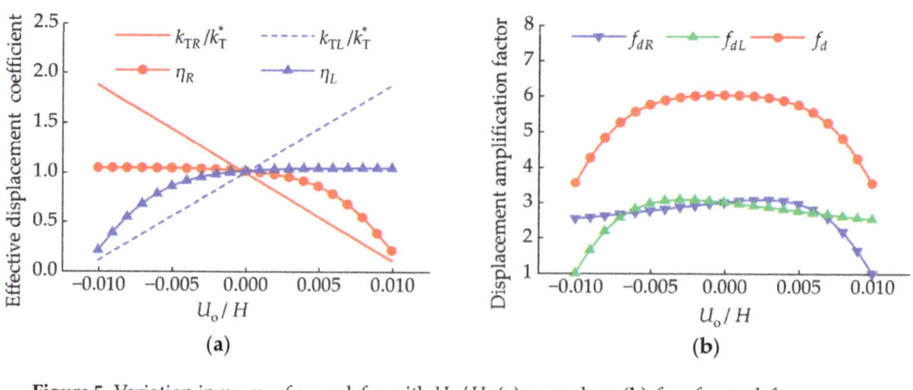

Figure 5. Variation in η_R, η_L, f_{dR}, and f_{dL} with U_o/H: (**a**) η_R and η_L; (**b**) f_{dR}, f_{dL}, and f_d.

Table 2. Parameters of TNVD components.

Device	$A_1(\mathrm{mm}^2)$	$E(\mathrm{kN/mm}^2)$	$c\omega(\mathrm{kN/mm})$	$k_N(\mathrm{kN/mm})$
TNVD	7600	206	2π	−2

2.3. The Improved Symmetrical TNVD

2.3.1. Composition of the Improved TNVD

In order to greatly eliminate the impact of inter-story deformation on the displacement amplification ability of the TNVD, an improved TNVD is designed by adding a spring device between the left and right TNVDs illustrated in Figure 1a, which is revealed in Figure 6a.

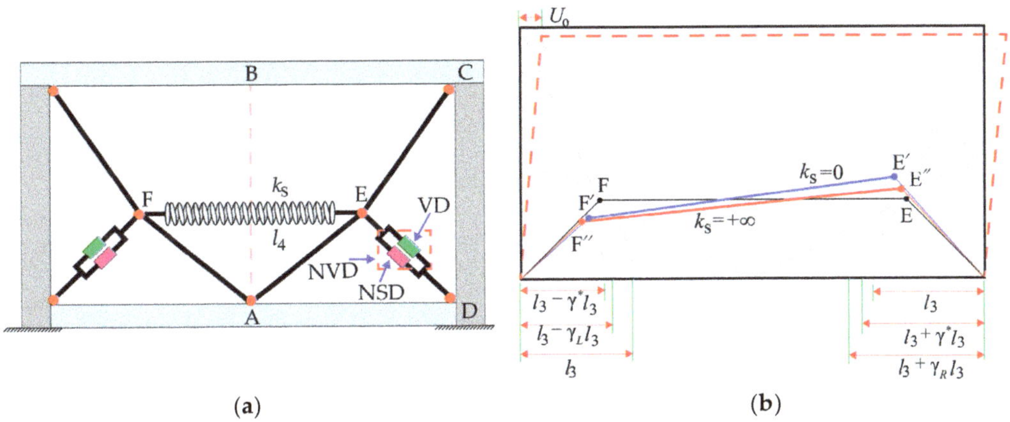

Figure 6. The improved TNVD and its deformation diagram: (**a**) the improved TNVD; (**b**) the deformation diagram.

The spring's stiffness is k_S. When $|U_o| > 0$, since $\gamma_R \neq \gamma_L$, the spring EF will be in a deformed state. The initial length of the spring is l_4, the spring after deformation is l_4'; then, the deformation Δ_S of the spring is as follows:

$$\Delta_S = l_4 - l_4' = \sqrt{2}l_1\left[\sqrt{1+\cos(\theta_{1R}'+\theta_{1L}')} - \sqrt{1+\cos(2\theta_1)}\right] \quad (15)$$

In this formula, θ_{1R}' is the horizontal inclination angle of the lower connecting rod after the right TNVD is deformed, and θ_{1L}' is the horizontal inclination angle of the lower connecting rod after the left TNVD is deformed.

It can be inferred from the calculation of the first-order partial derivative that there exists a unique zero point in its partial derivative function $\partial\Delta_S/\partial U_o$, which is obtained at $U_o = 0$. By taking the second-order partial derivative of Δ_S on U_o, we can determine that its partial derivative function is $\partial(\partial\Delta_S/\partial U_o)/\partial U_o < 0$. Therefore, there exists a unique maximum point, $\Delta_S|_{U_o=0} = 0$. In this way, when inter-story deformation occurs, the spring EF is always in a compressed state. The compressed spring EF exerts a thrust on the left and right TNVDs, respectively. This thrust causes γ_R and γ_L to approach γ^*, and k_{TR} and k_{TL} to approach k_T^*.

The improved symmetric TNVD's deformation is illustrated in Figure 6b. The spring EF is in its initial state, and the length of the left and right VDs and their supports is l_3. When the inter-story deformation is U_o and the viscous damping coefficient, negative stiffness, and spring stiffness k_S are all 0, the spring EF moves to E′F′, and the geometric amplification factor of the left and right TNVDs are γ_L and γ_R, respectively. When the spring stiffness is $k_S > 0$, the reaction of the compressed spring E′F′ on the TNVD causes the spring E′F′

to move to E″F″, and the left VD further increases the compression deformation, while the right VD reduces the tensile deformation. Meanwhile, due to the reaction force of the spring E″F″, it decreases k_{TL} and increases k_{TR}. When the spring stiffness is high, it can be considered that $\gamma_R \approx \gamma_L \approx \gamma^*$ and $k_{TR} \approx k_{TL} \approx k_T^*$.

In short, the spring plays a role in balancing the difference in displacement amplification ability between the left and right TNVDs. The greater the spring stiffness is, the stronger the balance ability is, until the left and right TNVDs' displacement amplification abilities are equal. The displacement amplification factor of the improved TNVD can be calculated without considering the influence of inter-story deformation as follows:

$$f_{dR} = \gamma^* \eta^* \tag{16}$$

$$f_d = f_{dL} + f_{dR} = 2\gamma^* \eta^* \tag{17}$$

2.3.2. Theoretical Formula Verification

In order to verify the displacement amplification factor of the improved TNVD, Equation (17), based on the assumption of small deformation, can be applied for calculation. A single-span steel frame model is established by means of finite element software SAP2000 (version 23.3.0), as shown in Figure 6a. The geometric parameters of this improved TNVD model are displayed in Table 1, and the remaining parameters are shown in Table 3. The section of the steel frame beam is hot-rolled H-section HN800 × 300 × 14 × 26 (unit: mm), and the frame column is a box section of 500 × 500 × 18 × 18 (unit: mm). The beams, columns, and the TNVD's connecting rods are all made of Q345 steel. Firstly, we verify the geometric amplification factor, and adopt displacement control loading to induce frame deformation of different amplitudes. Secondly, we verify the effective displacement coefficient and displacement amplification factor. The ground acceleration excitation adopts the sine excitation $\sin(10\pi t)\text{mm/s}^2$, and the main vibration frequency of the frame is $\omega = 10\pi$ rad/s. The sine wave amplitude is adjusted to cause the frame to deform in a different amplitude. The limit value of frame deformation U_o is $H/100$.

Table 3. Parameters of the improved TNVD components.

Device	$A_1(\text{mm}^2)$	$E(\text{kN/mm}^2)$	$c(\text{kN·s/mm})$	$k_N(\text{kN/mm})$	k_S/k_T^*
TNVD	7600	206	0.2	−2	[0, 10]

Figure 7 displays the effect of spring stiffness on the TNVD displacement amplification factor. From Figure 7a to Figure 7b, it can be seen that as the spring stiffness k_S increases, γ_R and γ_L tend to approach γ^*. When $k_S/k_T^* \geq 4$, the finite element calculation results of γ_R and γ_L are in good agreement with γ^* (the error is within 10%). From Figure 7b to Figure 7d, it can be seen that when $k_S/k_T^* = 10$, the finite element calculation results of the geometric amplification factor, effective displacement coefficient, and displacement amplification factor of the TNVD fluctuate less with inter-story deformation, and the consistency with the calculation results based on the assumption of small deformation is high, with a difference of less than 10%. Therefore, when $k_S/k_T^* \geq 10$, the displacement amplification factor of the improved TNVD can ignore the influence of inter-story deformation, and can be calculated by Equation (17).

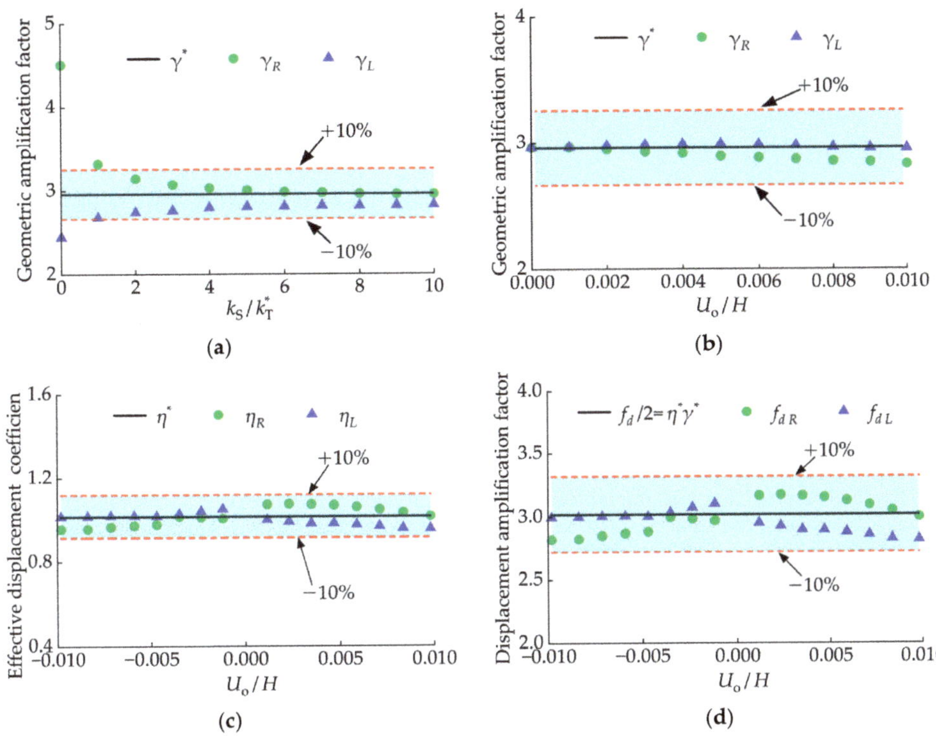

Figure 7. Verification of geometric amplification factor, effective displacement coefficient, and displacement amplification factor: (**a**) $U_o = H/100$, γ_R, and γ_L; (**b**) $k_S/k_T^* = 10$, γ_R, and γ_L; (**c**) $k_S/k_T^* = 10$, η_R, and η_L; (**d**) $k_S/k_T^* = 10$, f_{dR}, and f_{dL}.

3. Parameter Analysis and Optimization Design

The displacement amplification effect of the improved TNVD is jointly affected by the geometric properties of the TNVD, the equivalent stiffness of the upper and lower connecting rods, the damping coefficient of the VD, the negative stiffness of the NSD, etc. The total displacement amplification factor f_d can be obtained by combining Equations (5), (14), and (17) as follows:

$$f_d = \frac{2\sin(\theta_1+\theta_3)\cos(\theta_2)}{\sin(\theta_2-\theta_1)} \frac{1}{\left[(1+k_N/k_T^*)^2+(c\omega/k_T^*)^2\right]^{0.5}} \tag{18}$$

In Equation (18), the equivalent stiffness k_T^* based on small deformation is calculated from Equation (11), and ω is the main vibration frequency of the structure.

f_d describes the displacement amplification ability of the improved TNVD, where an energy dissipation factor, f_E, is introduced to describe the energy dissipation ability of the improved TNVD. The energy consumption factor f_E is the ratio of the sum of the maximum energy consumption of the left and right TNVDs in a single cycle to the maximum energy consumption of the structural stiffness proportional damping C_k in a single cycle (see Equation (19)).

$$f_E = \frac{2\pi c u_{do} \dot{u}_{do}}{\pi C_k U_o \dot{U}_o} = f_d^2 \frac{c}{2C_k} \tag{19}$$

In Equation (19), C_k represents the structural stiffness proportional damping; U_o and $\dot{U}_o$ represent the maximum inter-story displacement and maximum velocity; u_{do} and $\dot{u}_{do}$ represent the maximum displacement and maximum velocity of the VD.

f_d and f_E are relatively complex and influenced by multiple parameters. In order to facilitate direct discussion of the variation patterns with different parameters (such as θ_1, l_1, A_1, c, k_N, etc.), the structure shown in Figure 8 is analyzed. This structure has a total of eight spans, with TNVDs arranged on the side spans. The section of the steel frame beam is hot-rolled H-section HN800 × 300 × 14 × 26 (unit: mm), and the section of the column is hot-rolled H-section HW400 × 400 × 13 × 21 (unit: mm). The geometric parameters of the TNVD are displayed in Table 1, and the parameters of each component are shown in Table 3. $k_S/k_T^* = 10$; the main vibration frequency of the frame is that $\omega = 10\pi$ rad/s; the lateral stiffness is that $k_f = 305$ kN/mm, and the stiffness ratio damping is that $C_k = 0.78$kN/(mm/s). In this section, based on Equations (18) and (19), the impacts of the length of the lower connecting rod, the horizontal inclination angle of the lower connecting rod, the inter-story deformation limit, the damping coefficient, and the negative stiffness on f_d and f_E are studied.

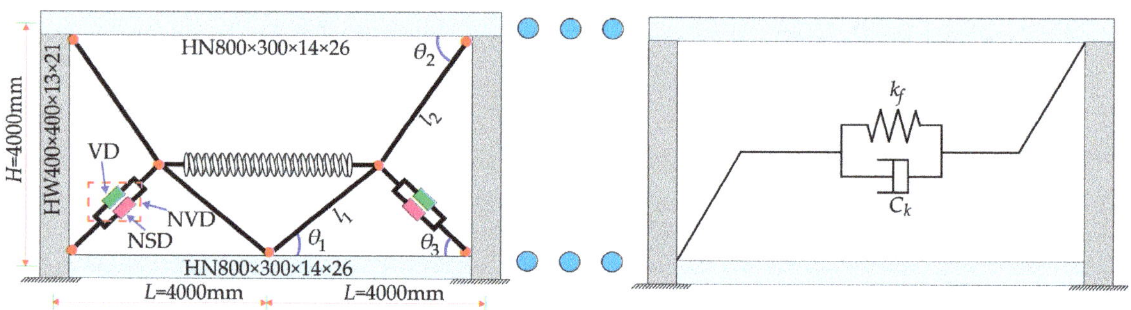

Figure 8. Schematic diagram of structure with TNVDs.

3.1. Parameter Analysis

3.1.1. Impacts of the Length l_1 and the Horizontal Inclination Angle θ_1 of the Lower Connecting Rod

The geometric shape of the TNVD can be fixed by the length and horizontal inclination angle of the lower connecting rod. l_1 and θ_1 are taken as the basic parameters to study the impact on the TNVD's f_d and f_E. The value range of l_1 is $0 < l_1 < L$, and the value range of θ_1 is $0 < \theta_1 < \arctan(H/L)$. Figure 9 illustrates the variation in f_d and f_E within a larger value range of l_1 and θ_1. From Figure 9a,b, it can be seen that for different lengths l_1, f_d and f_E first increase and then decrease as the horizontal inclination angle θ_1 increases. The reasons for this trend are shown in Figure 9c. From Figure 9c, it can be seen that when l_1 is fixed, as θ_1 increases, γ^* increases but η^* decreases, which causes f_d and f_E to increase first and then decrease. Meanwhile, as l_1 increases, there is a trend that f_d's and f_E's peak f_{dP} and f_{EP} decrease first and then increase. When θ_1 is fixed, f_d and f_E present different change laws as l_1 increases. As is revealed in Figure 9d, when θ_1 is small, f_d and f_E decrease as l_1 increases. When θ_1 approaches the limit value $\arctan(H/L)$, f_d and f_E increase with the increase in l_1.

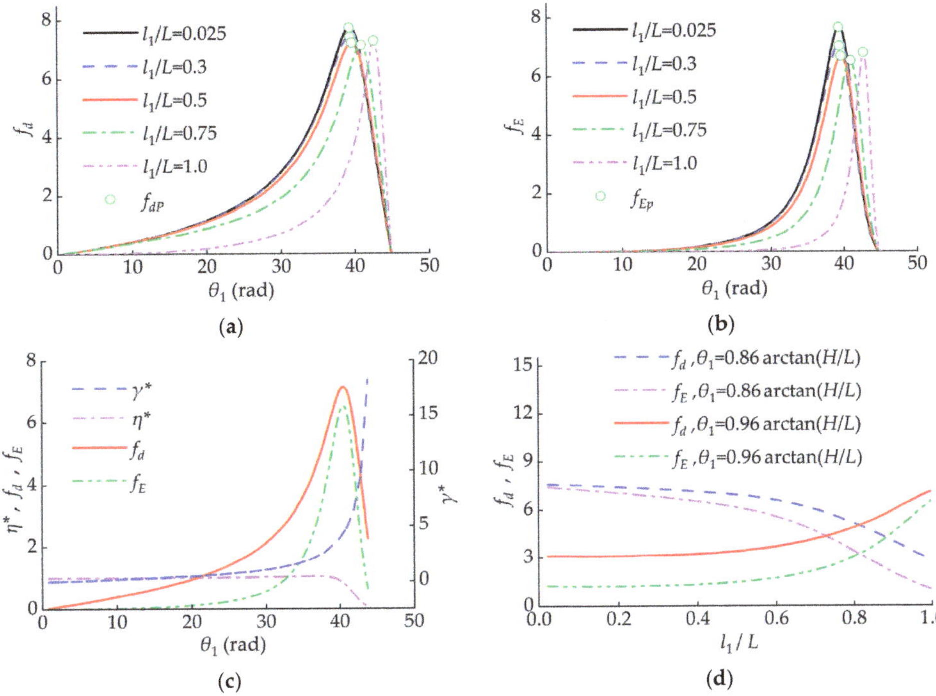

Figure 9. The impact of θ_1 and l_1 on f_d and f_E: (**a**) impact of θ_1 on f_d; (**b**) impact of θ_1 on f_E; (**c**) impact of θ_1 on f_d and f_E when $l_1/L = 0.7$; (**d**) impact of l_1 on f_d and f_E.

3.1.2. Impact of l_1 When the Inter-Story Deformation Limit $U_{o\ \mathrm{Max}}$ Is Fixed

Under normal working conditions of the TNVD, the inter-story deformation limit $U_{o\ \mathrm{Max}}$ will change the lower limit of the sum of the lengths of the upper and lower connecting rods, which is likely to affect the geometric shape of the TNVD. Equation (9) takes the equal sign to determine the sum of the lengths of the upper and lower connecting rods. Figure 10 reveals the variation in γ^* with $l_1/(l_1+l_2)$ under different inter-story displacement limits. As is revealed in Figure 10a, the larger $U_{o\ \mathrm{Max}}$ is, the smaller γ^* becomes. When $U_{o\ \mathrm{Max}}$ is unchanged, there exists a lower connecting rod's length to enable γ^* to reach optimal results. In Figure 10, γ_P^* represents the peak points of each curve, and γ_{PS}^* represents the γ^* of each curve when $l_1/(l_1+l_2) = L/(L+H)$. According to Figure 10a,b, it can be seen that under different inter-story displacement limits and different ratios of span to height, the approximate maximum value of γ^* is obtained when $l_1/(l_1+l_2) = L/(L+H)$. f_d consists of γ^* and η^*, and η^* is affected by γ^*. In order to enable f_d to reach a better value, the optimal value of γ^* can be obtained first, and then the internal damper parameters of the TNVD can be adjusted to ensure that η^* is within a better value range.

Meanwhile, Reference [17] studied the effect of the length of the connecting rod on the geometric amplification factor when the inter-story deformation is fixed. The results reveal that when the ratio of the length of the connecting rod to the floor height is 0.7, the geometric amplification factor reaches its maximum value. When the ratio of span to floor height is large, the geometric amplification factor obtained using the research results of this paper is similar to that obtained in Reference [17]. However, when the ratio of span to floor height is small, the length of the lower connecting rod that meets the requirements proposed in this paper will result in a larger geometric amplification factor for the TNVD.

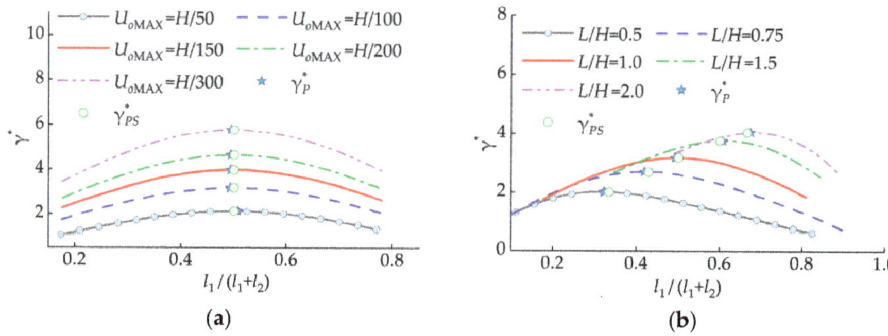

Figure 10. Impact of $l_1/(l_1 + l_2)$ on γ^* when $U_{o\,\mathrm{Max}}$ is fixed: (**a**) $L/H = 1.0$, γ^*; (**b**) $U_{o\,\mathrm{Max}} = H/100$, γ^*.

3.1.3. Impact of the Connecting Rod's Cross-Sectional Area A_1 and Damping Coefficient c

The cross-sectional area A_1 and damping coefficient c affect f_d and f_E by changing the η^*. As is shown in Figure 11a, with the increase in A_1, f_d and f_E first rapidly increase and then stabilize. Increasing A_1 is beneficial to improve η^*. As is revealed in Figure 11b, with the increase in the damping coefficient c, η^* and f_d decrease, while f_E displays a trend of first increasing and then decreasing. Hence, there exists an optimal c to achieve f_E to reach the maximum value. Figure 11c illustrates the variation law of f_E with $(c\omega - k_N)/k_T^*$ under different cross-sectional areas A_1 of connecting rods. f_E represents the peak points of each curve. From Figure 11c, it can be seen that f_E obtains the optimal result when $(c\omega - k_N)/k_T^* = 1$.

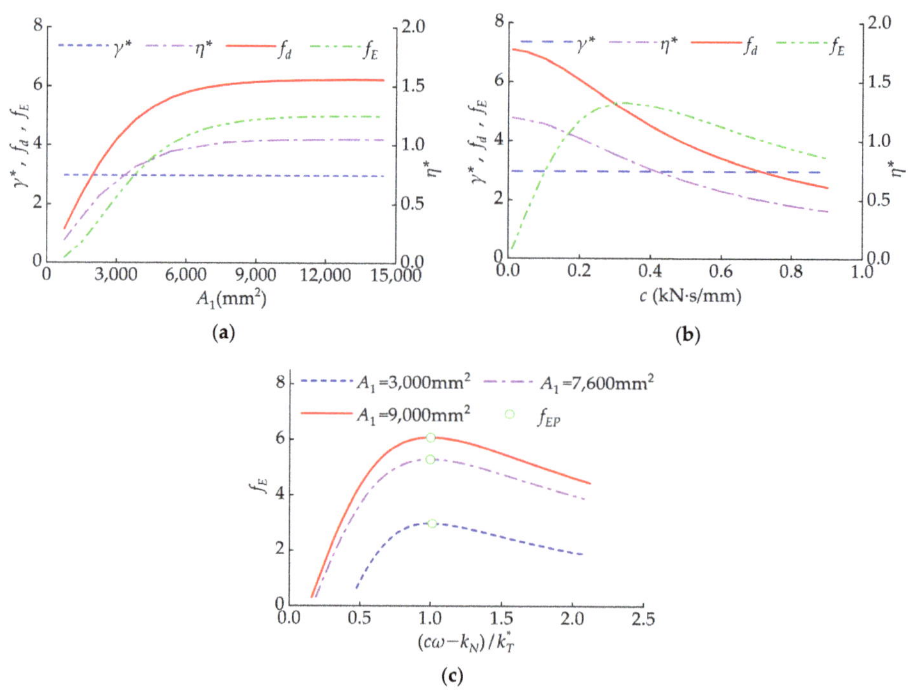

Figure 11. Impact of A_1 and c on f_d and f_E: (**a**) impact of A_1 on f_d and f_E; (**b**) impact of c on f_d and f_E; (**c**) impact of $(c\omega - k_N)/k_T^*$ on f_E.

3.1.4. Impact of NSD's Negative Stiffness k_N

This section explores the impact of changing the negative stiffness k_N of NSD on f_d and f_E. From Figure 12a,b, it can be seen that as the negative stiffness k_N decreases, η^*, f_d, and f_E gradually increase, while k_{eq} presents a trend of first decreasing and then increasing. Meanwhile, when the damping coefficient c is small, the TNVD will provide negative stiffness to the structure. Thus, the value of k_N should ensure that the TNVD achieves a high value of f_d while avoiding providing too much negative stiffness k_{eq}, which significantly reduces the total stiffness of the structure. "Technical specification for seismic energy dissipation of buildings" [43] requires that the support stiffness should be greater than three times the loss stiffness of the viscous damper. This regulation is still applied here. When $f_d = 1/(1+1/3^2)^{0.5}\gamma^* = 0.95\gamma^*$, $f_d = \gamma^*$, or $f_d = 2\gamma^*$, the relationship between k_N/k and $c\omega/k_T^*$ is as illustrated in Equation (20).

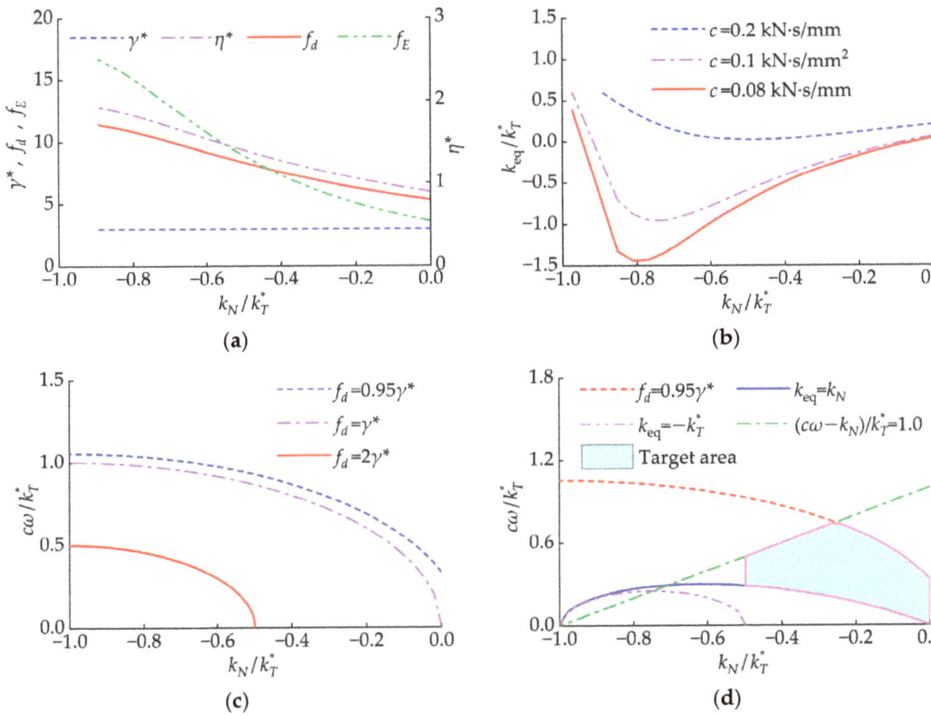

Figure 12. Impact of k_N/k_T^* on f_d and f_E and the target value region of k_N/k_T^* and $c\omega/k_T^*$: (**a**) impact of k_N/k_T^* on f_d and f_E; (**b**) impact of k_N/k_T^* on k_{eq}/k_T^*; (**c**) corresponding k_N/k_T^* and $c\omega/k_T^*$ of different f_d; (**d**) target value region of k_N/k_T^* and $c\omega/k_T^*$.

$$\begin{cases} k_N/k_T^* = \left[1+1/3^2-(c\omega/k_T^*)^2\right]^{0.5}-1 \ , f_d = 0.95\gamma^* \\ k_N/k_T^* = \left[1-(c\omega/k_T^*)^2\right]^{0.5}-1 \ , f_d = \gamma^* \\ k_N/k_T^* = \left[(1/2)^2-(c\omega/k_T^*)^2\right]^{0.5}-1 \ , f_d = 2\gamma^* \end{cases} \tag{20}$$

when $k_{\text{eq}} = k_N$ or $k_{\text{eq}} = -k_T^*$, the relationship between k_N/k_T^* and $c\omega/k_T^*$ is as shown in Equation (21).

$$\begin{cases}(1+k_N/k_T^*)(k_N/k_T^*)^2/(1-k_N/k_T^*) = (c\omega/k_T^*)^2 & , k_{\text{eq}} = k_N \\ (1+k_N/k_T^*)(1+2k_N/k_T^*) = -2(c\omega/k_T^*)^2 & , k_{\text{eq}} = -k_T^* \end{cases} \tag{21}$$

According to Equation (20) and Figure 12c, it can be seen that when $k_N/k_T^* \leq \left[1 + 1/3^2 - (c\omega/k_T^*)^2\right]^{0.5} - 1$, $f_d \geq 0.95\gamma^*$. According to Equation (21) and Figure 12d, it can be seen that when $(1+k_N/k_T^*)(k_N/k_T^*)^2/(1-k_N/k_T^*) \geq (c\omega/k_T^*)^2$, $k_{\text{eq}} \geq k_N$. By Equation (13), it can be concluded that when $k_N/k_T^* \geq -0.5$, it can be avoided that $k_{\text{eq}} \leq -k_T^*$ in any situation. According to Figure 11c, it can be seen that when $(c\omega - k_N)/k_T^* \leq 1$, increasing the damping coefficient c results in an increase in f_E. In order to achieve high displacement amplification capability of the TNVD while avoiding providing too much negative stiffness to the structure, it is recommended to select the target area shown in Figure 12d as the value range for k_N/k_T^* and $c\omega/k_T^*$. This target area possesses three properties, namely, $f_d \geq 0.95\gamma^*$, $k_{\text{eq}} \geq k_N$, and the fact that increasing the damping coefficient c leads to an increase in f_E.

3.2. *TNVD's Optimization Design*

3.2.1. TNVD's Optimization Strategies

To ensure the safe and effective operation of the improved TNVD with the ideal displacement amplification coefficient f_d and energy dissipation coefficient f_E, in combination with the discussion in Section 3.1, the following steps are taken to optimize the TNVD.

(1) The maximum inter-story deformation $U_{o\ \text{Max}}$ is determined based on the limit value of the inter-story displacement angle. $U_{o\ \text{Max}}$ is substituted into Equation (9) to obtain the minimum value of the $l_1 + l_2$.

(2) The length of the lower connecting rod $l_1 = (l_1 + l_2)L/(L + H)$ is selected and the length of the upper connecting rod l_2 is calculated. The horizontal inclination angle of the upper and lower connecting rods and damping components is calculated in a non-deformable state according to Equation (6). Subsequently, the optimal geometric amplification factor γ^* is calculated based on Equation (5). It is judged if $0.95\gamma^*$ is greater than the pre-set target of the displacement amplification factor f_d. If it is not satisfied, the following methods can be adopted: to keep θ_1 unchanged and to reduce the length l_1, or to keep the length l_1 unchanged and increase θ_1. At this moment, $l_1 + l_2$ changes, and it is necessary to substitute the inter-story displacement of the structural time history analysis into Equation (9) to verify whether the lengths of the upper and lower connecting rods of the TNVD are reasonable.

(3) The cross-sectional area A_1 of the connecting rod, damping coefficient c, and negative stiffness k_N are selected to make sure the values of k_N/k_T^* and $c\omega/k_T^*$ are within the target area shown in Figure 12d, and we employ Equation (14) to calculate the effective displacement coefficient η^*.

(4) γ^* and η^* are substituted into Equations (18) and (19) to calculate the displacement amplification factor f_d and the energy dissipation factor f_E.

(5) It is judged whether f_E satisfies the predetermined goals. If the goal is not met, the following methods should be adopted for further optimization: to increase damping coefficient c or decrease the negative stiffness k_N of the NSD. After that, it is verified again whether the values of k_N/k_T^* and $c\omega/k_T^*$ are within the target area shown in Figure 12d. Step (4) and Step (5) are repeated. If the value of $c\omega/k_T^*$ is on the upper boundary of the target area shown in Figure 12d, and f_E still cannot meet the set target, then increase the cross section of the upper and lower connecting rods of the TNVD, and return to Step (4) and Step (5).

3.2.2. Vibration Reduction Design Method for Structures with the Improved TNVD

For energy dissipation and seismic reduction structures with additional viscous dampers, the current design methods mostly focus on the additional damping ratio and inter-story displacement angle limits as control objectives. For a structure with an improved TNVD, this paper proposes a multi-objective control vibration reduction scheme design, including displacement amplification factor f_d, energy dissipation factor f_E, additional damping ratio, and inter-story displacement angle limit. f_d evaluates the displacement amplification ability of the improved TNVD; f_E intuitively evaluates the energy dissipation ability of the TNVD; the additional damping ratio evaluates the damping contribution of the TNVD to the overall structure, and the inter-story displacement angle limit controls the overall deformation of the structure. The design flowchart of this method is shown in Figure 13. The design process is as follows:

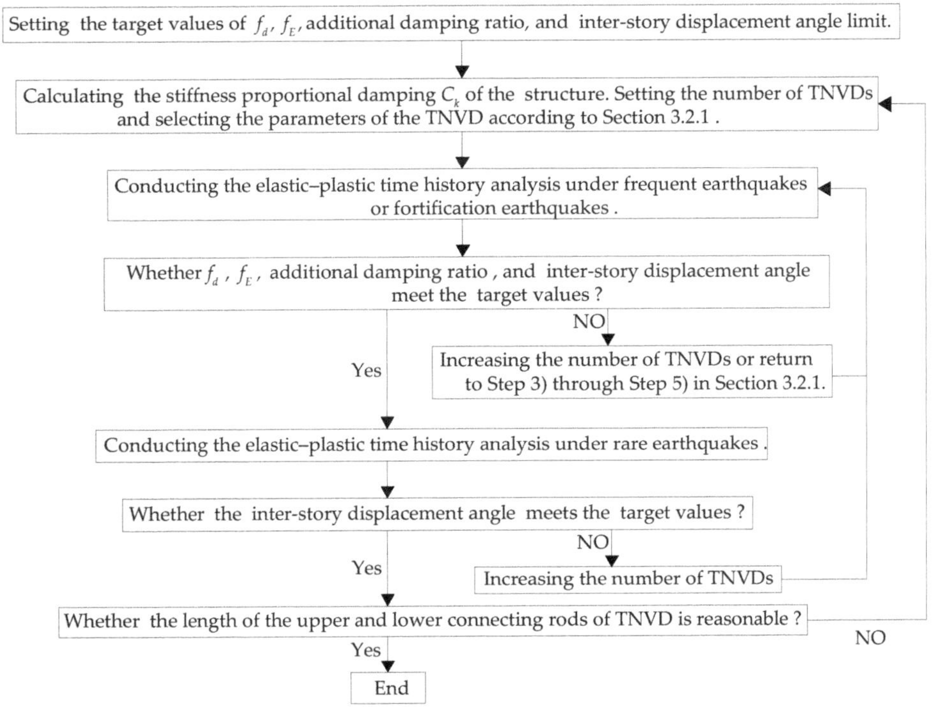

Figure 13. Damping design flowchart of structures with the improved TNVD.

(1) The target values of f_d, f_E, additional damping ratio, and inter-story displacement angle limit are set.

(2) The stiffness proportional damping C_k of the structure is calculated. The number of TNVDs is set up, and TNVD parameters are selected according to Section 3.2.1.

(3) The elastic–plastic time history analysis for frequent earthquakes or fortification earthquakes is conducted. It is judged whether f_d, f_E, and the additional damping ratio and inter-story displacement angle meet the set goals. If the goal is not achieved, increase the number of TNVDs or return to Step (3) through Step (5) in Section 3.2.1.

(4) The elastic–plastic time history analysis for rare earthquakes is conducted, and it is concluded whether the inter-story displacement angle meets the set target. If the goal is not achieved, increase the number of TNVDs and return to Step (3). According to the final rare earthquake inter-story deformation, Equation (9) is employed to recheck whether

the length of the upper and lower connecting rods of the TNVD is reasonable. If it is not reasonable, return to Step (2) to optimize the TNVD again.

4. Finite Element Analysis of Engineering Example

4.1. Project Overview

In order to verify the vibration reduction effect of structures with the improved TNVD and the effectiveness of the optimization method, this section takes a nine-story steel frame as an example. The height of the steel frame is 36 m; the seismic fortification intensity is 8 degrees (0.2 g); the site characteristic period is 0.45 s, and the self-weight is 7712 tons. The mass distribution on each floor is relatively uniform and 856 tons. The fundamental periods are 1.66 s (E–W), 1.60 s (N–S), and 1.36 s (rotation), respectively. The inherent damping ratio of the structure is 0.04, using the Rayleigh damping model with a mass-proportional coefficient of 0.314 and a stiffness-proportional coefficient of 1.83×10^{-3}. The steel frame adopts Q345 steel. The main beam is hot-rolled H-section HN630 × 200 × 15 × 20 (unit: mm), and the secondary beam is hot-rolled H-section HN500 × 200 × 10 × 16 (unit: mm). The beam connected to the TNVD is hot-rolled H-section HN800 × 300 × 14 × 26 (unit: mm). The frame columns have a box-shaped cross section, with the first floor measuring 700 × 700 × 20 × 20 (unit: mm); the second to the third floor measures 600 × 600 × 20 × 20 (unit: mm); the fourth floor to the sixth floor measures 550 × 550 × 18 × 18 (unit: mm), and the seventh to the ninth floor measures 500 × 500 × 18 × 18 (unit: mm). The elastic–plastic deformation of beams and columns is simulated using plastic hinges. The expected goals of the vibration reduction design are as follows: the damping ratio added to the structure by the TNVD under fortification earthquakes (peak acceleration PGA is 0.2 g) is 15%, and the limit value of the inter-story displacement angle of the structure is 1/300; the limit value of the inter-story displacement angle of the structure under rare earthquakes (peak acceleration PGA is 0.4 g) is 1/150; the TNVD displacement amplification factor is $f_d > 7$ and energy dissipation factor is $f_E > 25$.

4.2. Scheme Design

This section shows a total of three design schemes. According to the optimization strategy in Section 3.2.1 and the vibration reduction design method in Section 3.2.2, Scheme 1 was obtained. Scheme 2 was not optimized according to the TNVD optimization strategy. Scheme 3 has removed the NSD on the basis of Scheme 1. The installation positions of these three schemes in floors are shown in Figure 14, and the geometric parameters and internal component parameters of the TNVD are revealed in Table 4. According to the Chinese Code [44], five natural waves and two artificial waves are selected, and the normalized spectra of the selected seismic waves and the normalized standard response spectra are shown in Figure 15. SAP2000 software (version 23.3.0) was adopted to model, and elastic–plastic time history analysis for fortification earthquakes and rare earthquakes was conducted. All analysis results in this section are taken as the average of seven seismic wave calculation results. This section only elaborates on the application of the improved TNVD in the north–south direction of structures.

Table 4. Geometric parameters and various component parameters of the improved TNVD.

Scheme	Story	l_1 (mm)	l_2 (mm)	A_1 (mm)	c (kN · s/mm)	k_N (kN/mm)	k_S (kN/mm)
Scheme 1	6–8	2900	2811.3	7600	0.6	−2.2	353
	2–5	2900	2811.3	7600	0.8	−2.2	353
Scheme 2	6–8	3400	2349.5	7600	0.6	−2.2	292
	2–5	3400	2349.5	7600	0.8	−2.2	292
Scheme 3	6–8	2900	2811.3	7600	0.6	0	353
	2–5	2900	2811.3	7600	0.8	0	353

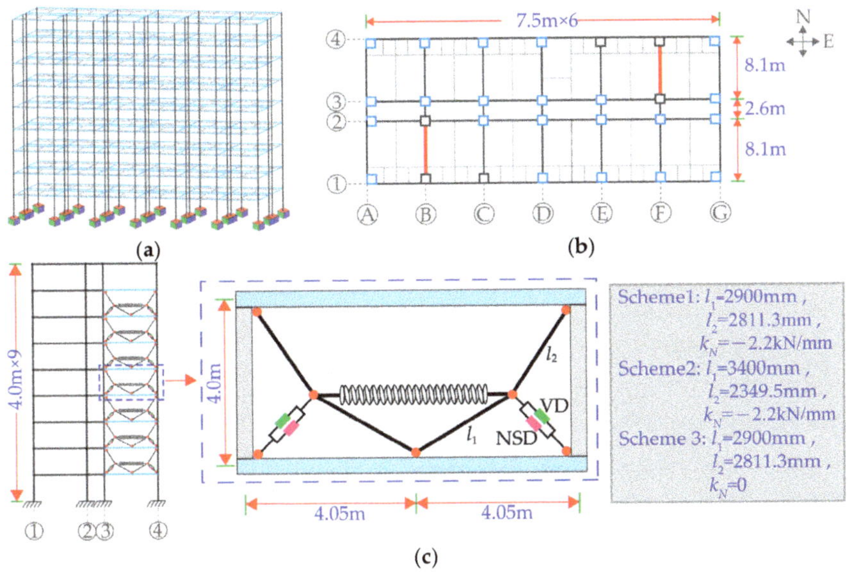

Figure 14. 3D Structural schematic diagram and layout of the improved TNVD: (**a**) structural 3D schematic diagram; (**b**) plane layout of the improved TNVD; (**c**) F-axis facade layout of the improved TNVD.

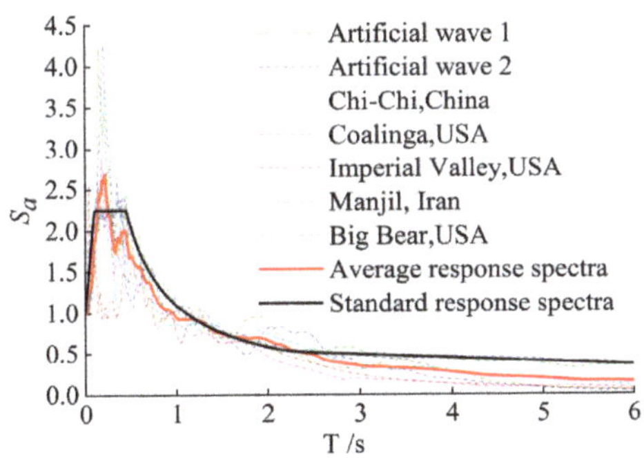

Figure 15. Normalized time history response spectra and standard response spectra.

4.3. Vibration Reduction Effect

In order to verify again that the displacement amplification ability of the improved TNVD can ignore the influence of inter-story deformation, the displacement and energy dissipation of the viscous dampers on the left side and right side of the same span in Scheme 1 under rare earthquakes and Big Bear seismic waves are compared. The viscous dampers are located on the F-axis of the fifth floor. According to Figure 16a, it can be seen that due to the spring effect in the improved TNVD, the right damper's displacement u_{dR} and left damper's displacement u_{dL} have a high degree of agreement throughout the entire duration of the earthquake. The peak values of u_{dR} and u_{dL} are 61.1 mm and 63.8 mm, respectively, with a difference of within 5%. As shown in Figure 16b, the energy dissipation of the left dampers and right dampers almost coincides throughout the entire time history. Therefore, the displacement amplification factor and energy dissipation factor of the improved TNVD can be calculated by applying Equations (18) and (19) based on the assumption of small structural deformation.

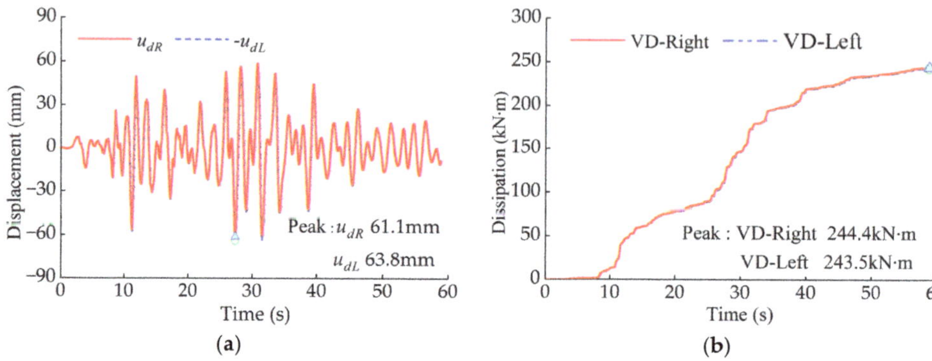

Figure 16. Displacement and energy dissipation of viscous dampers under rare earthquakes and Big Bear wave: (**a**) displacement comparison; (**b**) energy dissipation comparison.

The comparison of NVD hysteresis curves among the three schemes in this paper is revealed in Figure 17a. The NVD is located on the right side of the F-axis on the fifth floor. As is displayed in Figure 17a, due to the presence of negative stiffness, the hysteresis curves of the NVD for Scheme 1 and Scheme 2 exhibit a tilted state. In contrast to other schemes, the damper in Scheme 1 consumes more energy. Scheme 1 has been optimized and designed to achieve the optimal geometric amplification factor, while k_N/k_T^* and $c\omega/k_T^*$ are also within the target area shown in Figure 12d, as shown in Figure 17b.

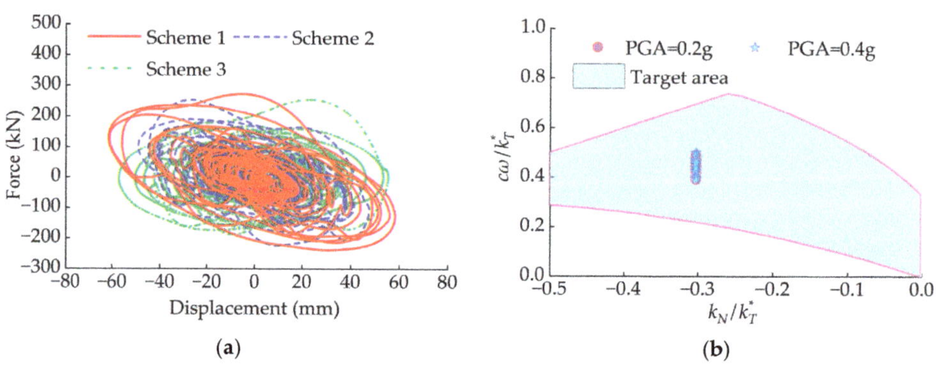

Figure 17. NVD's hysteresis curves of various schemes and combinations of k_N/k_T^* and $c\omega/k_T^*$ in Scheme 1: (**a**) NVD's hysteresis curves; (**b**) values of k_N/k_T^* and $c\omega/k_T^*$ in Scheme 1.

The comparison of the TNVD's f_d and f_E under three schemes is shown in Figure 18. Under fortification earthquakes (0.2 g) and rare earthquakes (0.4 g), Scheme 1 is optimal, followed by Scheme 3, and Scheme 2 is the worst. Scheme 1 meets the expected goals, namely, the TNVD's $f_d > 7.0$ and $f_E > 25$. In Scheme 1, the energy consumption factor is higher on the middle floor, and lower on the top and bottom floors. Although the lower floor has a large inter-story deformation and a large damping coefficient is adopted, the proportional stiffness damping of the structure is large, so that the energy dissipation factor of the lower floor is smaller than that of the middle floor. By comparing Scheme 1 and Scheme 2, we can find that the optimization method proposed in Section 3.2 can significantly improve the TNVD's displacement and energy consumption. Meanwhile, by comparing Scheme 1 and Scheme 3, it can be seen that the negative stiffness of the NSD is equally effective in increasing the displacement and energy dissipation of the TNVD.

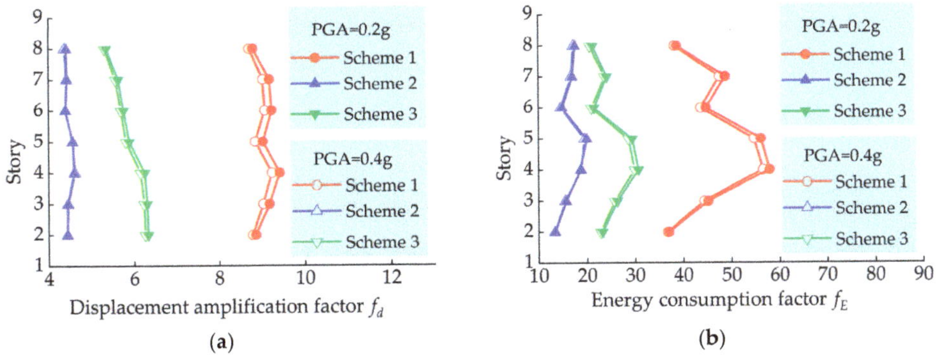

Figure 18. The TNVD's f_d and f_E under fortification earthquakes and rare earthquakes: (**a**) f_d; (**b**) f_E.

Table 5 summarizes the additional damping ratios of the TNVD under different schemes. The calculation of the additional damping ratio is based on the traditional specification method in Reference [38]. According to Table 5, under fortification earthquakes, the additional damping ratios of Scheme 1, Scheme 2, and Scheme 3 are 19.3%, 7.5%, and 10.0%, respectively. Only Scheme 1 achieves the expected target of an additional damping ratio of 15%.

Table 5. Average additional damping ratios of three schemes under fortification earthquakes.

Earthquake Wave	Scheme 1	Scheme 2	Scheme 3
Artificial wave 1	18.4%	7.0%	9.3%
Artificial wave 2	18.9%	7.6%	9.8%
Chi-Chi	20.5%	7.2%	9.6%
Coalinga	19.4%	6.9%	9.4%
Imperial Valley	21.0%	8.3%	9.2%
Manjil	18.1%	7.6%	8.9%
Big Bear	19.0%	8.2%	10.2%
Average of multiple waves	19.3%	7.5%	9.5%

The control effects of the TNVD on inter-story displacement and floor shear under different schemes are revealed in Figures 19 and 20. Whereas Scheme 2 cannot, both Scheme 1 and Scheme 3 can satisfy the target limit requirements for the inter-story displacement angle under fortification earthquakes and rare earthquakes, while Scheme 1 exhibits a smaller floor shear force. Under rare earthquakes, the maximum inter-story displacement angle of Scheme 1 is 1/200, which is much smaller than the limit of 1/150. Therefore, although the geometric parameters of the TNVD in Scheme 1 are designed according to 1/150, they still possess a sufficient safety margin.

In summary, the optimization design method proposed in this paper can effectively control the seismic response of structures with the improved TNVD. Designers can apply this optimization design program to easily design the improved TNVD vibration reduction schemes with high displacement amplification ability and strong energy consumption ability. Simultaneously, the displacement amplification factor and the energy dissipation factor of the improved TNVD can both use the small deformation assumption, which further brings convenience to the design of vibration reduction schemes.

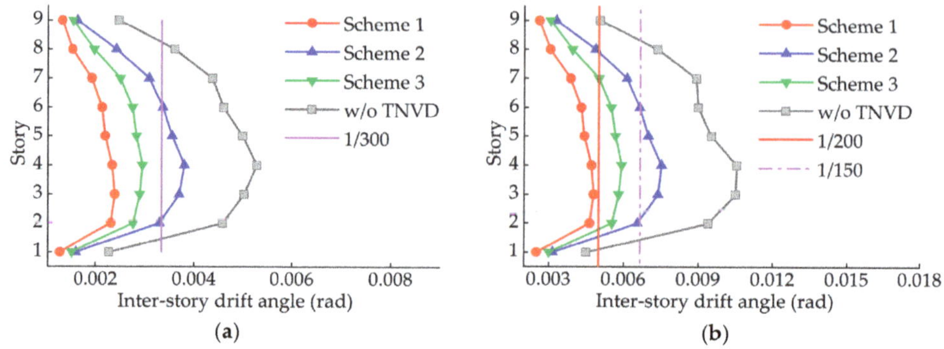

Figure 19. Inter-story displacement angle under fortification earthquakes and rare earthquakes: (**a**) fortification earthquakes; (**b**) rare earthquakes.

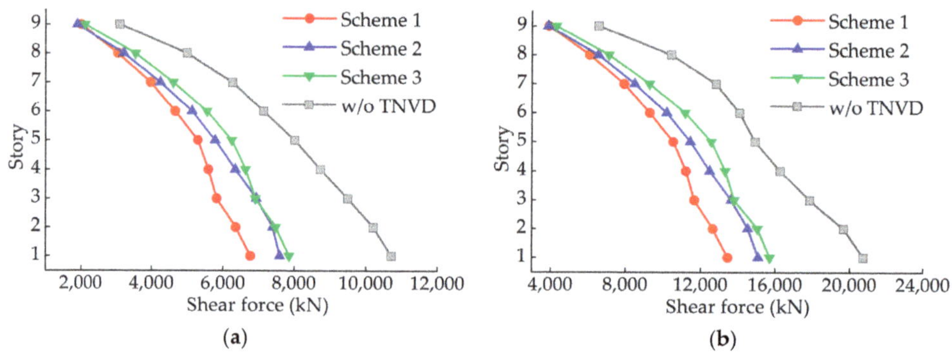

Figure 20. Story shear forces under fortification earthquakes and rare earthquakes: (**a**) fortification earthquakes; (**b**) rare earthquakes.

The TNVD can not only be applied to frame structures, but also to other structures with high stiffness. The inter-story deformation of buildings adopting shear walls or frame core tubes is relatively small, and the application of the TNVD can achieve larger damper deformation, providing greater energy dissipation for the structure. Transmission towers can also be equipped with the TNVD to control deformation under earthquake or wind loads. The application of the TNVD on offshore drilling platforms can also increase their ability to withstand wave loads, wind loads, or earthquakes.

It should be noted that the length of the TNVD connecting rod cannot be arbitrarily set. To enable the TNVD to have greater displacement amplification capability, the length of the connecting rod usually varies within a certain range. The inter-story displacement limit results in a lower limit for the sum of the lengths of the upper and lower connecting rods. Meanwhile, there are multiple factors that affect the performance of the TNVD, and the impact of these factors on the amplification ability of the TNVD is not consistent. Therefore, designers need to accurately understand these rules in order to effectively apply them to engineering.

In order to ensure the normal operation of the TNVD, in addition to meeting the specifications of Equation (9), the bearing capacity of the beams and columns around the TNVD should also be checked to ensure that they can still maintain elasticity under rare earthquakes. Due to the significant displacement amplification effect of the TNVD, it is necessary to ensure that the damper has a large deformation capacity to prevent the damper from failing when the structure encounters seismic intensity exceeding expectations. The improved TNVD has multiple connection nodes. The specific implementation form of the node will affect the length and horizontal inclination angle of the TNVD's connecting rod, thereby affecting the displacement amplification ability. Meanwhile, the improved TNVD's greater displacement amplification ability is more sensitive to the length of the TNVD's connecting rod. A slight change in the length of the connecting rod is likely to cause significant changes in the displacement amplification factor and energy dissipation factor. The above situations require further research.

5. Conclusions

The TNVD is a combination of a symmetrical toggle-style amplification device and a negative stiffness device, which further enhances the displacement amplification effect of the toggle-style device. This paper proposes to comprehensively evaluate the performance of the TNVD based on its displacement amplification factor f_d and the energy dissipation factor f_E. Aiming at the influence of inter-story deformation on the displacement amplification ability of the TNVD, this paper proposes an improved TNVD, which can use the assumption of small structural deformation. Furthermore, the impacts of the lower connecting rod's length, the lower connecting rod's horizontal inclination angle, the limit value of inter-story deformation, the damping coefficient, and the negative stiffness on the TNVD's f_d and f_E are expounded. On this basis, an optimization strategy for f_d and f_E is proposed, and a multi-objective control design method for a structure with the improved TNVD is proposed. The main conclusions are as follows:

(1) The displacement amplification factor f_d describes the displacement amplification ability of the TNVD, and energy dissipation factor f_E describes the energy dissipation ability. f_d and f_E can be used to evaluate the comprehensive performance of the TNVD.

(2) In comparison with the traditional toggle support damping systems, the TNVD adds a negative stiffness device to further increase the displacement of the damper. The correlation between the TNVD's displacement amplification ability and inter-story deformation is reflected in the influence of inter-story deformation on both the geometric amplification factor and the effective displacement coefficient. The improved TNVD adds a more rigid spring between the left-side TNVD and the right-side TNVD. This spring balances the difference of the displacement amplification ability between the left and right TNVDs. Therefore, the improved TNVD's displacement amplification factor f_d and energy dissipation coefficient f_E can both use the small deformation assumption.

(3) When the lower connecting rod's length l_1 of the TNVD remains unchanged, with the increase in the lower connecting rod's horizontal inclination angle θ_1, the geometric amplification factor γ^* increases while the effective displacement coefficient η^* decreases, making f_d and f_E present a trend of first increasing and then decreasing.

(4) When the inter-story deformation limit $U_{o\,\mathrm{Max}}$ is fixed, the lower limit of the length sum $(l_1 + l_2)$ of the upper and lower connecting rods is subsequently fixed. At this moment, there is an optimal length l_1 of the lower connecting rod to enable γ^* to obtain the maximum value. The optimal length of the lower connecting rod needs to satisfy the following relationship: $l_1/(l_1 + l_2) = L/(L + H)$.

(5) η^*, f_d, and f_E gradually increase with the decrease in negative stiffness k_N. In order to ensure that η^* is within a better value range, it is recommended to use the target area in Section 3.1.4 as the value range for k_N/k_T^* and $c\omega/k_T^*$. This target area possesses three properties, namely, $f_d \geq 0.95\gamma^*$, $k_{eq} \geq k_N$, and increasing the damping coefficient c leads f_E to increase.

(6) When the TNVD is actually installed in the structure, it is necessary to ensure that the length and horizontal inclination angle of the toggle's connecting rods have a high degree of conformity with the theoretical design value, avoiding f_d and f_E's having significant changes. Meanwhile, it is necessary to ensure the ratio $k_N/k_T^* > -0.5$, avoiding the TNVD's providing a large negative stiffness, which would significantly reduce the overall stiffness of the structure.

(7) The improved TNVD can have good control effects on the inter-layer displacement and story shear force of the structure through the optimization design in Section 3.2. This

design method utilizes the optimization strategy of the improved TNVD, which can easily design an improved TNVD damping scheme with larger displacement amplification ability and stronger energy consumption ability. Meanwhile, it can also provide a reference for the seismic reduction design of the other types of amplified negative stiffness damping systems.

Author Contributions: Conceptualization, Q.Z. and W.P.; software, Q.Z. and X.L.; validation, Q.Z. and X.L.; formal analysis, Q.Z.; data curation, Q.Z.; writing—original draft preparation, Q.Z.; writing—review and editing, Q.Z. and W.P.; visualization, X.L.; project administration, W.P.; funding acquisition, W.P. All authors have read and agreed to the published version of the manuscript.

Funding: This research was funded by the National Key Research and Development Program of China (Grant No. 2020YFD1100703-04),and Yunnan Provincial Basic Research Special Project (Grant No. 202101AU070059).

Data Availability Statement: The original contributions presented in the study are included in the article, further inquiries can be directed to the corresponding authors.

Conflicts of Interest: The authors declare no conflicts of interest.

References

1. Ou, J.P.; Li, H. Design approaches for active, semi-active and passive control systems based on analysis of characteristics of active control force. *Earthq. Eng. Eng. Vib.* **2009**, *8*, 493–506. [CrossRef]
2. Bhowmik, K.; Debnath, N. Semi-active Vibration Control of Soft-Storey Building with Magnetorheological Damper Under Seismic Excitation. *J. Vib. Eng. Technol.* **2024**, *12*, 6943–6961. [CrossRef]
3. Dong, B.; Sause, R.; Ricles, J.M. Seismic Response and Damage of Reduced-Strength Steel MRF Structures with Nonlinear Viscous Dampers. *J. Struct. Eng.* **2018**, *144*, 04018221. [CrossRef]
4. Zhang, C.; Li, H.N.; Gao, W.H.; Li, C. Experimental and analytical investigations on new viscoelastic damped joints for seismic mitigation of structures with precast shear walls. *Struct. Control Health Monit.* **2020**, *27*, e2485. [CrossRef]
5. Jaisee, S.; Yue, F.; Ooi, Y.H. A state-of-the-art review on passive friction dampers and their applications. *Eng. Struct.* **2021**, *235*, 112022. [CrossRef]
6. Jamkhaneh, M.E.; Ebrahimi, A.H.; Amiri, M.S. Experimental and numerical investigation of steel moment resisting frame with U-shaped metallic yielding damper. *Int. J. Steel Struct.* **2019**, *19*, 806–818. [CrossRef]
7. Sun, L.; Wang, G.; Zhang, C.W. Experimental investigation of a novel high performance multi-walled carbon nano-polyvinylpyrrolidone/silicon-based shear thickening fluid damper. *J. Intell. Mater. Syst. Struct.* **2024**, *35*, 661–672. [CrossRef]
8. Yang, L.Y.; Ye, M.; Huang, Y.H.; Dong, J.K. Study on Mechanical Properties of Displacement-Amplified Mild Steel Bar Joint Damper. *Iran. J. Sci. Technol. Trans. Civ. Eng.* **2023**, *7*, 01268. [CrossRef]
9. Constantinou, M.C.; Tsopelas, P.; Hammel, W.; Sigaher, A.N. New configurations of fluid viscous dampers for improved performance. In Proceedings of the Passive Structural Control Symposium; Tokyo Institute of Technology: Yokohama, Japan, 2000.
10. Constantinou, M.C.; Tsopelas, P.; Hammel, W.; Sigaher, A.N. Toggle-brace-damper seismic energy dissipation systems. *J. Struct. Eng.* **2001**, *127*, 105–112. [CrossRef]
11. Taylor, D.P. Toggle brace dampers: A new concept for structural control. In Proceedings of the Structures Congress, Philadelphia, PA, USA, 8–10 May 2000; The American Society of Civil Engineers: Reston, VA, USA, 2000.
12. Sigaher, A.N.; Constantinou, M.C. Scissor-jack-damper energy dissipation system. *Earthq. Spectra.* **2003**, *19*, 133–158. [CrossRef]
13. Hwang, J.S.; Huang, Y.N.; Hung, Y.H. Analytical and experimental study of toggle-brace-damper systems. *J. Struct. Eng.* **2005**, *131*, 1035–1043. [CrossRef]
14. Huang, H.C. Efficiency of the motion amplification device with viscous dampers and its application in high-rise buildings. *Earthq. Eng. Eng. Vib.* **2009**, *8*, 521–536. [CrossRef]
15. Zhang, R.; He, H.; Weng, D. Theoretical analysis and experimental research on toggle-brace-damper system considering different installation modes. *Sci. Iran.* **2012**, *19*, 1379–1390. [CrossRef]
16. Polat, E.; Constantinou, M.C. Open-space damping system description, theory, and verification. *J. Struct. Eng.* **2017**, *143*, 04016201. [CrossRef]
17. Lan, B.J.; He, H.X.; Wu, S.; Cheng, S.T.; Chen, Y.F. Multi-performance objectives optimum design method and vibration control application of generalized toggle–brace–damper system. *Earthq. Eng. Struct. Dyn.* **2022**, *51*, 1410–1435. [CrossRef]
18. She, A.Q.; Wang, L.J.; Peng, Y.L.; Li, J.H. Structural reliability analysis based on improved wolf pack algorithm AK-SS. *Structures* **2023**, *57*, 105289. [CrossRef]
19. Wang, M.; Sun, F.; Yang, J.Q.; Nagarajaiah, S. Seismic protection of SDOF systems with a negative stiffness amplifying damper. *Eng. Struct.* **2019**, *190*, 128–141. [CrossRef]
20. Javanbakht, M.; Cheng, S.; Ghrib, F. Impact of support stiffness on the performance of negative stiffness dampers for vibration control of stay cables. *Struct. Control Health Monit.* **2020**, *27*, e2610. [CrossRef]

21. Nagarajaiah, S.; Reinhorn, A.M. Applicability of Pseudo-Force Method to Highly Nonlinear Dynamic Problems. In Proceedings of the Structures Congress 94 ASCE, Atlanta, GA, USA, 24–28 April 1994; pp. 165–172.
22. Nagarajaiah, S.; Reinhorn, A.M.; Constantinou, M.C. True Adaptive Negative Stiffness: A New Structural Modification Approach for Seismic Protection. In Proceedings of the 5th World Conference on Structural Control and Monitoring, Tokyo, Japan, 12–14 July 2010.
23. Pasala, D.T.R.; Sarlis, A.A.; Nagarajaiah, S.; Reinhorn, A.M.; Constantinou, M.C. Adaptive negative stiffness: New structural modification approach for seismic protection. *J. Struct. Eng.* **2013**, *139*, 1112–1123. [CrossRef]
24. Sarlis, A.A.; Pasala, D.T.R.; Constantinou, M.C.; Reinhorn, A.M.; Nagarajaiah, S. Negative stiffness device for seismic protection of structures. *J. Struct. Eng.* **2013**, *139*, 1124–1133. [CrossRef]
25. Pasala, D.T.R.; Sarlis, A.A.; Reinhorn, A.M.; Nagarajaiah, S.; Constantinou, M.C. Simulated bilinear-elastic behavior in a SDOF elastic structure using negative stiffness device: Experimental and analytical study. *J. Struct. Eng.* **2014**, *140*, 04013049. [CrossRef]
26. Chen, L.; Sun, L.M.; Nagarajaiah, S. Cable with discrete negative stiffness device and viscous damper: Passive realization and general characteristics. *Smart Struct. Syst.* **2015**, *15*, 627–643. [CrossRef]
27. Kwon, I.Y.; Yang, H.T.; Hansma, P.K.; Randall, C.J. Implementable Bio-Inspired Passive Negative Spring Actuator for Full-Scale Structural Control under Seismic Excitation. *J. Struct. Eng.* **2016**, *142*, 04015079. [CrossRef]
28. Liu, M.; Zhou, P.; Li, H. Novel self-centering negative stiffness damper based on combination of shape memory alloy and pre-pressed springs. *J. Aerosp. Eng.* **2018**, *31*, 04018100. [CrossRef]
29. Shi, X.; Zhu, S.Y. Magnetic negative stiffness dampers. *Smart Mater Struct.* **2015**, *24*, 072002. [CrossRef]
30. Shi, X.; Zhu, S.Y.; Spencer, B.F. Experimental study on passive negative stiffness damper for cable vibration mitigation. *J. Eng. Mech.* **2017**, *143*, 04017070. [CrossRef]
31. Shi, X.; Zhu, S.Y.; Nagarajaiah, S. Performance comparison between passive negative stiffness damper and active control in cable vibration mitigation. *J. Bridge Eng.* **2017**, *22*, 04017054. [CrossRef]
32. Tan, Y.Q.; Dang, J.; Igarashi, A.; Himeno, T.; Hamada, Y. Hybrid simulation tests evaluating the seismic performance of lead high-damping rubber bearings at low temperatures. *Struct. Infrastruct. Eng.* **2023**, 2292757. [CrossRef]
33. Attary, N.; Symans, M.; Nagarajaiah, S.; Reinhorn, A.M.; Constantinou, M.C. Performance evaluation of negative stiffness devices for seismic response control of bridge structures via experimental shake table tests. *J. Earthq. Eng.* **2015**, *19*, 249–276. [CrossRef]
34. Attary, N.; Symans, M.; Nagarajaiah, S. Numerical simulations of a highway bridge structure employing passive negative stiffness device for seismic protection. *Earthq. Eng. Struct. Dyn.* **2015**, *44*, 973–995. [CrossRef]
35. Sun, T.W.; Peng, L.Y.; Li, X.; Guan, Y.X. Developing and Applying a Double Triangular Damping Device with Equivalent Negative Stiffness for Base-Isolated Buildings. *Buildings* **2023**, *13*, 3008. [CrossRef]
36. Chalarca, B.; Filiatrault, A.; Perrone, D. Influence of Fluid Viscous Damper Stiffness on the Floor Acceleration Response of Steel MomentResisting Frames Under Far-Field Ground Motions. *J. Eng. Struct.* **2024**, *28*, 2158–2187.
37. Chen, Y.; Chai, Y.H. Effects of brace stiffness on performance of structures with supplemental Maxwell model-based brace-damper systems. *Earthq. Eng. Struct. Dyn.* **2011**, *40*, 75–92. [CrossRef]
38. Fournier, J.A.; Cheng, S.H. Impact of damper stiffness and damper support stiffness on the efficiency of a linear viscous damper in controlling stay cable vibrations. *J. Bridge Eng.* **2014**, *19*, 04013022. [CrossRef]
39. Losanno, D.; Londono, J.M.; Zinno, S. Effective damping and frequencies of viscous damper braced structures considering the supports flexibility. *Comput. Struct.* **2017**, *207*, 121–131. [CrossRef]
40. Li, S.Y.; Chen, Y.T.; Chai, Y.H. Effects of Brace Stiffness and Nonlinearity of Viscous Dampers on Seismic Performance of Structures. *Int. J. Struct. Stab. Dyn.* **2021**, *21*, 2150188. [CrossRef]
41. Zhou, P.; Liu, M.; Li, H. Optimized negative stiffness damper with flexible support for stay cables. *Struct. Control Health Monit.* **2021**, *28*, e2717. [CrossRef]
42. *RISN-TG046-2023*; Chinese Standard, Technical Guideline for Maintaining Normal Functionality of Buildings in Earthquakes. Chinese Architecture and Building Press: Beijing, China, 2023.
43. *JGJ 297-2013*; Chinese Standard, Technical Specification for Seismic Energy Dissipation of Buildings. Chinese Architecture and Building Press: Beijing, China, 2013.
44. *GB 50011-2016*; Chinese Standard, Code for Seismic Design of Buildings. Chinese Architecture and Building Press: Beijing, China, 2016.

buildings

MDPI

Article

Study on Vibration Reduction Effect of the Building Structure Equipped with Intermediate Column–Lever Viscous Damper

Qiang Zhou [1,2,*], Wen Pan [1,2,*] and Xiang Lan [3]

[1] Faculty of Civil Engineering and Mechanics, Kunming University of Science and Technology, Kunming 650500, China
[2] Yunnan Seismic Engineering Technology Research Center, Kunming 650500, China
[3] College of Architecture and Civil Engineering, Kunming University, Kunming 650214, China; lx523947015@kmu.edu.cn
* Correspondence: qzhou19920120@163.com (Q.Z.); 13312039@kust.edu.cn (W.P.)

Abstract: Generally speaking, the traditional lever amplification damping system is installed between adjacent columns in a building, which occupies a significant amount of space in the building. In contrast to amplification devices in different forms, the damper displacement of the intermediate column damper system is smaller, and the vibration reduction efficiency is lower. In light of these drawbacks, this study proposes a new amplification device for energy dissipation and vibration reduction, which is based on an intermediate column–lever mechanism with a viscous damper (CLVD). Initially, a specific simplified mechanical model of CLVD is derived. Subsequently, an equivalent Kelvin mechanical model of CLVD is derived to intuitively reflect CLVD's damping and stiffness effect. The damping ratio added by CLVDs to the structure is calculated according to that model; the additional damping ratio and additional stiffness are utilized to calculate the displacement ratio R_d and shear force ratio R_v of the structure with CLVDs to the structure without CLVDs. R_d and R_v are introduced to evaluate the vibration reduction effect of the structure with CLVDs, and the effects of various parameters (such as intermediate column position, beam's bending line stiffness, lever amplification factor, damping coefficient, and earthquake intensity) on R_d and R_v are analyzed. The results indicate that when the ratio of the distance from the intermediate column to the edge column to the span of the beam is 0.5, CLVD owns the optimal vibration reduction effect. Increasing the beam's bending line stiffness is beneficial for CLVD to control structural displacement and shear force; when the leverage amplification factor is too large, the CLVD provides the structure with stiffness as the main factor, followed by damping. Additionally, when the ratio of the displacement amplification factor to the geometric amplification factor satisfies $f_d/\gamma = 1/2^{1-0.5\alpha}$, the CLVD has the optimal displacement control effect on the structure. After that, measures are provided to optimize the CLVD in different situations in order to effectively control the inter-story displacement and the story shear force of the structure. Consequently, a nine-story frame is taken as an example to elaborate the application of CLVDs in the design for energy dissipation and vibration reduction. The results reveal that the CLVD scheme adopting the proposed optimization method can effectively enhance the displacement amplification ability of CLVDs, resulting in an additional damping ratio of up to 12%. At the same time, the inter-story displacement was reduced by almost 40% under fortification earthquakes. Through the research in this study, designers can obtain a new choice in structural vibration reduction design.

Keywords: energy dissipation structure; intermediate column–lever viscous damper; simplified mechanical model of CLVD; displacement amplification factor; vibration reduction effect

1. Introduction

The passive control technology [1–4] for energy dissipation and vibration reduction is designed to control the structural seismic response by installing dampers in the structure to consume the seismic input energy. Compared to the active control system and semi-active control system [5,6], the passive control system has the advantage of not requiring external energy, low overall cost, and simple maintenance in the later stage. These dampers can be divided into displacement-related dampers [7,8] and velocity-related dampers according to different energy dissipation principles. Currently, installing velocity-dependent viscous dampers [9,10] in structures has become a typical example of energy dissipation and vibration reduction technology. Under frequent or fortification earthquakes, when the structural deformation is small or the connecting components of dampers deform significantly, the insufficient deformation of dampers makes it difficult to effectively control the seismic response of the structure [11–14]. Therefore, it is essential to propose an efficient viscous damping energy dissipation system to significantly control the seismic response of structures.

In recent years, numerous researchers have proposed various amplification dampers to effectively control the seismic response of structures. Yang [15] proposed the displacement-amplified mild steel bar joint damper using a lever and analyzed the performance of each component of the damper using ABAQUS finite element software. The analysis results indicated that the energy consumption of dampers is almost three-times that of ordinary dampers. Naeem [7] proposed a rotational friction damper with restoring force characteristics. The research results indicated that the use of torsion springs in rotational friction dampers is more effective in improving the seismic performance of structures than traditional dampers. Ribakov et al. [16–18] studied different forms of the inclined brace lever damping system and proposed an optimal design program to realize the design of the lever damping system, demonstrating that this system possessed a good control effect on the inter-story displacement of the structure through numerical simulation. Combining a lever amplification device with a viscous damping wall, He et al. [19,20] conducted model and vibration table experiments, revealing that this new amplification viscous damping wall had a significant energy dissipation ability. The research of Fournier et al. [21,22] indicated that the flexibility of the support would significantly reduce the effect of the viscous damper. Thus, to ensure significant stiffness in the lever amplification device, the traditional lever amplification devices in various forms require a considerable cross-sectional size due to the longer axial length of the support or lever arm. Meanwhile, the traditional lever amplification device within the entire frame is likely to affect the building's functionality at its location. Learning from the crank and connecting rod technology in the mechanical field, Constantinou et al. [23,24] proposed toggle-brace dampers in different forms and conducted theoretical analyses and vibration table tests. The results revealed that the toggle-brace damping device can multiplicatively magnify the inter-story displacement to make the damper fully dissipate energy. Sigaher et al. [25] proposed a scissor-type damping system and analyzed the impact of different scissor-type arrangements on the damper energy dissipation and structural seismic response. Hwang et al. [26] proposed a toggle-brace device that directly connected the dampers to beam column nodes, and established a program for the relationship among displacement amplification factors, toggle geometry, and maximum inter-story displacement limits. Huang et al. [27] analyzed the impact of the toggle's geometry and brace stiffness on the displacement amplification of the damper and demonstrated that the toggle-brace stiffness has a significant influence on the structure's displacement response. Jiang et al. [28] applied a toggle damper to a cable bridge system. In order to improve the issue of traditional toggle devices occupying too much space in the building, Polat et al. [29] proposed an open damping amplification device. Lan et al. [30] suggested a local toggle-brace damper device and designed the local toggle amplifier using the geometric optimization method. Xu [31] proposed a new calculation method for the displacement amplification effect of the toggle-brace damper and suggested installing the toggle-brace damper in a frame with lower floor height and large span. Zhou [32] studied the displacement amplification effect and energy dissipation effect of the toggle-style negative stiffness dampers. The research results indicated that compared to the traditional toggle-brace damper, toggle-style negative stiffness dampers add a negative stiffness device, which further increases the displacement of the dampers. The premise for achieving high displacement amplification capability in the above-mentioned toggle-brace damper is to have high installation accuracy, which increases the difficulty for practical application.

The traditional intermediate column damper [33,34] links viscous dampers or viscous damping walls to the upper and lower intermediate columns, and then arranges them within the frame span. Due to the advantages of having a low impact on the building function and its convenient construction, this system is widely used in engineering. A number of researchers have adopted different methods of studying the optimization of dampers to achieve optimal control of the seismic response. Lan [35] conducted shaking table experiments on intermediate column damper systems. The results indicated that when the intermediate column is located in the middle of the span, the displacement of the viscous damper is relatively large, and the inter-story displacement of the structure can be effectively controlled. Sonmez et al. [36] adopted the artificial bee colony algorithm to determine the optimal placement of dampers in the structure; Bishop et al. [37,38] adopted the genetic algorithm; and Lan et al. [39] employed the response surface method. However, the increase in damper displacement caused by optimization is significantly smaller than the amplification effect caused by the amplification device. In addition, the amplification effect reduces the demand for the number of dampers, resulting in a decrease in cost. Therefore, when comparing different forms of amplification devices, it is particularly important to significantly increase the deformation in the intermediate column damper.

In order to make the lever damping system have the characteristics of a large displacement method effect, convenient construction, and small impact on building space, this study proposes a new intermediate column–lever viscous damping system (CLVD). Firstly, the displacement amplification mechanism of the CLVD is developed. Subsequently, the simplified mechanical model of the CLVD is derived. In order to visually display the damping and stiffness effects of CLVD, an equivalent Kelvin mechanical model of CLVD is provided. On the basis of the Kelvin mechanical model, theoretical expressions for displacement ratio R_d and shear force ratio R_v are established, and these two indicators are utilized to comprehensively evaluate the vibration reduction effect of the structure with a CLVD. After that, the effects of the intermediate column's position, the beam's bending line stiffness, the lever amplification factor, the damping coefficient, the damping index, and the earthquake intensity on the additional CLVD structure's total damping ratio ζ, the stiffness ratio k_a'/k_f, R_d, and R_v are analyzed. On this basis, specific measures for optimizing additional CLVD structure's damping effects under different situations are also provided to guide the damping design. Consequently, a nine-story frame structure is taken as an example to illustrate the application of CLVDs in vibration reduction structures. Through the research in this paper, designers can obtain a new choice in structural vibration reduction design.

2. Composition and Mechanical Model of CLVD

2.1. Composition and Characteristics of CLVD

The intermediate column–lever viscous damper (CLVD) consists of an intermediate column, a lever amplification device, and a viscous damper (VD). In Figure 1, a comparison is provided between the traditional intermediate column viscous damper and the intermediate column–lever viscous damper. The CLVD can be flexibly arranged within the span of beams according to the limitations of the building space. During the construction process of the building, the intermediate column can be integrated with the beam for pouring, and the lever amplification device and viscous damper can be installed after the structural construction is completed.

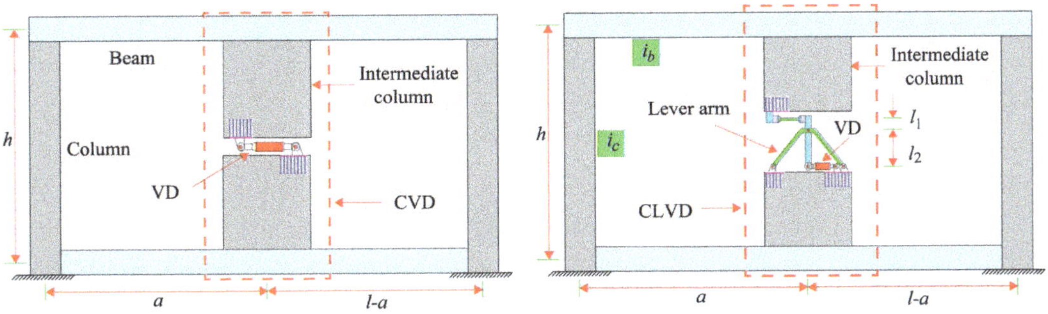

Figure 1. Traditional intermediate column viscous damper (CVD) and intermediate column–lever viscous damper (CLVD): (**a**) CVD; (**b**) CLVD.

For the CLVD, the viscous damper (VD) is connected to the lower intermediate column and a lever. The lever amplification device is connected to the upper intermediate column and the lower intermediate column, and both the upper intermediate column and lower intermediate column are connected to the beam. When the structure undergoes horizontal deformation, the deformation of the beam causes the upper intermediate column to rotate. The horizontal movement and rotation of the upper intermediate column cause the significant deformation of the VD through the rotation of the lever arm. Therefore, the displacement amplification effect of the viscous damper in the CLVD damping system originates from the deformation of the beam and the rotation of the lever arm. The deformations in the CVD and CLVD are shown in Figure 2.

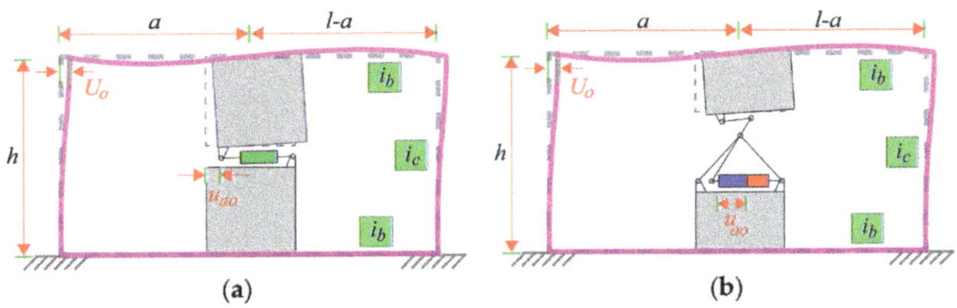

Figure 2. Schematic diagram of structural deformation and damper deformation: (**a**) CVD; (**b**) CLVD.

2.2. Simplified Mechanical Model of CLVD

To derive the mechanical model of CLVD, the following provisions are made for the CLVD damping system and are revealed in either Figure 1b or Figure 2b. The span length where the CLVD is located is l; the floor height is h; the distance from the intermediate column to the left column is a; l_1 and l_2 are the lengths of the upper and lower ends of the lever arm; i_b is the bending line stiffness of the beam; and i_c is the bending line stiffness of the column. The damping force of VD is $F = c\text{sgn}(\dot{u}_d)\left|\dot{u}_d\right|^{\alpha}$, where c and α represent the VD's damping coefficient and damping index, respectively; $\dot{u}_d$ is the velocity of the damper; ρ represents the line stiffness ratio of the beam to column; β represents the position of the intermediate column; and χ represents the leverage amplification factor, as shown in Equation (1):

$$\rho = i_b/i_c\,,\beta = a\,/l\,,\chi = l_2/l_1 \tag{1}$$

Under earthquake action, the maximum inter-story displacement of the structure is U_o, and the maximum displacement of the VD is u_{do}. When U_0 remains constant and the damping coefficient $c = 0$, the maximum displacement of the VD is u_{ao}. This study refers to the ratio of u_{ao} to U_o as the geometric amplification factor γ, which represents the geometric amplification property of the CLVD. In concept, this amplification property is consistent with the damper's displacement amplification factor cos θ (θ is the angle between the slant brace and the horizontal direction) in the classic slant brace connection. Through analysis, γ can be calculated as follows:

$$\gamma = u_{ao}/U_o = \chi - 3\chi\left[6(\beta^2 - \beta) + 1\right]\Big/[2(3\rho + 2)] \tag{2}$$

Figure 3 depicts the changes in γ with respect to ρ and β. $\beta_A = 1/2 - 1/\sqrt{12}$ and $\beta_B = 1/2 + 1/\sqrt{12}$. Whether it is the CVD or CLVD, with the increase in β, the geometric amplification factor γ first increases and then decreases and reaches its maximum value when $\beta = 0.5$. When $\beta_A < \beta < \beta_B$, γ decreases as ρ increases. Meanwhile, due to the amplification effect of the leverage, the geometric amplification factor t of the CLVD is much greater than that of CVD.

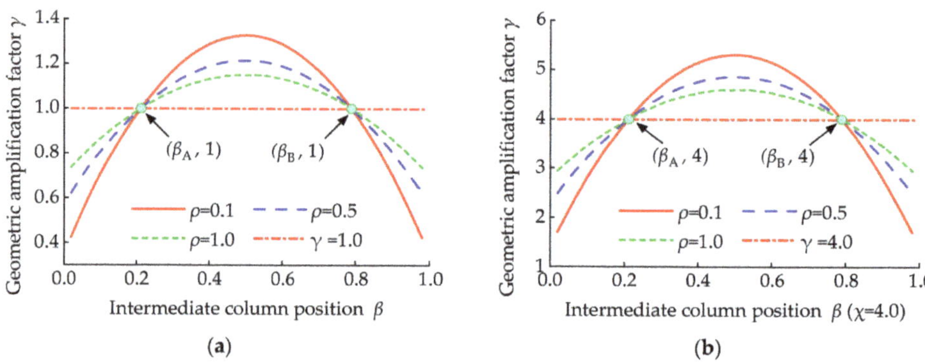

Figure 3. Geometric amplification factor γ: (**a**) CVD; (**b**) CLVD (χ = 4.0).

The simplified mechanical model of the CLVD is shown in Figure 4a. In this figure, the leverage amplification factor in the mechanical model is γ. k is the equivalent stiffness for connecting the intermediate column and the lever in series. The equivalent stiffness is composed of the following three types of the equivalent stiffness in series: the equivalent stiffness of the upper intermediate column k_{mc-up}, the equivalent stiffness of the lower intermediate column k_{mc-low}, and the lever's equivalent stiffness k_{le}. Schematic diagrams of these three equivalent stiffness types are displayed in Figure 4b–d. The equivalent stiffness k is calculated using Equation (3):

$$k = \frac{k_{le}k_{mc-up}}{2k_{le} + k_{mc-up}} \tag{3}$$

$$k_{le} = \frac{1}{\dfrac{(l_1 + l_2)^2\chi^2}{3i_{le}(\chi + 1)^2} + \dfrac{(\chi + 1)^2 l_b}{2EA_b\cos^2\theta}} \tag{4}$$

$$k_{mc-up} = \frac{1}{\dfrac{\chi^2[\beta f_1 + (1-\beta)f_2]h^2}{2f_3 i_b} + \dfrac{\chi^2(12i_{mc} + i_{vmc}h^2)}{12i_{mc}i_{vmc}}} \tag{5}$$

In Equation (5), $f_1 = (1+\frac{\beta}{\rho})[\beta(2-3\beta) - 2(1+\frac{1}{\rho})(1-\beta)(3\beta-1)]$, $f_2 = (1+\frac{1-\beta}{\rho})[(1-\beta)(3\beta-1) - 2\beta(1+\frac{1}{\rho})(2-3\beta)]$, $f_3 = 2 - 8(1+\frac{1}{\rho})^2$.

In Equation (3), k_{le} is calculated using Equation (4) and k_{mc-up} is calculated using Equation (5). In Equation (4), l_b is the length of the support; θ is the horizontal angle of the support; A_b is the cross-sectional area of the support; i_{mc} and i_{vmc} are the bending line stiffness and shear line stiffness of the intermediate column, respectively; and i_{le} is the bending line stiffness of the lever arm. In Equation (5), due to the assumption that the height of the upper intermediate column is 1/2 of the floor height, it can be considered safe to assume that k_{mc-low} is approximately equal to k_{mc-up}.

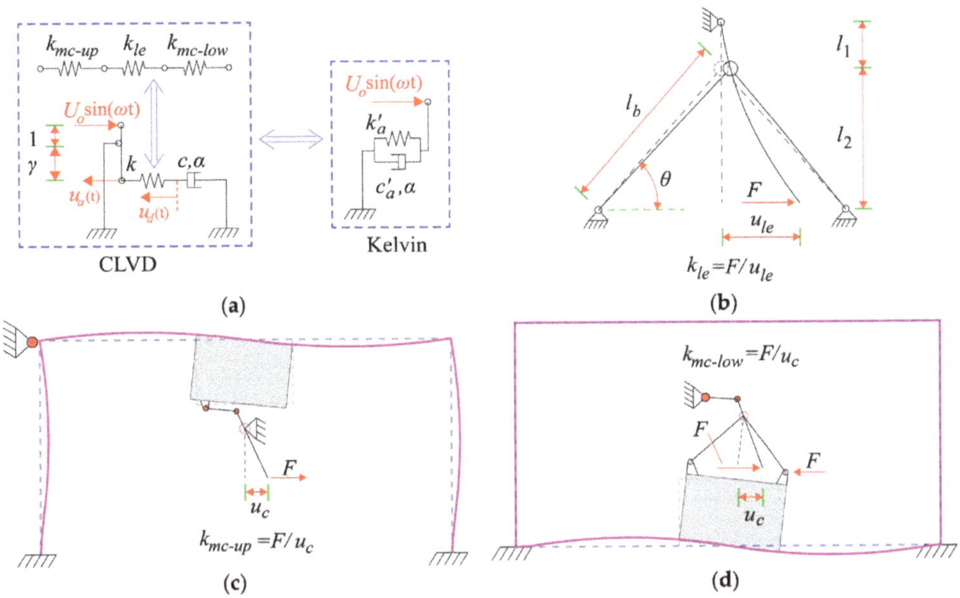

Figure 4. Schematic diagram of the CLVD's mechanical model and equivalent stiffness: (**a**) simplified mechanical model and Kelvin model; (**b**) k_{le}; (**c**) k_{mc-up}; (**d**) k_{mc-low}.

According to Equations (2) and (3), the position of the intermediate column will simultaneously affect the geometric amplification factor γ and the intermediate column–lever equivalent stiffness k. Therefore, the CLVDs located at different positions within the span will provide different vibration reduction effects to the structure.

2.3. Kelvin Model of CLVD

The vibration reduction effect of the energy dissipation structure originates from the stiffness and damping effects that are added by the relevant damping system. The Kelvin mechanical model directly adds stiffness and damping to the structure. In order to facilitate the analysis of the vibration reduction effect of the CLVD damping structures, it is necessary to transform the simplified mechanical model of the CLVD into a Kelvin mechanical model. In the Kelvin model, the damping coefficient is c'_a, the damping index is α, and the spring stiffness is k'_a, as shown in Figure 4a.

When the VD is linear, the mechanical model of the CLVD can be transformed into the Kelvin model. Here, the transformation of the mechanical model is achieved through application of the following three conditions: equal deformation, equal force, and equal energy dissipation. The constitutive relationship of VD is that $F(t) = c\dot{u}_d(t)$, where $u_d(t)$ and $\dot{u}_d(t)$ are the deformation and velocity of VD, respectively. Assuming that the CLVD mechanical model is subjected to stable sinusoidal excitation, the system can then be studied in a steady state without involving dynamic analysis. We assume that the sinusoidal deformation generated by CLVDs is $U_0 \sin(\omega t)$. Since the mechanical model of the CLVD and the Kelvin model have equal deformation and force, the $u_d(t)$, $\dot{u}_d(t)$, and $F(t)$ of the VD are as follows:

$$\left\{\begin{array}{l} u_d(t) = \gamma U_0 \sin(\omega t) - [k'_a U_0 \sin(\omega t) + c'_a \omega U_0 \cos(\omega t)]/(\gamma k) \\ \dot{u}_d(t) = \gamma \omega U_0 \cos(\omega t) - [k'_a \omega U_0 \cos(\omega t) - c'_a \omega^2 U_0 \sin(\omega t)]/(\gamma k) \\ F(t) = c\dot{u}_d(t) = [k'_a U_0 \sin(\omega t) + c'_a \omega U_0 \cos(\omega t)]/\gamma \end{array}\right. \tag{6}$$

In this study, the ratio of VD's maximum displacement u_{do} to the inter-story maximum displacement U_0 is identified as CLVD's displacement amplification factor f_d, representing the CLVD's displacement amplification ability. The Laplace transform is conducted in Equation (6) to solve k'_a and c'_a. The fact that the energy dissipation of the CLVD mechanical model and Kelvin model is equal is utilized to solve f_d, as follows:

$$f_d = u_{do}/U_0 = \gamma \Big/ \left[1 + (c\omega/k)^2\right]^{0.5} \tag{7}$$

$$\left\{\begin{array}{l} k'_a = (\gamma^2 - f_d^2)k \\ c'_a = f_d^2 c \end{array}\right. \tag{8}$$

Equations (7) and (8) are linear for the VD. When the VD is nonlinear, its constitutive relationship is $F = c\mathrm{sgn}[\dot{u}_d(t)]|\dot{u}_d(t)|^{\alpha}$. When the damping index $\alpha = 0$, $f_d = \gamma/[1 + c/(u_{do}k)]$, $k'_a = \gamma^2(1 - f_d/\gamma)k$, and $c'_a = f_d\, c$. When $0 < \alpha \leq 1$, f_d, k'_a and c'_a are as follows:

$$f_d = \gamma \Big/ \left\{1 + [c\omega^{\alpha}/(u_{do}^{1-\alpha}k)]^{1+\alpha}\right\}^{1-0.5\alpha}, \; 0 \leq \alpha \leq 1 \tag{9}$$

$$\left\{\begin{array}{ll} k'_a = \gamma^2[1 - (f_d/\gamma)^{\frac{1}{1-0.5\alpha}}]k & , 0 \leq \alpha \leq 1 \\ c'_a = f_d^{1+\alpha} c & , 0 \leq \alpha \leq 1 \end{array}\right. \tag{10}$$

3. Parameter Analysis of CLVD's Vibration Reduction Effect and Optimization

3.1. Evaluation Indicators of CLVD's Vibration Reduction Effect

The principle of setting a damper to reduce the seismic response of a structure is based on the effect achieved by increasing the damping due to both the energy dissipation of the damper and the stiffness effect of decreasing the period of the structure due to the additional stiffness. The damping effect enables the structure's displacement and shear force to decrease simultaneously, and the stiffness effect makes the displacement decrease and the shear force increase. According to the discussion in Sections 2.2 and 2.3, the single-layer structure with a CLVD is illustrated in Figure 1b, the simplified mechanical model is shown in Figure 5a, and the equivalent Kelvin mechanical model is revealed in Figure 5b. According to the Chinese Code [40], the total damping of the damping structure can be obtained as follows:

$$\zeta = \zeta_0 + \lambda c'_a \omega^{\alpha} U_0^{\alpha-1}/[2\pi\left(1 + k'_a/k_f\right)k_f] \tag{11}$$

where ζ_0 is the inherent damping ratio of the structure; ω can be approximated as the natural vibration period of the structure; k_f is the lateral stiffness of the structure; and k'_a/k_f represents the additional stiffness characteristics of CLVD. λ is a function of the damping index, whose value is referenced in the Chinese Code [40].

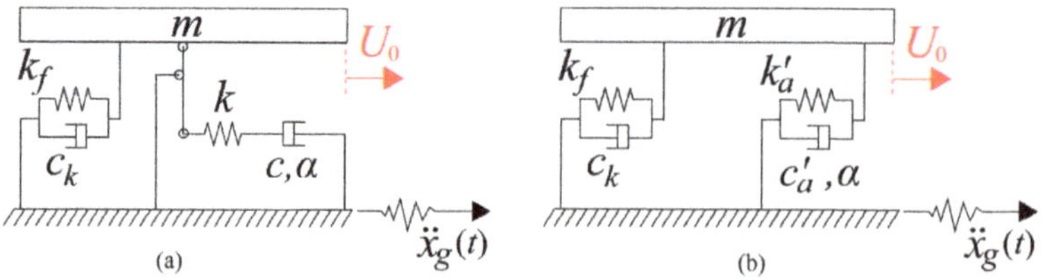

Figure 5. Mechanical model of the single-story structure with CLVD: (**a**) simplified mechanical model; (**b**) equivalent Kelvin model.

The displacement ratio R_d is the ratio of the damping structure's displacement U_0 to the non-damping structure's displacement U_0'. The shear force ratio R_v is the ratio of the damping structure's shear force V_0 to the non-damping structure's shear force V_0'. For most frame structures with additional viscous damping systems, the natural vibration period of the structure is in the medium and long period of the quasi-velocity response spectrum with a constant value. The authors of reference [41] identified that for the structure of the additional energy dissipation devices with medium and long periods, the damping effect of the energy dissipation device can be represented with $[(1+25\zeta_0)/(1+25\zeta)]^{0.5}$, and the stiffness effect can be represented with $[1/(1+k_a'/k_f)]^{0.5}$. Therefore, R_d and R_v are as follows:

$$R_d = U_0/U_0' = [(1+25\zeta_0)/(1+25\zeta) \times 1/(1+k_a'/k_f)]^{0.5} \quad (12)$$

$$R_v = V_0/V_0' = [(1+25\zeta_0)/(1+25\zeta) \times (1+k_a'/k_f)]^{0.5} \quad (13)$$

When the damping index $\alpha < 1$, the calculation cannot directly use Equation (9) due to the unknown displacement u_{do} of the damper. Since the non-damping structure's inter-story displacement is known, the following iterative method can be applied for calculation:

(1) It is assumed that the displacement of the damper u_{do} is equal to the inter-story displacement U_0' of the non-damping structure.
(2) u_{do} is substituted into Equation (9) for calculation f_d, and k_a', c_a', ζ, R_d, and R_d are calculated through Equations (10)–(13), respectively.
(3) The error $\varepsilon = u_{do}/(f_d R_d U_0') - 1$ is calculated. When $\varepsilon > 1\%$, u_{do} is made to be equal to $f_d R_d U_0'$. Steps 2 and 3 are repeated. When $\varepsilon < 1\%$, the iteration ends, and the final calculation values R_d and R_v are obtained.

3.2. Parameter Analysis

There are many factors that influence R_d and R_v, including the position of the intermediate column β, the beam's bending line stiffness i_b, the leverage amplification factor χ, the damping coefficient c, and the damping index α. In this section, the impacts of the above various factors on R_d and R_v are successively discussed.

For the convenience of discussion, the following assumptions are made regarding the single-story structure in Figure 1b. The structural floor height is h = 4000 mm; the span length is l = 8100 mm; the beam section is 400 mm × 800 mm; the column section is 600 mm × 600 mm; the intermediate column section is 1600 mm × 250 mm; and the concrete material grade is C35. The box section corresponding to the lever arm is 300 mm × 140 mm × 20 mm × 20 mm, the section of the inclined strut box is 200 mm × 200 mm × 25 mm × 25 mm, and the material grade is Q235. The lever length is 1000 mm, and the lever amplification factor is $\chi = 4$. The intermediate column position is $\beta = 0.5$. The parameter of the viscous damper is $c = 5\text{kN}/(\text{mm/s})^{0.5}$. The structural modal frequency is that $\omega = 8.1$ rad/s; the lateral stiffness is that $k_f = 120$ kN/ mm; and the structural inherent damping ratio is $\zeta_0 = 0.05$. Under fortification earthquakes (PGA = 2 m/s^2), the non-damping structure's displacement is $U_0' = h/200$.

3.2.1. Impact of the Position of Intermediate Column

This section explores the impact of changing the intermediate column position β on R_d and R_v. Figure 6 reveals the curves of changes in k_a'/k_f, ζ, R_d, and R_v with the intermediate column position β. As shown in Figure 6, when the intermediate column is in the middle of the span, the stiffness k_a' and total damping ratio ζ provided by the CLVD to the structure reach their maximum values, while R_d and R_v reach their minimum values. Therefore, in order to ensure the control effect of the CLVD on structural displacement and shear force, the CLVD should be located at the mid-span as long as the building function is available.

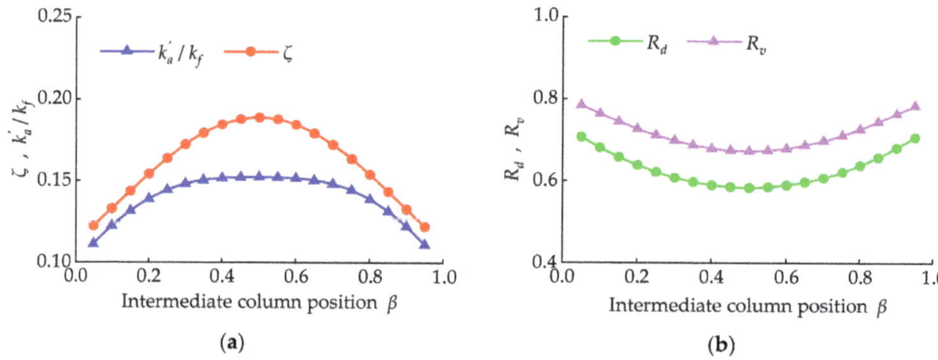

Figure 6. Changes in k_a'/k_f, ζ, R_d, and R_v with β: (**a**) changes in ζ and k_a'/k_f with β; (**b**) changes in R_d and R_v with β.

3.2.2. Impact of Beam's Bending Line Stiffness

This section explores the impact of changing the beam's bending line stiffness i_b on R_d and R_v. In Figure 7, i_{b1} represents the beam's bending line stiffness with a cross-section of 400 mm × 800 mm. When the intermediate column is located in different positions, the impact of the change in i_b on k_a'/k_f and ζ is different. As illustrated in Figure 7a, when the intermediate column is located at the end of the span, the stiffness k_a' provided by the CLVD to the structure increases with the increase in i_b.

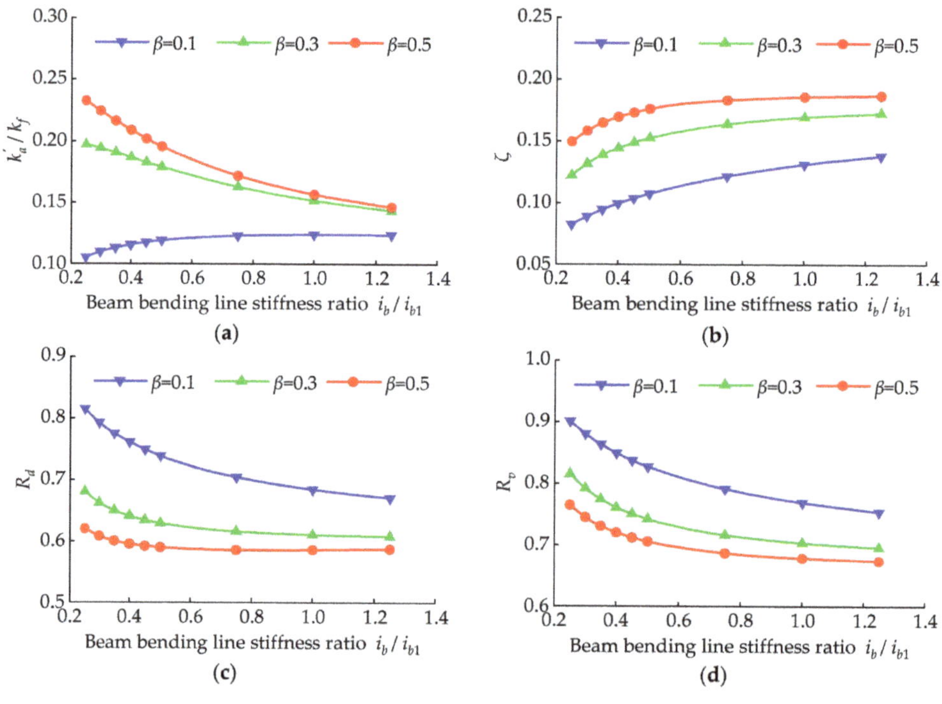

Figure 7. Changes in k_a'/k_f, ζ, R_d, and R_v with i_b/i_{b1}: (**a**) changes in k_a'/k_f with i_b/i_{b1}; (**b**) changes in ζ with i_b/i_{b1}; (**c**) changes in R_d with i_b/i_{b1}; (**d**) changes in R_v with i_b/i_{b1}.

When the intermediate column is located in the middle of the span, the stiffness k_a' provided by the CLVD to the structure decreases as i_b increases. In Figure 7b, the total damping ratio ζ is shown increasing with the increase in i_b. From Figure 7c,d, it can be seen that no matter where the intermediate column lies within the span, R_d and R_v decrease with the increase in i_b. In contrast to the CLVD located in the middle of the span, the CLVD located at the end of the span requires a larger beam section to control the structure's displacement and shear force.

3.2.3. Impact of the Leverage Amplification Factor

This section explores the impact of changing the leverage amplification factor χ on R_d and R_v. In Figure 8a, it can be seen that with the increase in the leverage amplification factor χ, k_a'/k_f increases, and ζ first increases and then decreases. Therefore, an optimal leverage amplification factor χ exists that maximizes the total damping ratio ζ of the structure. In Figure 8b, with the increase in χ, R_d and R_v are shown to first decrease and then increase. When the leverage amplification factor is too large, the CLVD provides the structure with stiffness as the main factor, followed by damping. This enables the CLVD to have an improved displacement control effect on the structure, while it has a worse control effect on the shear force.

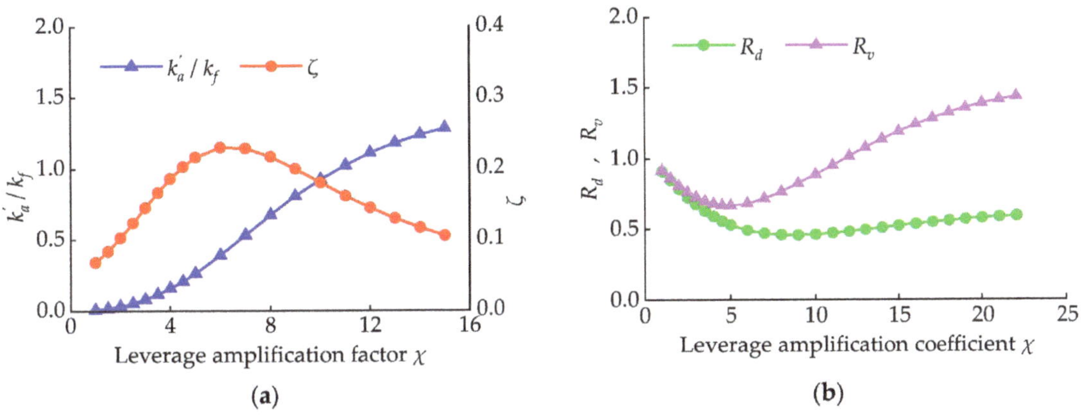

Figure 8. Changes in k_a'/k_f, ζ, R_d, and R_v with χ: (**a**) changes in ζ and k_a'/k_f with χ; (**b**) changes in R_d and R_v with χ.

3.2.4. Impact of the Damping Coefficient and Damping Index

This section explores the impact of changing the damping coefficient c and damping index α on the R_d and R_v. In Figure 9a,c, it can be seen that as c or α increases, k_a'/k_f gradually increases, and ζ first increases and then decreases. From Figure 9b,d, it can be seen that as c or α increase, R_d and R_v first decrease and then increase. Therefore, an optimal combination exists of c and α to achieve optimal control of the structural displacement. In Figure 9e, R_{dP} represents the peak points of each curve, and R_{dP}^S represents R_d of each curve when $(\gamma/f_d)^{1/(1-0.5\alpha)} = 2$. In Figure 9e, it can be seen that the CLVD has the optimal effect on structural displacement control when c and α make $(\gamma/f_d)^{1/(1-0.5\alpha)} = 2$.

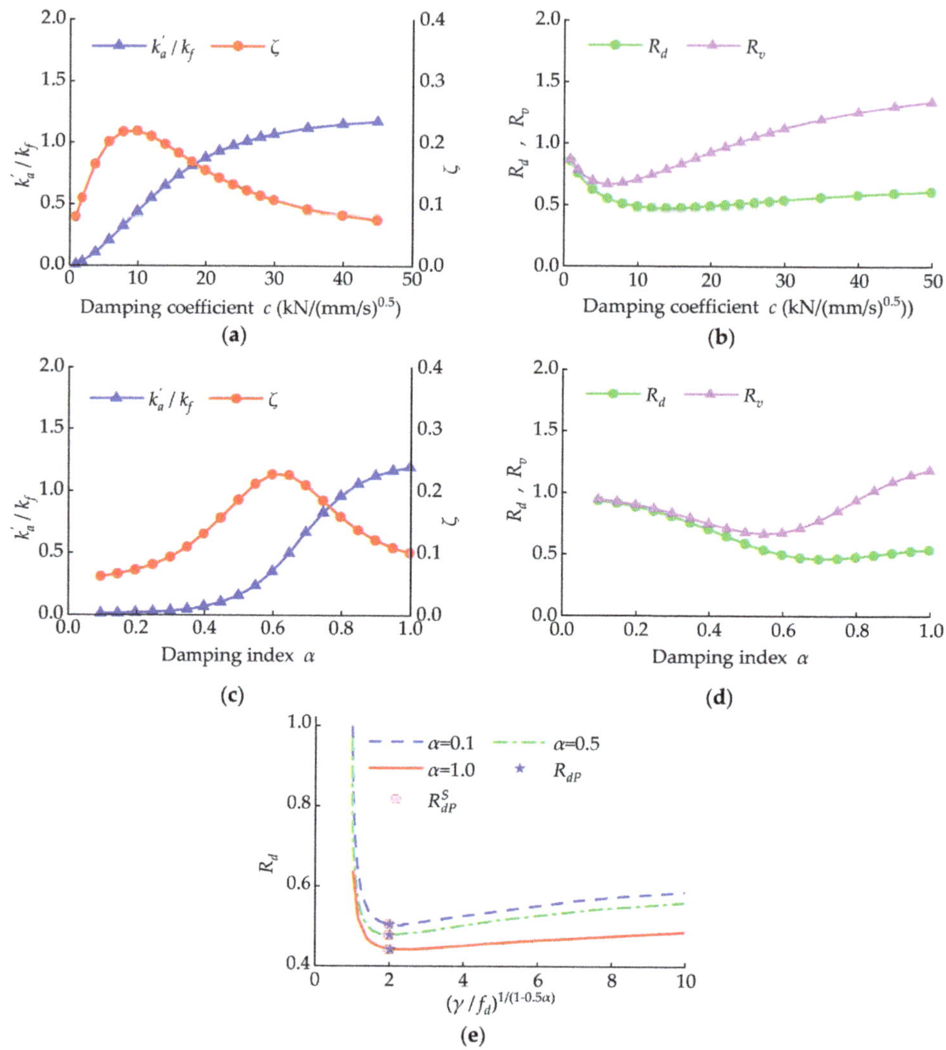

Figure 9. Changes in k_a'/k_f, ζ, R_d, and R_v with c and α: (**a**) changes in k_a'/k_f and ζ with c; (**b**) changes in R_d and R_a with c; (**c**) changes in k_a'/k_f and ζ with α; (**d**) changes in R_d and R_v with α; (**e**) changes in R_d with $(\gamma/f_d)^{1/(1-0.5\alpha)}$.

3.2.5. Impact of Earthquake Intensity

This section analyzes the impact of changing the earthquake intensity on R_d and R_v. When the VD is nonlinear, the displacement amplification factor f_d of the CLVD is related to the displacement of the damper. Therefore, under different earthquake intensities, R_d and R_v are different. In Figure 10a,b, it can be seen that as the earthquake intensity increases, k_a'/k_f gradually decreases, while ζ first increases and then decreases. In the initial stages of the increasing earthquake intensity, the CLVD mainly provides stiffness to the structure. This provides the CLVD with a better displacement control effect on the structure and a poorer control effect on the shear force. When the earthquake intensity is high, the CLVD mainly provides damping to the structure.

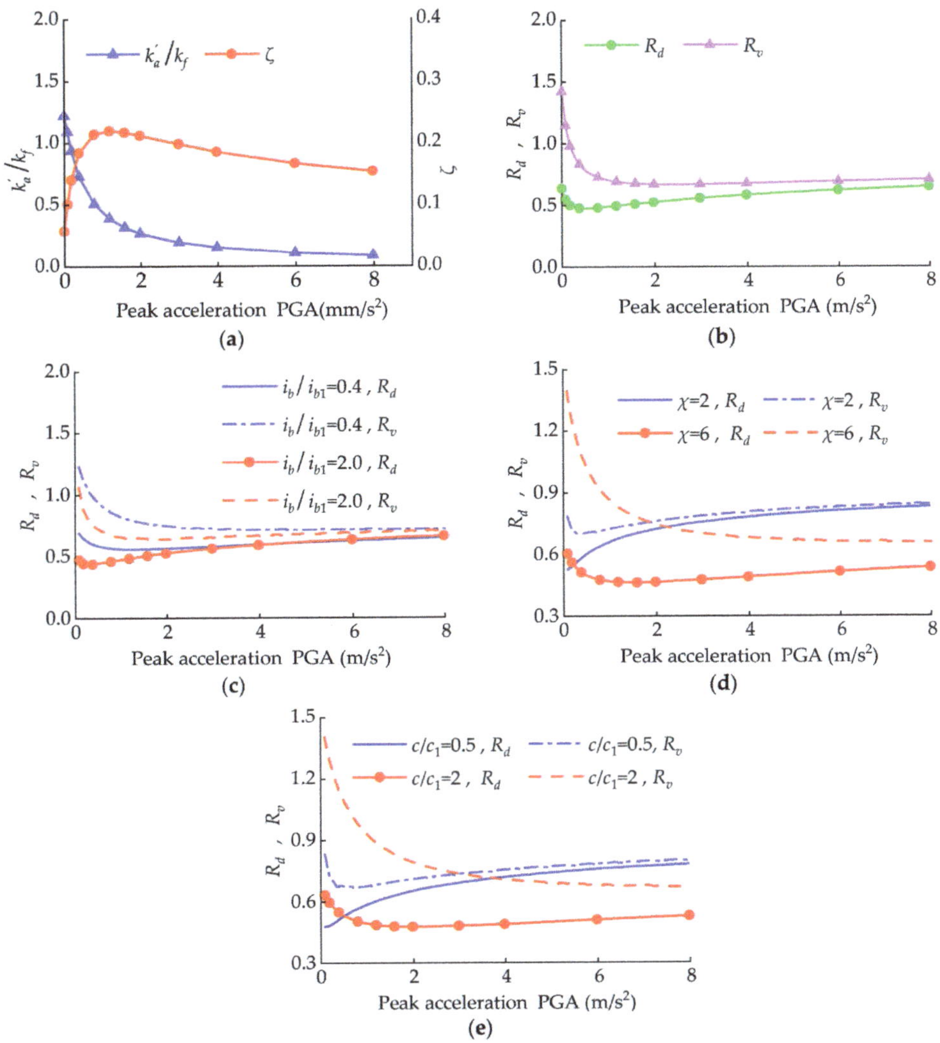

Figure 10. Changes in k_a'/k_f, ζ, R_d, and R_v with PGA: (**a**) changes in k_a'/k_f and ζ with the PGA; (**b**) changes in R_d and R_v with the PGA; (**c**) changes in R_d and R_v with the PGA and i_b; (**d**) changes in R_d and R_v with the PGA and χ; (**e**) changes in R_d and R_v with PGA and c.

In Figure 10c, i_{b1} represents the bending line stiffness of a beam with a cross-section of 400 mm × 800 mm. In Figure 10c, it is revealed that when the earthquake intensity is low, increasing i_b can significantly reduce both R_d and R_v. However, when the earthquake intensity is high, increasing i_b has a worse effect on decreasing R_d and R_v.

In Figure 10d, when the earthquake intensity is high, it is illustrated that increasing the leverage amplification factor χ has a significant reducing effect on R_d and R_v. However, when the earthquake intensity is low, increasing χ has an insignificant effect on reducing R_d. Meanwhile, R_v will be significantly increased.

In Figure 10e, $c_1 = 5(\text{kN}/(\text{mm/s})^{0.5})$. In Figure 10e, when the earthquake intensity is large, it is shown that an increase in the damping coefficient c can significantly reduce R_d. However, when the earthquake intensity is low, increasing the damping coefficient c will result in an increase in R_d and R_v.

In order to ensure that the CLVD demonstrates significant control effects on the structure's displacement and shear force under different-intensity earthquakes, in Figure 10c–e, three methods are illustrated for achieving this goal. The first method is to simultaneously increase the leverage amplification factor χ and the beam's bending line stiffness i_b; the second one is to simultaneously increase the damping coefficient c and the beam's bending line stiffness i_b; and the third one is to increase the leverage amplification factor χ while reducing the damping coefficient c.

In summary, under the allowable conditions of construction engineering, the position of the intermediate column should be $\beta = 0.5$, and the beam connected to the intermediate column should have high stiffness. Meanwhile, an excessive leverage amplification factor should be avoided, and the recommended value is between 2 and 5, and $f_d/\gamma < 1/2^{1-0.5\alpha}$ should also be avoided. When the displacement and shear control effect of CLVD on the structure is poor, the suggestions proposed in Section 3.2.5 should be selected for improvement. When CLVD is applied to real-world buildings, the universal ball node shown in Figure 1 should be applied to reduce the adverse effect of CLVD's out-of-plane deformation on the CLVD displacement amplification ability. Due to the large force transmitted by the lever amplification device to the intermediate column, it is necessary to strengthen the node connecting the lever amplification device and the intermediate column.

3.3. *Optimization Strategy for CLVD*

The design of the energy dissipation and vibration reduction structures generally adopts an elastic section design when encountering frequent earthquakes (or fortification earthquakes) and an elastic–plastic deformation verification calculation when encountering rare earthquakes. In the process of the vibration reduction design, continuous trial and error are required for optimizing the relevant damping system. The discussion in Section 3.2 dealt with the effective reduction in the displacement and shear force of the structure with a CLVD; consequently, this section provides optimization strategies for the CLVD in different situations to guide the seismic design of buildings, as illustrated in the following scenarios:

(1) As long as permitted by the building space, efforts should be made to position the CLVD at the mid-span.
(2) If both R_d and R_v are very large in frequent earthquakes (or fortification earthquakes), the optimization measures should be as follows: try to increase the damping coefficient c while increasing the beam section, or try to increase the lever amplification factor χ while increasing the beam section.
(3) If R_d is relatively small while R_v is relatively large under frequent earthquakes (or fortification earthquakes), and both R_d and R_v are relatively small under rare earthquakes, the optimization measures are as follows: try to increase the beam section, or reduce the damping coefficient c, or reduce the leverage amplification factor χ.
(4) If R_d is relatively small while R_v is relatively large under frequent earthquakes (or fortification earthquakes), and R_d is relatively large under rare earthquakes, the optimization measures are as follows: try to increase the beam section while increasing the quantity of CLVDs.
(5) If both R_d and R_v are relatively small under frequent earthquakes (or fortification earthquakes), and R_d is relatively large under rare earthquakes, the optimization measures are as follows: try to increase the damping coefficient c, or increase the leverage amplification factor χ, or increase the quantity of CLVDs.

The above-mentioned optimization scenarios are applicable to both single and different-intensity earthquakes. In the optimization process of the CLVD, it may be necessary to repeatedly utilize the improvement measures of the above five different scenarios. For instance, after adopting the improvement measures in Scenario 2, the CLVD is likely to shift from Scenario 2 to Scenario 5. At this point, it is necessary to continue adopting the improvement measures of Scenario 5 until the expected goals of displacement reduction and shear force reduction are satisfied.

4. Finite Element Analysis of Engineering Example

4.1. Project Overview

This section takes a nine-story building as the research object and elaborates on the application of CLVD dampers in energy dissipation and vibration reduction structures. This concrete frame has a fortification intensity of 8 degrees (0.2 g), a building height of 36 m, a self-weight of 9784 tons, and a site characteristic period of 0.45 s. The fundamental periods of the structure are 1.61 s (E-W), 1.53 s (N-S), and 1.38 s (rotation). The structural inherent damping ratio is 5.0%. The column section of the first floor is 700 mm × 700 mm; the second floor is 600 mm × 600 mm; from the third floor to the fourth floor is 550 mm × 550 mm; and from the fifth floor to the ninth floor is 500 mm × 500 mm. The main beam section is 300 mm × 600 mm, and the secondary beam section is 200 mm × 400 mm. The beams and columns from the first floor to the fourth floor are made of concrete material with a grade of C40, while the beams and columns on the fifth floor to the ninth floor are made of concrete material with a grade of C35. The expected goals for the vibration reduction design are as follows: an additional damping ratio of 10% and a limit of 1/300 of the inter-story displacement angle of the structure under fortification earthquakes (with a peak earthquake acceleration of 0.2 g); the limit of the inter-story displacement angle of the structure under rare earthquakes (with a peak earthquake acceleration of 0.4 g) is 1/150.

4.2. Scheme Design

This section adopts the optimization strategy set out in Section 3.3 to conduct multiple trial calculations on the CLVD, consequently obtaining Scheme 1. Firstly, We set the number of CLVDs based on the total area of the building. According to the engineering experience, when the ratio of the total building area to the number of viscous dampers is between 100 and 200, it is relatively reasonable. This study takes 56 sets of CLVDs, including 28 sets in the north–south direction. Subsequently, according to the inter-story deformation of the structure without the CLVDs under fortification earthquakes, we choose to arrange CLVDs from the second floor to the eighth floor. According to engineering experience, the parameters of the intermediate column section, beam section, lever, and damper for CLVD are preliminarily selected. We select the position of the intermediate column $\beta = 0.2$ and obtain Scheme 3 through preliminary calculation. As Scheme 3 does not meet the set goals and belongs to Scenario 3 proposed in Section 3.3, increasing the beam section will result in Scheme 2. Since Scheme 2 also cannot meet the set goals and belongs to Scenario 1 proposed in Section 3.3, we further optimize it to make the position of the intermediate column $\beta = 0.5$ and obtain Scheme 1. After calculation, Scheme 1 meets the set goal.

The intermediate column of Scheme 1 is located in the middle of the span, namely, $\beta = 0.5$. The cross-section of the intermediate column is 1700 mm × 350 mm. The lever arm and support are made of Q235 steel. The box section corresponding to the lever arm is 300 mm × 140 mm × 20 mm × 20 mm, and the box section corresponding to the support is 200 mm × 200 mm × 25 mm × 25 mm. The horizontal angle of the support is 48°, and the total length of the lever arm is 1000 mm. The lever amplification factor is $\chi = 4$. The parameters of the viscous dampers from the second floor to the fifth floor are $c = 10\text{kN}/(\text{mm}/\text{s})^{0.5}$; the parameters of the viscous dampers from the sixth floor to the eighth floor are $c = 6\text{kN}/(\text{mm}/\text{s})^{0.5}$. The position of the intermediate column in Scheme 2 is $\beta = 0.2$, and the beam section connected to the intermediate column in Scheme 3 is 300 mm × 600 mm. The parameters of CLVD for Scheme 1, Scheme 2, and Scheme 3 are shown in Table 1. The arrangements of the CLVDs in the structure are illustrated in Figure 11.

Table 1. Parameters of CLVD.

Scheme	Story	β	Beam Section (mm × mm)	Column Section (mm × mm)	χ	c $(\mathrm{kN}\cdot(\mathrm{s/mm})^{0.5})$
Scheme 1	6–8	0.5	400 × 800	1700 × 350	4	6
	2–5	0.5	400 × 800	1700 × 350	4	10
Scheme 2	6–8	0.2	400 × 800	1700 × 350	4	6
	2–5	0.2	400 × 800	1700 × 350	4	10
Scheme 3	6–8	0.2	300 × 600	1700 × 350	4	6
	2–5	0.2	300 × 600	1700 × 350	4	10

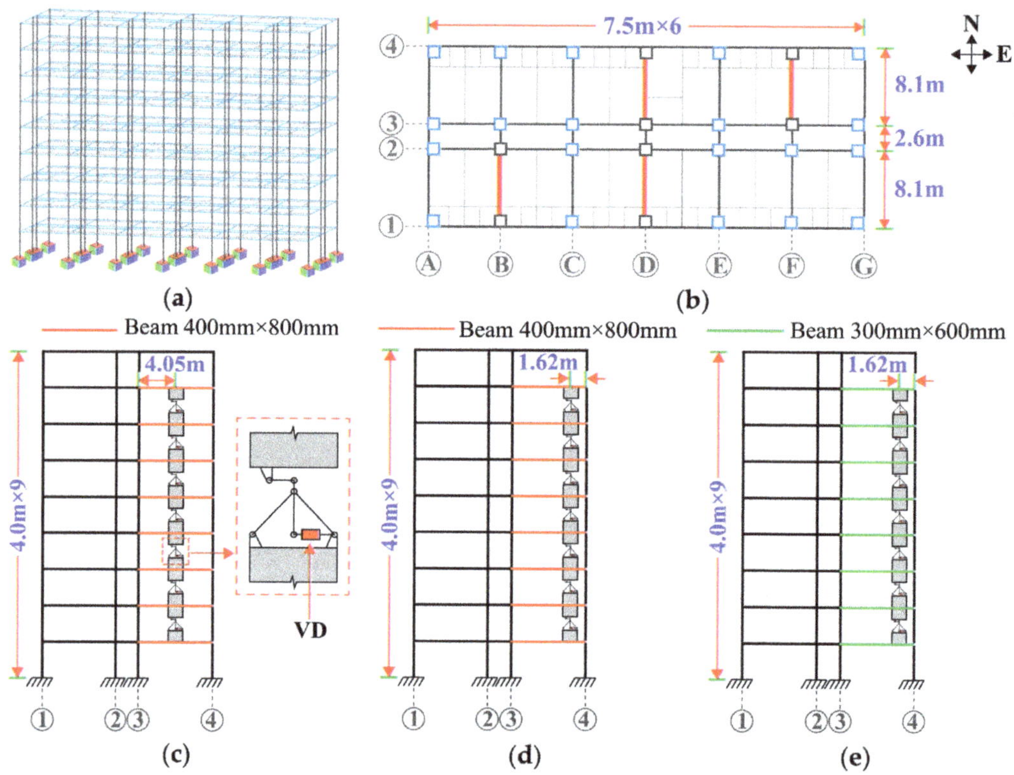

Figure 11. Three-dimensional schematic diagram of the structure and layout of dampers in different schemes: (**a**) a 3D schematic diagram of the structure; (**b**) the layout of CLVD; (**c**) Scheme 1 (F-axis); (**d**) Scheme 2 (F-axis); (**e**) Scheme 3 (F-axis).

The elastic section design of the additional CLVD structure is based on the response spectrum design under fortification earthquakes according to the Chinese Code [42]. In the response spectrum design, the effect of CLVDs on the structure is replaced by the equivalent damping ratio and the equivalent stiffness of Scheme 1. The seismic elastic dynamic analysis is conducted utilizing SAP2000 software (version 23.3.0), while the elastic–plastic dynamic analysis for rare earthquakes is conducted utilizing Perform-3d software (version 7.0). In the elastic–plastic analysis, fiber section models are adopted for beam and column components, and the steel reinforcement adopts a double-line model; the unconstrained concrete adopts a three-line model; and the constrained concrete adopts the Mander model. The viscous damper adopts a fluid damper element. In the parameter analysis of the vibration reduction efficient of the CLVD in Section 3.2, it is assumed that the structure without a CLVD maintains elasticity. Therefore, the structure without an additional CLVD is forced to maintain elasticity. Subsequently, the dynamic calculations of rare earthquakes are performed. This can avoid the plastic deformation of the structure without a CLVD, which increases the complexity of the vibration reduction effect analysis. According to the Chinese Code [42], five natural waves and two artificial waves are selected. A comparison between the normalized spectrum of the selected seismic wave and the normalized standard response spectrum is shown in Figure 12. All analysis results in this section are taken as the average of seven seismic wave calculation results. Due to space limitations, this section only elaborates on the application of CLVD damping systems in the north–south direction of structures.

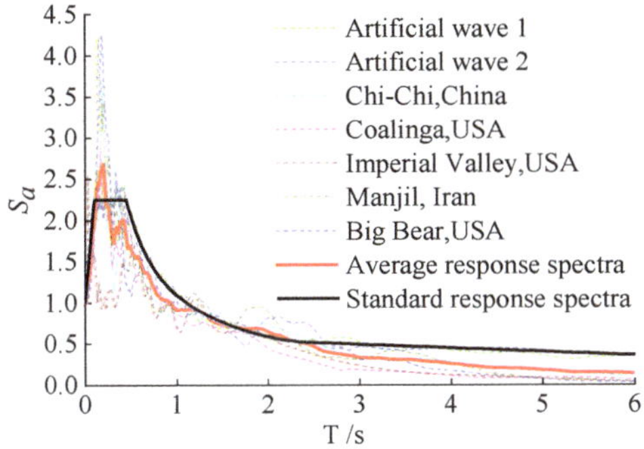

Figure 12. Normalized time–history response spectra and standard response spectra.

4.3. Vibration Reduction Effect

Table 2 summarizes the damping ratios, which are added to structures under different damper schemes. The calculation of the additional damping ratio is based on the traditional specification method in [40]. $\zeta_d = W_c/(4\pi W_s)$. ζ_d is the additional damping ratio, W_c is the total energy dissipation of the damper, and W_s is the total potential energy of the structure. According to Table 2, under fortification earthquakes, the additional damping ratios from Scheme 1 to Scheme 3 are 12.1%, 7.1%, and 4.4%, respectively. Scheme 1 also meets the expected additional damping ratio of 10%.

Table 2. Additional damping ratios for three schemes under fortification earthquakes.

Earthquake Wave	Scheme 1	Scheme 2	Scheme 3
Artificial wave 1	11.2%	6.8%	3.9%
Artificial wave 2	12.2%	6.9%	4.3%
Chi-Chi	12.6%	7.3%	4.9%
Coalinga	12.8%	7.1%	4.1%
Imperial Valley	12.6%	7.9%	4.8%
Manjil	11.6%	6.6%	4.3%
Big Bear	11.9%	7.1%	4.2%
Average of multiple waves	12.1%	7.1%	4.4%

Figure 13 reveals the CLVD's displacement amplification factor f_d under different schemes. On comparing the three schemes, we observe that under fortification earthquakes and rare earthquakes, Scheme 1 is the best, while Scheme 2 and Scheme 3 are relatively poor. On comparing Scheme 1 and Scheme 2, we discover that the CLVD mid-span layout is better than the edge-span layout. On comparing Scheme 2 and Scheme 3, we demonstrate that increasing the beam's bending line stiffness can significantly improve the displacement of the damper.

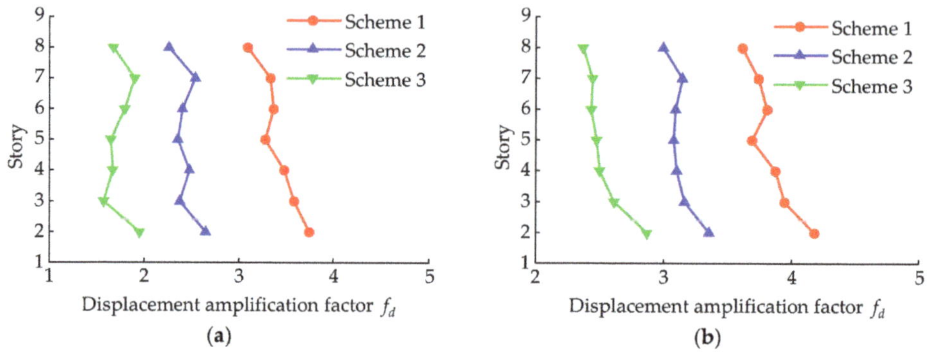

Figure 13. Displacement amplification factors f_d under fortification earthquakes and rare earthquakes: (**a**) f_d under fortification earthquakes; (**b**) f_d under rare earthquakes.

Figure 14a illustrates the hysteresis curves of the damper at position F-(3-4) on the fifth floor under the Big Bear wave in different schemes. In Figure 14a, due to the good displacement amplification ability of the CLVD in Scheme 1, it is illustrated that the hysteresis curve of its damper is also increased in size. In Figure 14b, under rare earthquakes, the CLVD energy dissipation of Scheme 1 is significantly greater than the inherent damping energy dissipation of the structure, which suppresses the plastic deformation of the structure.

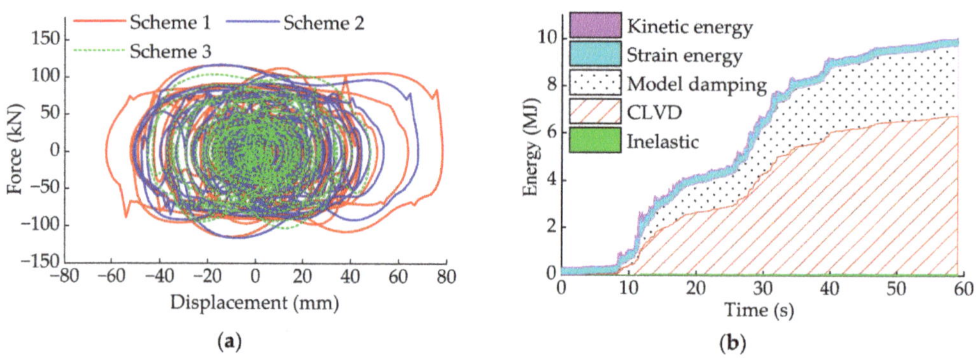

Figure 14. VD hysteresis curve and energy dissipation under rare earthquake and the Big Bear wave: (**a**) damper hysteresis curve at position 5-F-(3-4); (**b**) energy dissipation in Scheme 1.

The control effects of the CLVD damping system for the inter-story displacement and floor shear force from Schemes 1 to Scheme 3 are illustrated in Figures 15 and 16. According to Figure 15, Scheme 1 has a maximum inter-story displacement angle of less than 1/300 under fortification earthquakes and less than 1/150 under rare earthquakes. For all the floors except for the first floor, the inter-story displacement of Scheme 1 has been reduced by almost 40% under fortification earthquakes and by almost 30% under rare earthquakes. In comparison with Scheme 1, Scheme 2 and Scheme 3 have worse effects. The story shear forces of Scheme 1 are reduced to a certain extent, while the reduction is smaller than the reduction in inter-story displacement. This is because the CLVD adds both damping and stiffness to the structure. For Scheme 2 and Scheme 3, the reduction in the story shear forces is greater than the reduction in inter-story drift. This is because the structures with CLVDs in Schemes 2 and Schemes 3 undergo significant plastic deformation during rare earthquakes, while the structure without CLVDs is forced to maintain its elasticity.

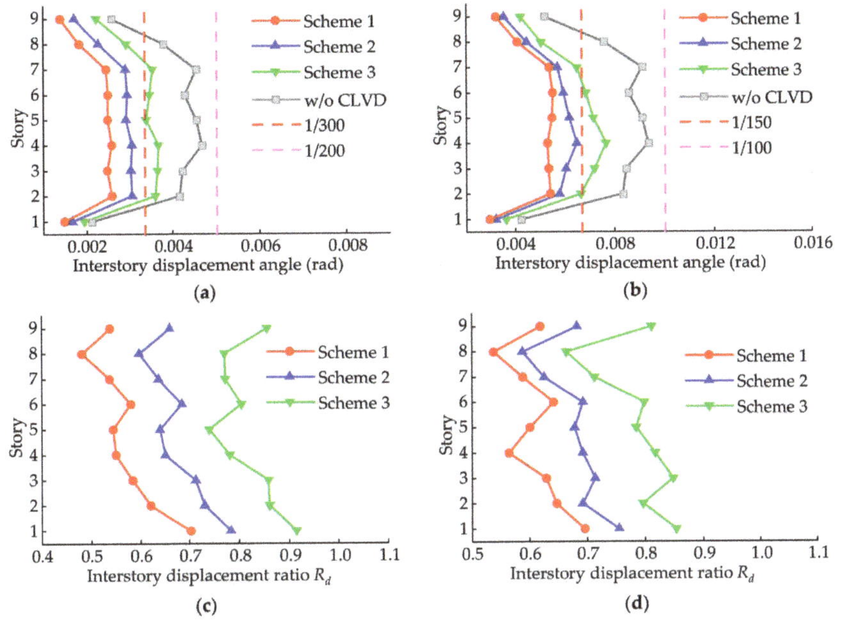

Figure 15. Inter-story displacement angle and R_d under fortification earthquakes and rare earthquakes: (**a**) inter-story displacement angle under fortification earthquakes; (**b**) inter-story displacement angle under rare earthquakes; (**c**) R_d under fortification earthquakes; (**d**) R_d under rare earthquakes.

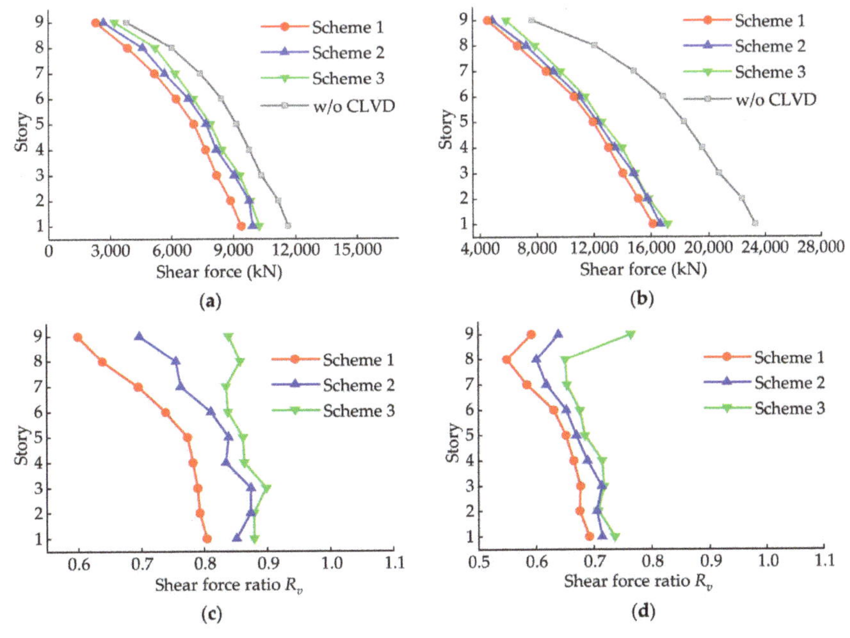

Figure 16. Story shear force and R_v under fortification earthquakes and rare earthquakes: (**a**) story shear force under fortification earthquakes; (**b**) story shear force under rare earthquakes; (**c**) R_v under fortification earthquakes; (**d**) R_v under rare earthquakes.

The research on the vibration reduction efficiency of CLVD in this study mainly considers the frame structure. For frame shear wall structures and shear wall structures, the proportion of inter-story bending deformation to the total inter-story deformation is relatively large. Therefore, the control effect of CLVD on inter-story displacement and shear force of such structures will deteriorate, which is not covered in this paper. CLVD has a high demand for the stiffness of beams, and applying negative stiffness devices to CLVD can further increase the deformation of dampers and reduce the stiffness demand of beams. We will further explore the displacement amplification effect and vibration reduction efficiency of CLVD dampers containing negative stiffness devices in the future.

5. Conclusions

CLVDs are a combination of an intermediate column, lever, and viscous damper, which are mainly designed to overcome the drawbacks of the low-energy-consumption effect of the intermediate column damper and the high cost and significant occupation of building space of the lever damper. Specifically, in this study, the displacement ratio R_d and the shear force ratio R_v of the structure with CLVDs to the structure without CLVDs are proposed as a means of comprehensively evaluating the vibration reduction effect of the structure with CLVDs; the effects of the position of the intermediate column, the beam's bending line stiffness, the lever amplification factor, the damping coefficient, and the earthquake intensity on the additional CLVD structure's total damping ratio ζ, the stiffness ratio k_a'/k_f, R_d, and R_v were studied; countermeasures to improve the damping effect of the structure with CLVDs under different conditions were provided. The main conclusions are as follows:

(1) In contrast to the intermediate column located at both ends of the span, when the ratio of the distance from the intermediate column to the edge column to the span of the beam is 0.5, the stiffness k_a' and total damping ratio ζ provided by CLVDs to the structure reach the maximum value, and the CLVD owns the optimal vibration reduction effect.
(2) When the intermediate column is located at the end of the span, k_a'/k_f increases with the increase in the beam's bending line stiffness i_b. When the intermediate column is located in the middle of the span, k_a'/k_f decreases as i_b increases. Regardless of the position of the intermediate column, ζ increases as i_b increases. Meanwhile, increasing the beam's bending line stiffness is beneficial for CLVD to control structural displacement and shear force.
(3) There exists an optimal lever amplification factor χ, viscous damping coefficient c, and exponent α to achieve the optimal values of R_d and R_v. When the leverage amplification factor is too large, the CLVD provides the structure with stiffness as the main factor, followed by damping. When the ratio of the displacement amplification factor to the geometric amplification factor satisfies $f_d/\gamma = 1/2^{1-0.5\alpha}$, the CLVD possesses the optimal effect for controlling structural displacement.
(4) In order to ensure that the CLVD has positive control effects on the structure's displacement and shear force under different intensity earthquakes, there are three approaches for achieving this goal. The first approach is to simultaneously increase the leverage amplification factor χ and the beam's bending stiffness i_b; the second one is to simultaneously increase the damping coefficient c and the beam's bending line stiffness i_b; and the third one is to increase the leverage amplification factor χ while reducing the damping coefficient c.
(5) In the design of energy dissipation and vibration reduction, it is recommended to adopt the CLVD optimization strategy proposed in Section 3.3 to guide the design analysis. The CLVD with optimized parameters possesses a good control effect on the structure's inter-story displacement and story shear force. Scheme 1 using optimization methods has an additional damping ratio of up to 12% under fortification earthquakes. At the same time, the inter-story displacement was reduced by almost

40% under fortification earthquakes, and the inter-story displacement was reduced by almost 30% under rare earthquakes.

Author Contributions: Conceptualization, Q.Z. and W.P.; software, Q.Z.; validation, Q.Z., W.P. and X.L.; formal analysis, Q.Z.; data curation, Q.Z.; writing—original draft preparation, Q.Z.; writing—review and editing, Q.Z.; visualization, X.L.; project administration, W.P.; funding acquisition, W.P. All authors have read and agreed to the published version of the manuscript.

Funding: This research was funded by the National Key Research and Development Program of China (Grant No. 2020YFD1100703-04), and Yunnan Provincial Basic Research Special Project (Grant No. 202101AU070059).

Data Availability Statement: The original contributions presented in the study are included in the article, further inquiries can be directed to the corresponding authors.

Conflicts of Interest: The authors declare no conflicts of interest.

References

1. Symans, M.D.; Charney, F.A.; Whittaker, A.S.; Constantinou, M.C. Energy dissipation systems for seismic applications: Current practice and recent developments. *J. Struct. Eng.* **2008**, *134*, 3–21. [CrossRef]
2. Gioiella, L.; Tubaldi, E.; Gara, F.; Dezi, L.; Dall'Asta, A. Modal properties and seismic behaviour of buildings equipped with external dissipative pinned rocking braced frames. *Eng. Struct.* **2018**, *172*, 807–819. [CrossRef]
3. Mcnamara, R.J.; Taylor, D.P. Fluid viscous dampers for high-rise buildings. *Struct. Des. Tall Spec. Build.* **2003**, *12*, 145–154. [CrossRef]
4. Pavia, A.; Scozzese, F.; Petrucci, E.; Zona, A. Seismic upgrading of a historical masonry bell tower through an internal dissipative steel structure. *Buildings* **2021**, *11*, 24. [CrossRef]
5. Ou, J.P.; Li, H. Design approaches for active, semi-active and passive control systems based on analysis of characteristics of active control force. *Earthq. Eng. Eng. Vib.* **2009**, *8*, 493–506. [CrossRef]
6. Bhowmik, K.; Debnath, N. Semi-active Vibration Control of Soft-Storey Building with Magnetorheological Damper under Seismic Excitation. *J. Vib. Eng. Technol.* **2024**, *12*, 6943–6961. [CrossRef]
7. Naeem, A.; Kim, J. Seismic retrofit of structures using rotational friction dampers with restoring force. *Adv. Struct. Eng.* **2020**, *23*, 16. [CrossRef]
8. Rayegani, A.; Soureshjani, O.K.; Alaee, S.A.M.; Mualla, I.H.; Nemati, F. Seismic Performance of Buildings Equipped with Four-Joint Rotational Friction Dampers in Mainshock–Aftershock Sequences. *J. Struct. Eng.* **2024**, *150*, 04023235. [CrossRef]
9. Dong, B.; Sause, R.; Ricles, J.M. Seismic Response and Damage of Reduced–Strength Steel MRF Structures with Nonlinear Viscous Dampers. *J. Struct. Eng.* **2018**, *144*, 04018221. [CrossRef]
10. Kang, J.D.; Tagawa, H. Seismic performance of steel structures with seesaw energy dissipation system using fluid viscous dampers. *Eng. Struct.* **2013**, *56*, 431–442. [CrossRef]
11. Pollini, N.; Lavan, O.; Amir, O. Minimum–cost optimization of nonlinear fluid viscous dampers and their supporting members for seismic retrofitting. *Earthq. Eng. Struct. Dyn.* **2017**, *46*, 1941–1961. [CrossRef]
12. Chen, Y.T.; Chai, Y.H. Effects of brace stiffness on performance of structures with supplemental Maxwell model–based brace–damper systems. *Earthq. Eng. Struct. Dyn.* **2011**, *40*, 75–92. [CrossRef]
13. Londono, J.M.; Neild, S.A.; Wagg, D.J. A noniterative design procedure for supplemental brace–damper systems in single–degree–of–freedom systems. *Earthq. Eng. Struct. Dyn.* **2013**, *42*, 2361–2367. [CrossRef]
14. Li, S.Y.; Chen, Y.T.; Chai, Y.H. Effects of Brace Stiffness and Nonlinearity of Viscous Dampers on Seismic Performance of Structures. *Int. J. Struct. Stab. Dyn.* **2021**, *21*, 2150188. [CrossRef]
15. Yang, L.Y.; Ye, M.; Huang, Y.H.; Dong, J.K. Study on Mechanical Properties of Displacement-Amplified Mild Steel Bar Joint Damper. *Iran. J. Sci. Technol. Trans. Civ. Eng.* **2023**, *7*, 01268. [CrossRef]
16. Ribakov, Y.; Gluck, J.; Gluck, N. Practical design of MDOF structures with supplemental viscous dampers using mechanical levers. In Proceedings of the 8th ASCE Specialty Conference on Probabilistic Mechanics and Structural Reliability, Notre Dame, IN, USA, 24–26 July 2000.
17. Ribakov, Y.; Reinhorn, A.M. Design of Amplified Structural Damping Using Optimal Considerations. *J. Struct. Eng.* **2003**, *129*, 1422–1427. [CrossRef]
18. Aly, S.A.; Zasso, A.; Resta, F. Proposed Configurations for the Use of Smart Dampers with Bracings in Tall Buildings. *Smart Mater. Res.* **2012**, *2012*, 251543. [CrossRef]
19. He, W.; Xu, H.; Xu, Z.; Liu, W. Experimental investigation and earthquake response analysis of a multilayer viscous damping wall with amplified deformation. *Eng. Struct.* **2022**, *251*, 113427. [CrossRef]
20. Shang, F.; Liu, W.G.; Zhang, Q.; Xu, H.; Guo, Y. Shaking table test and seismic behaviour evaluation of viscous damping wall with amplification mechanism. *Eng. Struct.* **2023**, *284*, 115974. [CrossRef]

21. Fournier, J.A.; Cheng, S.H. Impact of damper stiffness and damper support stiffness on the efficiency of a linear viscous damper in controlling stay cable vibrations. *J. Bridge Eng.* **2014**, *19*, 04013022. [CrossRef]
22. Losanno, D.; Londono, J.M.; Zinno, S.; Serino, G. Effective damping and frequencies of viscous damper braced structures considering the supports flexibility. *Comput. Struct.* **2017**, *207*, 121–131. [CrossRef]
23. Constantinou, M.C.; Tsopelas, P.; Hammel, W.; Sigaher, A.N. New configurations of fluid viscous dampers for improved performance. In *Proceedings of the Passive Structural Control Symposium*; Tokyo Institute of Technology: Yokohama, Japan, 2000.
24. Constantinou, M.C.; Tsopelas, P.; Hammel, W.; Sigaher, A.N. Toggle–brace–damper seismic energy dissipation systems. *J. Struct. Eng.* **2001**, *127*, 105–112. [CrossRef]
25. Sigaher, A.N.; Constantinou, M.C. Scissor–jack–damper energy dissipation system. *Earthq. Spectra.* **2003**, *19*, 133–158. [CrossRef]
26. Hwang, J.S.; Huang, Y.N.; Hung, Y.H. Analytical and experimental study of toggle–brace–damper systems. *J. Struct. Eng.* **2005**, *131*, 1035–1043. [CrossRef]
27. Huang, H.C. Efficiency of the motion amplification device with viscous dampers and its application in high–rise buildings. *Earthq. Eng. Eng. Vib.* **2009**, *8*, 521–536. [CrossRef]
28. Jiang, H.J.; Wu, S.Y.; He, L.S. Theoretical analysis and optimization of toggle–brace damper for cable tray system. *J. Constr. Steel Res.* **2021**, *187*, 106936. [CrossRef]
29. Polat, E.; Constantinou, M.C. Open–space damping system description, theory, and verification. *J. Struct. Eng.* **2017**, *143*, 04016201. [CrossRef]
30. Lan, B.J.; He, H.X.; Wu, S.; Cheng, S.T.; Chen, Y.F. Multi–performance objectives optimum design method and vibration control application of generalized toggle–brace–damper system. *Earthq. Eng. Struct. Dyn.* **2022**, *51*, 1410–1435. [CrossRef]
31. Xu, J.W.; Ma, P.F.; Hu, Y.k.; Yao, J.T. Discussion on Calculation Method of Magnification Factor of Toggle-Brace-Viscous Damper. *Buildings* **2024**, *13*, 2006. [CrossRef]
32. Zhou, Q.; Pan, W.; Lan, X. Analysis of Amplification Effect and Optimal Control of the Toggle-Style Negative Stiffness Viscous Damper. *Buildings* **2024**, *14*, 1625. [CrossRef]
33. Lan, X.; Pan, W.; Kuang, H.W. Correction additional damping of an energy–dissipation structure based on a story drifts utilization ratio method. *J. Vib. Shock.* **2017**, *36*, 64–71. (In Chinese)
34. Çerçevik, A.E.; Avşar, Ö.; Dilsiz, A. Optimal placement of viscous wall dampers in RC moment resisting frames using metaheuristic search methods. *Eng. Struct.* **2021**, *249*, 113108. [CrossRef]
35. Lan, X.; Zhang, L.F.; Sun, B.F.; Pan, W. Study on the Damping Efficiency of a Structure with Additional Viscous Dampers Based on the Shaking Table Test. *Buildings* **2024**, *14*, 1506. [CrossRef]
36. Sonmez, M.; Aydin, E.; Karabork, T. Using an artificial bee colony algorithm for the optimal placement of viscous dampers in planar building frames. *Struct. Multidiscip. Optim.* **2013**, *48*, 395–409. [CrossRef]
37. Bishop, J.A.; Striz, A.G. On using genetic algorithms for optimum damper placement in space trusses. *Struct. Multidiscip. Optim.* **2004**, *28*, 2–3. [CrossRef]
38. Sarcheshmehpour, M.; Estekanchi, H.E.; Ghannad, M.A. Optimum placement of supplementary viscous dampers for seismic rehabilitation of steel frames considering soil–structure interaction. *Struct. Des. Tall Spec. Build.* **2020**, *29*, e1682. [CrossRef]
39. Lan, X.; Wei, G.L.; Zhang, X.X. Study on the Influence and Optimization Design of Viscous Damper Parameters on the Damping Efficiency of Frame Shear Wall Structure. *Buildings* **2024**, *14*, 497. [CrossRef]
40. *JGJ 297–2013*; Chinese Standard, Technical Specification for Seismic Energy Dissipation of Buildings. Chinese Architecture and Building Press: Beijing, China, 2013.
41. Kasal, K.; Ito, H.; Watanabe, A. Peak response prediction rule for a SDOF elasto–plastic system based on equivalent linearization technique. *J. Struct. Constr. Eng.* **2003**, *571*, 53–62. (in Japanese).
42. *GB 50011–2016*; Chinese Standard, Code for Seismic Design of Buildings. Chinese Architecture and Building Press: Beijing, China, 2016.

buildings

MDPI

Article

Assessment of the Compound Damping of a System with Parallelly Coupled Anti-Seismic Devices

Polidor Bratu [1], Claudiu-Sorin Dragomir [2,3] and Daniela Dobre [2,4,*]

1 Solid Mechanics Institute of the Romanian Academy, 030167 Bucharest, Romania; icecon@icecon.ro
2 National Institute for Research and Development in Construction, Urban Planning and Sustainable Spatial Development "URBAN-INCERC", 021652 Bucharest, Romania; dragomirclaudiusorin@yahoo.com
3 Environment and Land Reclamation Department, Faculty of Land Reclamation and Environment Engineering, 012244 Bucharest, Romania
4 The Structural Mechanics Department, Technical University of Civil Engineering, 020396 Bucharest, Romania
* Correspondence: daniela.dobre@utcb.ro

Abstract: (1) Background: Romanian earthquakes caused severe damage over time to a significant number of constructions, and that is why efforts are being made to make structural systems safer. (2) Methods: For structural systems with protection against seismic actions or vibrational actions that have linear viscous dissipation devices, the requirement to assess the equivalent modal damping rate for the entire functional assembly related to the other dynamic parameters arises. (3) Results: This article presents the analytical development of formulas for the compound damping and circular frequency when anti-seismic devices have different dynamic characteristics and their application in order to solve some real engineering cases of bridges and viaducts in Romania with distinct viscoelastic supports. In support of this idea, some experimental tests on a beam system resting on two different anti-seismic elastic supports highlighted the fact that the compound damping of the system can be calculated with the relations established in this paper, provided that the displacements in the horizontal direction of excitation are in the linear domain. Also, we determined the seismic response considering the Vrancea 1977 accelerogram for critical damping ratios of 5% and 18.5%, and then we obtained the variation in the factor of transmissibility depending on the frequency, in order to highlight the optimized value of the equivalent amortization/damping. (4) Conclusions: In the specific context of Romanian seismicity, seismic isolation through the use of isolators with different characteristics represents an optimal technical solution, and it is also optimal from an economic point of view, with an appropriate level of dynamic isolation obtained.

Keywords: dissipated energy; vibrations; equivalent amortization; modal damping; elastomeric anti-seismic device

1. Introduction

Resilience is an important concept in seismic risk management and in any disaster reduction strategy. Because a society becomes even more vulnerable in a built environment more vulnerable to natural hazards, the general main objective to increase resilience involves the reduction in physical vulnerability and maintaining functionality in the case of several infrastructures. The Vrancea intermediate source is important in Romania, taking into account the energy, the extent of the macro-seismic effects, and that it can exert a direct and indirect impact at a national disaster level on a scale of more than 50% of the territory [1,2]. Romanian earthquakes have caused severe damage over time to a significant number of constructions, and in order to avoid the negative structural effects of future earthquakes, in important seismic areas, seismic isolation is implemented on a large scale, following the example of countries with high seismicity as well [3–7]. Passive control systems are applied to different structures, and the efficiency of these systems was demonstrated during exceptional natural events such as earthquakes [8,9]. In terms of the damping capacity, there are different methods for estimating the equivalent modal damping ratios of a bridge under strong ground motion [10–13]. As another alternative to the traditional design criteria, a system with parallelly coupled anti-seismic devices is considered in order to reduce the seismic actions that affect a structural system of construction.

Specifically, in the case of the rigid structural systems elastically supported with the help of anti-seismic devices with significant horizontal displacements in the elastic domain, the requirement of evaluating the degree of the global damping for the entire construction assembly rises. This analysis of the situation belongs to the field of system compound damping assessment, taking into account the possible and significant modal motions [14–16]. Research was carried out to solve some cases of building structures of bridges and viaducts with distinct viscoelastic supports either in the form of support devices according to EN 1337/2018 [17] or in the form of anti-seismic devices according to EN 15129/2010 [18]. In the ICECON-TEST laboratory in Bucharest (Romania), dynamic tests were carried out with excitation on the possible natural modes for a rigid beam system supported at the ends by two elastomeric devices with different individual values of rigidity and damping [19]. The concept and method of assessing the modal compound amortization for a system with several different viscoelastic devices in terms of rigidity and damping were defined both on this basis as well as by verifying the established calculated relations [14–16,19]. Based on a case study, for a 1200 m long viaduct, leaning on distinct elastomeric devices with k_i, ζ_i, $i = 1 \ldots 4$, characteristics of every support system using a pillar of viaduct, the value of the compound damping is determined using the established calculated relations [14–16,19].

2. Analytical Assessment of Compound Modal Amortization/Damping

In the case of an anti-seismic system composed of dissipation devices in the linear–elastic regime, the effective (equivalent) viscous damping ζ_{ef} of the system can be deduced based on the maximum energy [20–31]. When the system consists of n number of devices, for each device i, the fraction of the viscous damping $\zeta_i^{(r)}$ for the vibration mode r and the modal dissipation energy $W_{d,i}^{(r)} = E_{d,i}^{(r)}$ can be determined; the following relations can be formulated [14,20–22]:

$$W_{d,i}^{(r)} = E_{d,i}^{(r)} = \pi c_i^{(r)} \omega A_{ir}^2 \tag{1}$$

$$V_i^{(r)} = E_{el,i}^{max} = \frac{1}{2} k_i^{(r)} A_{ir}^2 \tag{2}$$

where corresponding to device i for the r mode of vibration, $W_{d,i}^{(r)} = E_{d,i}^{(r)}$ is the dissipated energy; $V_i^{(r)} = E_{el,i}^{max}$ is the maximum elastic energy; $c_i^{(r)} = 2\zeta_i^{(r)} m_i^{(r)} \omega_r$ is the equivalent viscous amortization coefficients; ω_r is the natural circular frequency pulsation of the structural system modeled as a system with 1 DoF (dynamic degree of freedom); m is the mass of the superstructure with rigid motion; k_i is the lateral (secant) rigidity/stiffness of a device; and A_i is the amplitude of the lateral displacement of device i knowing that the lateral displacement is sinusoidal as in $d(t) = Asin(\omega t)$. The equivalent viscous amortization is assessed in a resonance regime for the r mode of vibration, that is, when $\omega = \omega_r$ [14,22,23].

In this case, the energies formulated by Relations (1) and (2) can be explained as follows:

(a) The energy dissipated on mode r, for a complete cycle, corresponding to device i, is

$$W_{d,i}^{(r)} = 2\pi\zeta_i^{(r)} m_i^{(r)} \omega_r \omega A_{ir}^2$$

or

$$W_{d,i}^{(r)} = 2\pi\zeta_i^{(r)} \omega^2 m_i^{(r)} A_{ir}^2 \tag{3}$$

(b) The maximum elastic energy (of deformation) on r mode, for device i, can be written down as

$$V_i^{(r)} = \frac{1}{2}\omega^2 m_i^{(r)} A_{ir}^2 \quad (4)$$

Eliminating $\omega^2 m_i A_{ir}^2$ from the two Relations (3) and (4), we obtain

$$W_{d,i}^{(r)} = 4\pi\zeta_i^{(r)} V_i^{(r)} \quad (5)$$

For the entire system of devices $i \to n$, we have

$$W_{system} = 4\pi\sum_{i=1}^{n}\zeta_i^{(r)} V_i^{(r)} = 4\pi\zeta_{eq} V_{system}^{(r)}$$

Resulting in

$$\zeta_{eq}^{(r)} = \zeta_{ef}^{(r)} = \frac{\Sigma_i\zeta_i^{(r)} V_i^{(r)}}{\Sigma_i V_i^{(r)}} \quad (6)$$

Isolation system with parallelly connected anti-seismic devices

We consider an isolation system consisting of i, i = 1...n, devices connected in parallel, with a rigid structural system of mass m that gravitationally presses on a pile system that supports the viaduct structure. In this case, each anti-seismic device is actuated in compression vertically, with the same loading force, and horizontally, all devices have the same instantaneous displacement $x(t) = Asin(\omega t)$ at a sinusoidal excitation on the first order vibration mode (Figure 1).

$$\omega_r = \omega_n = \sqrt{\frac{k_{eq,parallel}}{m_{eq}}} \text{ or } \omega_r = \sqrt{\frac{\Sigma k_i}{\Sigma m_i}}$$

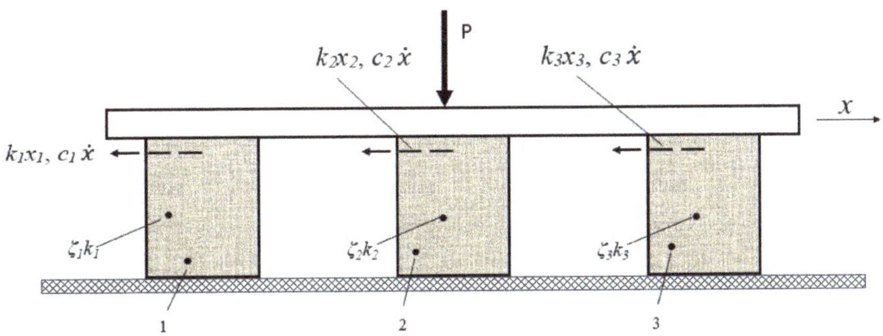

Figure 1. Dynamic model.

For each device i, we highlight the elastic forces $k_i x_i$ and the viscous amortization/damping forces $c_i\dot{x}_i$, reduced to the level of mass m_i, for the horizontal translation movement/displacement $x_i = x_i(t)$, as

$$\begin{cases} F_{el}^i = k_i x_i \\ F_v^i = c_i\dot{x}_i \end{cases} \quad (7)$$

As a result of the laws provided by Relation (7), the correlations that are established measure ζ_i and V_i because $x_i = x$ is the instantaneous translational lateral displacement common to all devices i = 1...n. In this case, the vibration mode is $r = 1$, which means that it corresponds to the fundamental mode with $\omega_n = \omega$, and $x(t) = A_1 sin(\omega_n t + \theta_1)$ with the maximum value $x_{max} = A_1$, and $V_i^{max} = \frac{1}{2}k_i A_1^2$. The fraction of the effective viscous amortization $\zeta_{ef} \equiv \zeta_{eq}$ of the system according to Relation (6) can be written down for the system assembly in Figure 1 as follows:

$$\zeta^{(1)}_{ef,parallel} = \frac{\frac{1}{2}A_1^2(\zeta_1 k_1 + \zeta_2 k_2 + \zeta_3 k_3)}{\frac{1}{2}A_1^2(k_1 + k_2 + k_3)}$$

where we insert $k_{eq,parallel} = k_1 + k_2 + k_3$ and obtain

$$\zeta^{(1)}_{ef,parallel} = \frac{\zeta_1 k_1 + \zeta_2 k_2 + \zeta_3 k_3}{k^{eq}_{paralel}} \tag{8}$$

Relation (8) can also be written down according to the ratios of stiffness (standard rigidities) as follows:

$$\zeta^{(1)}_{ef,parallel} = \frac{\zeta_1 + \zeta_2 \frac{k_2}{k_1} + \zeta_3 \frac{k_3}{k_1}}{1 + \frac{k_2}{k_1} + \frac{k_3}{k_1}} \tag{9}$$

In fact, a weighted average of the sum of viscous amortization/damping ratios that corresponds to device i for mode $r = 1$ with weights related to the normalized values of the stiffnesses can be calculated. In another form, ζ_{eq} can be written as the ratio between the sum of viscous damping coefficients with weights related to the normalized values of the circular frequencies ω_i and the sum of masses with weights related to the normalized values of the circular frequencies to the power of 2:

$$\zeta^{(1)}_{eq} = \frac{c_1 + c_2 \frac{\omega_2}{\omega_1} + c_3 \frac{\omega_3}{\omega_1}}{m_1 + m_2\left(\frac{\omega_2}{\omega_1}\right)^2 + m_3\left(\frac{\omega_3}{\omega_1}\right)^2} \cdot \frac{1}{2\omega_1} \tag{10}$$

By generalizing Relations (8) and (10), it can be written down as $\zeta^{(1)}_{ef,\ parallel} \equiv \zeta^{(1)}_{eq,\ parallel}$ for

$$\zeta^{(1)}_{ef,parallel} = \frac{\sum_{i=1}^{n=3} \zeta_i k_i}{\sum_{i=1}^{n=3} k_i} = \frac{\sum_{i=1}^{n=3} c_i \omega_i}{\sum_{i=1}^{n=3} m_i \omega_i^2} \cdot \frac{1}{2} \tag{11}$$

Relation (11) can be applied if it is easier to experimentally determine the circular frequencies of the isolators compared to their stiffnesses in a non-linear analysis, or in the case of non-linear behavior.

3. Results

3.1. Case Study

For a 1200 m long viaduct with a mass of m = 9600 t and 30 spans, the number of beams was 120 pieces resting on 480 pieces of elastomeric devices [19,24]. The sections were identical constructively and functionally, each with an opening a = 40 m and width b = 14 m on 16 elastomeric devices (Figure 2).

The elastomeric devices in the analyzed direction deform in shear and assure a natural frequency, for mode 1 of translation in the Ox and Oy directions, with f_n = 0.52 Hz. The elastic and amortization characteristics of the four types of elastomeric devices are provided in Table 1.

The elastomeric devices with the function of seismic isolators were made in four distinct variants of the elastomer mixture, respectively, with the elastic and damping characteristics shown in Table 1 and with the shear modules, G_i, i = 1. . .4, obtained from experimental data (which varied with the elastomer content). The constructive cylindrical shape with intermediate metal reinforcements (type sandwich) had the following dimensions: diameter D = 500 mm, total constructive height H=624 mm, and total thickness of the elastomeric layers T_q = 250 mm.

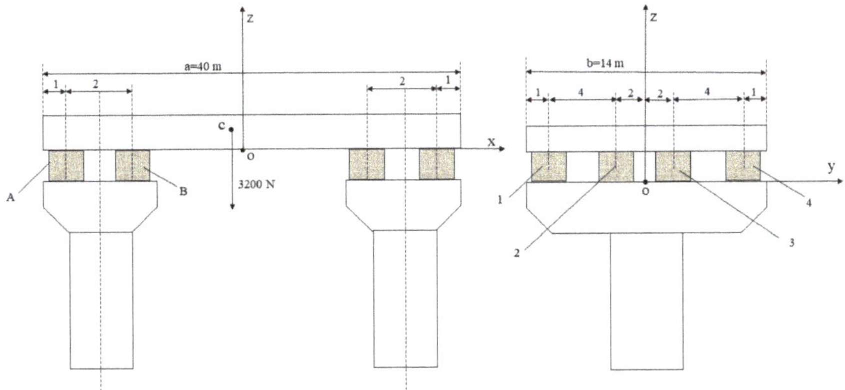

Figure 2. Support scheme for a viaduct section. OXYZ—system of rectangular axes, a—length of section, b—width of section; c—center of gravity of the assembly; 1. . .4—different elastomeric devices, placed in line and on two A and B rows.

Table 1. Elastic and damping characteristics for f_n= 0.52 Hz.

i-Type Elastomeric Device	Elastic G_i [MPa]	Elastic k_i [10^5 N/m]	Modal Viscous Amortization c_i [10^4 Ns/m]	Modal Amortization Ratio ζ_i [%]
1	0.15	1	0.3	10
2	0.25	1.5	1	12
3	0.40	2.5	3	18
4	0.55	3	5	25

During the tests on the dynamic stand, it was found that the hysteretic loop was a regular ellipse, and the force–deformation characteristic (±500 mm) was linear. On the other hand, the total deformation in the linear–elastic domain, in the horizontal direction, was obtained using the design relation provided by SR EN 15129 [17], in the form $x = \gamma T_q$, where $\gamma \leq 2$ is the angular deformation of the isolator during lateral displacement. In the case described in this article, $x_{max} = \gamma_{max} T_q$ = 2 × 250= 500 mm, so it is accepted that a force equivalent to the weight of the superstructure can be applied in the horizontal direction, i.e., $F_x = mg$, and then the static deformation can be determined using $\delta_{static} = \frac{F_x}{K}$. Taking into account that F_x = 96 × 10^6 N and $K = \sum K_i$ = 192 × 10^6 N/m, δ_{static} = 0.50 m is obtained. The linear–elastic function is ensured both by the parameters in dynamic mode with the elliptic hysteretic loop and by the engineering design system.

The variety of the four types of isolators was necessary due to the static equilibrium conditions of the bridge girder, as well as of the entire assembly. Thus, to ensure horizontality, the static equilibrium equations were applied to an elastically supported rigid body in the linear domain. This calculation method and related issues are not part of the objectives of this article. In this case, the basic assumptions of linear–elastic deformation, horizontality through the static equilibrium of the assembly and significant and dominant lateral displacements in relation to the vertical ones in case of an earthquake were specified.

The displacement amplitude of the system with one dynamic degree of freedom, for the model analyzed in the linear–elastic domain, corresponds to an excitation with an acceleration a =2.5 m/s^2 and frequency f = 0.52 Hz, having the maximum value of the elastic horizontal displacement d = 0.25 m < δ_{static}, which proves that in the dynamic regime, the hypothesis of elastic linearity is highlighted.

The assembly viaduct pile system consisting of elastomeric seismic isolation devices is presented in Figure 3. Considering the total stiffness K and the total damping C, the dynamic equation of motion for 1 DoF can be written, where the dynamic load $\ddot{x}_g(t)$ values represent Vrancea earthquake accelerations:

$$m\ddot{x}(t) + C\dot{x}(t) + Kx(t) = -m\ddot{x}_g(t) \tag{12}$$

$$\frac{1}{K} = \frac{1}{K_{str}} + \frac{1}{K_{eq}}$$

$$\frac{1}{C} = \frac{1}{C_{str}} + \frac{1}{C_{eq}}$$

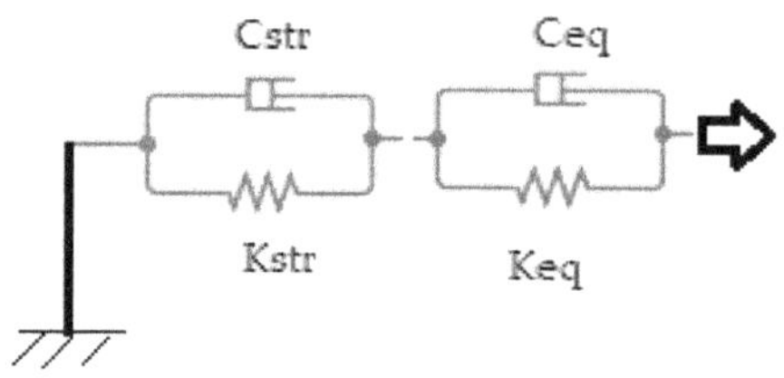

Figure 3. Model for the pile system (that supports the viaduct structure) and the isolation system with isolation devices (connected in parallel).

3.2. Specific Results

By applying Relations (9) and (10) to the scheme in Figure 2 and the data in Table 1, we have

$$\zeta_{ef}^{x,y} = \frac{2\left(\zeta_1 + \zeta_2\frac{k_2}{k_1} + \zeta_3\frac{k_3}{k_1} + \zeta_4\frac{k_4}{k_1}\right)}{2\left(1 + \frac{k_2}{k_1} + \frac{k_3}{k_1} + \frac{k_4}{k_1}\right)}$$

and/or

$$\zeta_{ef}^{x,y} = \frac{c_1 + c_2\frac{\omega_2}{\omega_1} + c_3\frac{\omega_3}{\omega_1} + c_4\frac{\omega_4}{\omega_1}}{m_1 + m_2\left(\frac{\omega_2}{\omega_1}\right)^2 + m_3\left(\frac{\omega_3}{\omega_1}\right)^2 + m_4\left(\frac{\omega_4}{\omega_1}\right)^2} \cdot \frac{1}{2\omega_1}$$

where using the previously established values as inputs, we obtain

$$\zeta_{ef}^{x,y} = \frac{2(0.1 + 0.12 \cdot 1.5 + 0.18 \cdot 2.5 + 0.25 \cdot 3)}{2(1 + 1.5 + 2.5 + 3)} = 0.185$$

$$\zeta_{ef}^{x,y} = \frac{0.3 \cdot 10^4 + 1 \cdot 10^4 \cdot \frac{3.6}{6.67} + 3 \cdot 10^4 \cdot \frac{2.99}{6.67} + 5 \cdot 10^4 \cdot \frac{3}{6.67}}{2250 + 11574.07 \cdot \left(\frac{3.6}{6.67}\right)^2 + 27777.78 \cdot \left(\frac{2.99}{6.67}\right)^2 + 33333.33 \cdot \left(\frac{3}{6.67}\right)^2} \cdot \frac{1}{2 \cdot 6.67} = 0.185$$

In order to find the near-optimal design ensemble for a specific situation, other values for ζ_{eq} can be obtained from permutations in the $\frac{k_i}{k_1} - \zeta_i$ relationship, as shown in Table 2.

For the viaduct structure with a previously given support scheme and configuration, a compound amortization of 18.5% is obtained.

B. For the frequency f_n = 0.52 Hz, $\zeta_{eq} = 18.5\%$ compared to $\zeta = 5\%$, and the 1977 Vrancea earthquake accelerations (accelerogram INCERC north–south, with time step Δt = 0.02 s, the peak ground acceleration being approx. 195 cm/s^2 occurring at t = 6.14 s), the instantaneous and maximum seismic responses of the system with one dynamic degree of freedom in the time domain, obtained by numerical integration of the equation of motion (12), are presented in Figure 4a–c. The applied approach is the linear acceleration method and the structural response is described in terms of absolute accelerations and relative displacements. As expected, the response is reduced more in cases of higher damping (with the elastomeric devices).

Table 2. Permutations in $\frac{k_i}{k_1} - \zeta_i$ relationship, i = 1…4.

$\frac{k_1}{k_1}$	$\frac{k_2}{k_1}$	$\frac{k_3}{k_1}$	$\frac{k_4}{k_1}$	ζ_1	ζ_2	ζ_3	ζ_4	ζ_{eq}
1	2.5	3	1.5	0.25	0.12	0.1	0.18	0.14
1	3	2.5	1.5	0.25	0.12	0.1	0.18	0.141
1	3	2.5	1.5	0.18	0.12	0.1	0.25	0.145
1	1.5	2.5	3	0.25	0.12	0.18	0.1	0.147
1	2.5	3	1.5	0.25	0.18	0.12	0.1	0.151
1	1.5	3	2.5	0.25	0.12	0.18	0.1	0.152
1	2.5	3	1.5	0.25	0.12	0.18	0.1	0.155
1	1.5	2.5	3	0.12	0.25	0.18	0.1	0.155
1	2.5	1.5	3	0.1	0.18	0.25	0.12	0.160
1	1.5	3	2.5	0.18	0.1	0.12	0.25	0.164
1	2.5	3	1.5	0.1	0.25	0.12	0.18	0.169
1	3	1.5	2.5	0.18	0.25	0.12	0.1	0.17
1	2.5	3	1.5	0.12	0.1	0.25	0.18	0.173
1	2.5	3	1.5	0.1	0.12	0.25	0.18	0.177
1	2.5	3	1.5	0.12	0.25	0.18	0.1	0.179
1	3	2.5	1.5	0.1	0.18	0.25	0.12	0.180
1	2.5	3	1.5	0.12	0.18	0.25	0.1	0.183
1	1.5	3	2.5	0.1	0.12	0.25	0.18	0.185

Analyzing the maximum of the graph obtained in displacements, shown in Figure 4c, while considering two values for damping, it can be seen that it is lower than the maximum value related to the elastic horizontal displacement for the model analyzed in the linear–elastic domain.

C. On the other hand, considering the existing correlations between the defining quantities (mass, stiffness, and damping) and the method of determining the equivalent amortization, values between 3.18 rad/s and 3.67 rad/s are determined for the equivalent circular frequency ω_{eq}, as presented in Table 3.

Table 3. Estimative values for the equivalent circular frequency ω_{eq}.

ω_{eq}	Related to $\frac{k_i}{m_i}$	Related to ζ_i	Related to $\frac{k_i}{k_1}$	Related to $\frac{c_i}{c_1}$
	$\sqrt{\frac{\Sigma k_i}{\Sigma m_i}}$	$\frac{\Sigma \zeta_i \omega_i}{\Sigma \zeta_i}$	$\frac{\Sigma \frac{k_i}{k_1} \omega_i}{\Sigma \frac{k_i}{k_1}}$	$\frac{\Sigma \frac{c_i}{c_1} \omega_i}{\Sigma \frac{c_i}{c_1}}$
[rad/s]	3.27	3.67	3.57	3.18
estimated value/ exact value	1	1.12	1.09	0.97

D. Finally, for the system formed by these four types of seismic isolators, the variation in the transmissibility factor T_i depending on the frequency is also of interest. Table 4 shows the values of the transmissibility factor for different frequencies of the vibration (f_{vibr}) and according to the individual characteristics of the isolators (ω_i,ζ_i) compared to the situation in which the frequency related to $\frac{k_i}{m_i}$ and the equivalent amortization is considered (ω_{system},ζ_{eq}).

$$T_i = \sqrt{\frac{1 + \left(2\zeta_i \frac{\omega_{vibr}}{\omega_i}\right)^2}{\left[1 - \left(\frac{\omega_{vibr}}{\omega_i}\right)^2\right]^2 + \left(2\zeta_i \frac{\omega_{vibr}}{\omega_i}\right)^2}} \quad (13)$$

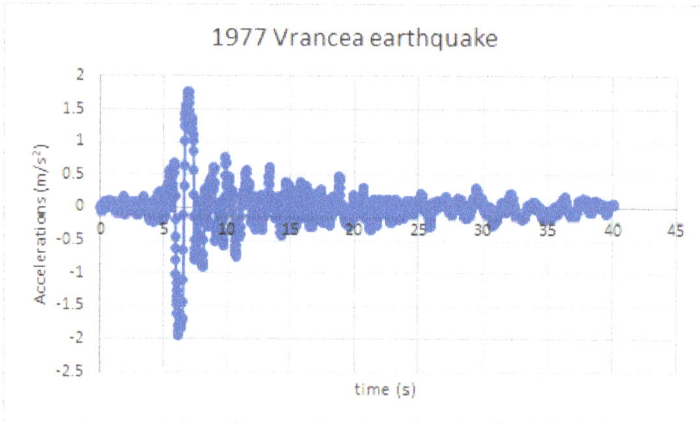

(a)

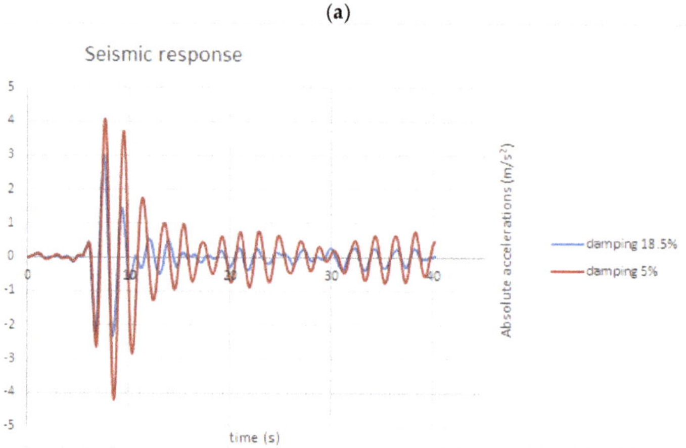

(b)

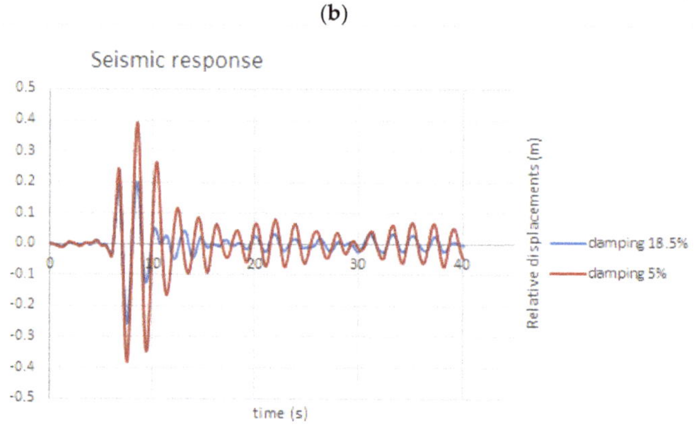

(c)

Figure 4. (**a**). Accelerogram 1977 Vrancea, INCERC north–south, with time step $\Delta t = 0.02$ s, the peak ground acceleration being approx. 195 cm/s^2, occurring at t = 6.14 s. (**b**). Dynamic analysis. Seismic response in the time domain in terms of absolute accelerations. (**c**). Dynamic analysis. Seismic response in the time domain in terms of relative displacements.

Table 4. Transmissibility factor related to the circular frequency, f_{vibr}, and to the equivalent parameters, ω_{system}, ζ_{eq}.

f_{vibr} [Hz]	$T\omega_1, \zeta_1$	$T\omega_2, \zeta_2$	$T\omega_3, \zeta_3$	$T\omega_4, \zeta_4$	$T\omega_{system}, \zeta_{eq}$
0	1	1	1	1	1
0.05	1.002	1.007	1.011	1.011	1.009
0.1	1.008	1.031	1.046	1.045	1.038
0.15	1.020	1.073	1.109	1.106	1.089
0.2	1.036	1.137	1.209	1.200	1.168
0.25	1.058	1.231	1.363	1.339	1.286
0.3	1.086	1.368	1.598	1.537	1.458
0.35	1.121	1.569	1.962	1.804	1.712
0.4	1.164	1.880	2.494	2.108	2.086
0.45	1.216	2.383	2.969	2.282	2.576
0.5	1.281	3.219	2.704	2.120	2.917
0.55	1.360	4.216	2.001	1.742	2.631
0.6	1.458	3.829	1.458	1.382	2.017
0.67	1.639	2.246	1.003	1.021	1.364
0.7	1.737	1.817	0.876	0.909	1.175
0.75	1.940	1.344	0.716	0.763	0.943
0.8	2.209	1.047	0.602	0.654	0.778
0.85	2.577	0.847	0.516	0.570	0.658
0.9	3.088	0.705	0.450	0.504	0.567
0.95	3.794	0.599	0.398	0.451	0.497
1	4.636	0.519	0.355	0.408	0.440
1.05	5.122	0.455	0.321	0.372	0.394
1.1	4.642	0.404	0.292	0.342	0.357
1.15	3.685	0.361	0.267	0.316	0.325
1.2	2.862	0.326	0.246	0.293	0.298
1.25	2.269	0.297	0.228	0.274	0.275
1.3	1.850	0.272	0.213	0.257	0.255
1.35	1.545	0.250	0.199	0.242	0.238
1.4	1.317	0.231	0.187	0.229	0.223
1.45	1.141	0.215	0.176	0.217	0.209
1.5	1.003	0.200	0.166	0.206	0.197
1.55	0.891	0.187	0.158	0.197	0.187
1.6	0.799	0.176	0.150	0.188	0.177
1.65	0.722	0.165	0.143	0.180	0.168
1.7	0.657	0.156	0.136	0.172	0.160
1.75	0.602	0.148	0.131	0.166	0.153
1.8	0.554	0.140	0.125	0.159	0.147
1.85	0.512	0.133	0.120	0.154	0.141
1.9	0.475	0.127	0.116	0.148	0.135
1.95	0.443	0.121	0.111	0.143	0.130
2	0.414	0.116	0.107	0.139	0.125

In Figure 5, the variation in the transmissibility factor related to the frequency f_{vibr} and to the equivalent parameters ω_{system} and ζ_{eq} is between the variation in values for high vibration frequencies and those for low frequencies.

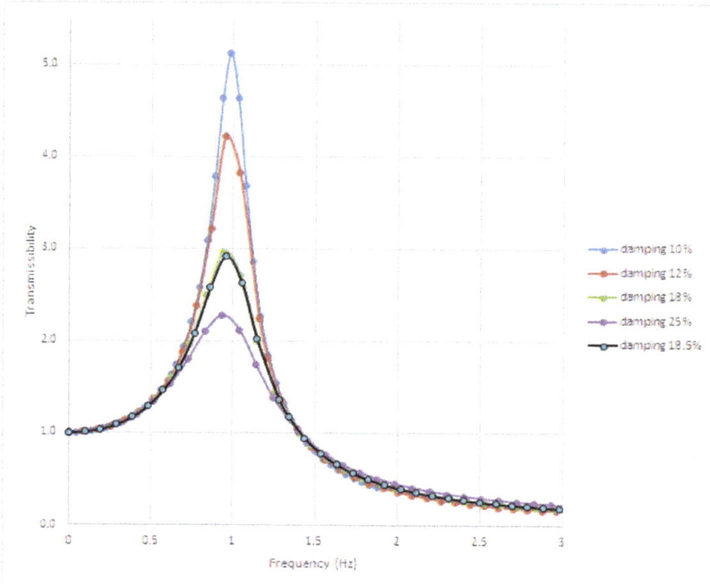

Figure 5. Transmissibility–vibration ratio relationship.

4. Conclusions

For structural systems with protection against seismic actions or vibrational actions that have linear viscous dissipation devices, the requirement to assess the equivalent modal amortization ratio arises. This article presents a novel method of assessment, computation and interpretation of the equivalent or effective modal amortization $\zeta_{\text{ef}}^{(\text{r})} \equiv \zeta_{\text{eq}}^{(\text{r})}$, called the compound damping of the structural assembly for the r mode of vibration. The calculated relations were verified on small-scale models in the laboratory at ICECON Bucharest and at the Solid Mechanics Institute of the Romanian Academy. Also, for a 1200 m viaduct constructed in Romania, dynamic analyses of behavior during seismic actions were performed, as well as the assessment of the modal compound damping for the fundamental mode at the natural frequency of 0.52 Hz [14,19,29]. The variety in the isolators was necessary due to the static equilibrium conditions of the bridge girder as well as the entire assembly in order to ensure horizontality, with the static equilibrium equations being applied to an elastically supported rigid body in the linear domain.

The maximum value of the elastic horizontal displacement smaller than the static deformation (δ_{static}) proves the hypothesis of elastic linearity in the dynamic regime. Consequently, the objective of this article to approach compound damping under the specified assumptions is fully achieved.

As a result of the symmetry studies of the laboratory tests and case study evaluations, the following conclusions can be formulated:

(a) For a structural system containing several anti-seismic or anti-vibratile devices distinct from one another, with elastic characteristics k_i and damping ratios ζ_i, where i is the order index, i = 1. . .n, based on Relation (6), the (effective) equivalent amortization/damping rate $\zeta_{\text{ef}}^{(\text{r})} \equiv \zeta_{\text{eq}}^{(\text{r})}$ can be determined for each r vibration mode;

(b) If the structural system has a dominant displacement in the instantaneous alternative translation in a certain direction and elastic devices and/or the devices are different, with parameters k_i,ζ_i, as a result of the parallel connection, then the calculation for $\zeta_{\text{eq}}^{(1)} \equiv \zeta_{\text{ef}}^{(1)}$ at a fundamental mode with $\omega_n = 2\pi\zeta_n$ can be performed using Relations (9) or (10);

(c) This case study shows that natural modal amortizations $\zeta_{\mathrm{i}}^{(1)}$, $i = 1\ldots4$, are significantly influenced by the stiffness ratios as $\frac{k_j}{k_1}$ with $j = 2\ldots4$;

(d) ζ_{eq} can be obtained from permutations in the $\frac{\mathrm{k_i}}{\mathrm{k_1}} - \zeta_i$ relationship for a specific situation;

(e) For large constructions, the use of only one type of seismic isolation device is not the optimal economical solution, and using more than one type could be more acceptable with a certain combination of technical characteristics;

(f) Some estimative values for the equivalent circular frequency ω_{eq} are determined considering the existing correlations between the defining quantities (mass, stiffness, and damping), and are of interest, especially in the case of eccentricities between the center of mass and center of rigidity;

(g) We determined the seismic response considering the Vrancea 1977 accelerogram for the critical damping ratios of 5% and 18.5%, and the results highlight a different dynamic behavior of the system with its natural vibration frequency of 0.52 Hz, in terms of absolute accelerations and relative displacements;

(h) The maximum value of the elastic horizontal displacement $d = 0.25$ m is smaller than $\delta_{static} = 0.50$ m, but larger than the maximum displacement equal to 0.20 m, considering a 0.185 value for damping;

(i) The performance of an isolation system was determined by the transmissibility factor of the system so that the variation in the T_i factor was found to be in correlation with the determined equivalent amortization.

In the situation where all anti-seismic devices are identical elastically and viscously, the amortization of a single device is obtained, that is, $\zeta_{\mathrm{eq}}^{(1)} = \zeta_{\mathrm{ef}}^{(1)} = \zeta_{1}^{(1)}$ for a fundamental vibration mode. Given the above, for complex structural systems that include distinct anti-seismic/anti-vibration devices in their dynamic scheme, the compound amortization rate can be assessed on the significant vibration modes. There are procedures in the specialized literature for defining the optimal values of the mechanical and geometrical parameters from device stiffness, maximum displacement, and vertical load, but this paper evaluates the degree of global amortization for the entire construction assembly (structural system and seismic devices), with a case study regarding an important construction. At the national level, the adoption of such effective measures using seismic isolation devices to record lower effects in the event of a future severe earthquake in Vrancea is a sustainable strategy.

As a future approach, the case of the non-linear regime should also be studied because the proposed approach is focused on the calculation of a global damping ratio in the linear–elastic regime. Also, the case of a system with several degrees of dynamic freedom subjected to a dynamic motion caused by an earthquake is certainly of interest. Monitoring the dynamic behavior using seismic instrumentation is another aspect to be addressed in the future (within a program like the one currently funded by the Romanian Ministry of Research, Innovation and Digitalization [32]).

Author Contributions: Conceptualization, P.B.; writing—original draft preparation, P.B.; writing—review and editing, P.B., D.D. and C.-S.D.; resources, P.B. and C.-S.D. All authors have read and agreed to the published version of the manuscript.

Funding: This research received external funding support from the project "Integrative concept for the digital analysis of data from the large-scale seismic monitoring of the national territory and built environment, aimed for rapid identification of the destructive potential of seismic events occurring in Romania and in the adjacent regions" (Nucleus Programme PN 23 35 01 01), financed by the Romanian Ministry of Research, Innovation and Digitalization.

Data Availability Statement: The original contributions presented in the study are included in the article, further inquiries can be directed to the corresponding author.

Conflicts of Interest: The authors declare no conflicts of interest.

References

1. Pavel, F.; Vacareanu, R.; Radulian, M.; Cioflan, C. Investigation on directional effects of Vrancea subcrustal earthquakes. *Earthq. Eng. Eng. Vib.* **2015**, *14*, 399–410. [CrossRef]
2. Pavel, F. An updated perspective of the impact of the 1940 Vrancea earthquake on design and construction practices in Romania. *Buildings* **2024**, *14*, 1152. [CrossRef]
3. Fujino, Y.; Siringoringo, D.; Kikuchi, M.; Kasai, K.; Kashima, T. Seismic Monitoring of Seismically Isolated Bridges and Buildings in Japan—Case Studies and Lessons Learned. In *Genetic Enhancement of Crops for Tolerance to a Biotic Stress: Mechanisms and Approaches*; Springer Tracts in Civil Engineering; Springer: Cham, Switzerland, 2019; Volume 1, pp. 407–447. [CrossRef]
4. Montes, K.; Dang, J. Seismic isolated bridge performance evaluation and design proposal using artificial neural network. *Earthq. Eng. Resil.* **2023**, *2*, 364–378. [CrossRef]
5. Makris, N.; Chang, S.P. *Effect of Damping Mechanisms on the Response of Seismically Isolated Structures*; Pacific Earthquake Engineering Research Center: Berkeley, CA, USA, 1998.
6. Makris, N.; Zhang, J. *Structural Characterization and Seismic Response Analysis of a Highway Overcrossing Equipped with Elastomeric Bearings and Fluid Dampers: A Case Study*; Pacific Earthquake Engineering Research Center: Berkeley, CA, USA, 2002.
7. Simionescu, C.S.; Leopa, A. The seismic response evaluation of a bridge deck, isolated by rubber bearings. *Acta Tech. Napoc. Ser. Appl. Math. Mech. Eng.* **2020**, *63*, 51–60.
8. Pastia, C.; Luca, S.G. Seismic Protection of a Realistic Bridge Structure Using Passive Friction Devices. *Bull. Polytech. Inst. Jassy Constr. Archit. Sect.* **2012**, *58*, 77–84.
9. Paulet-Crainiceanu, F.; Florea, V.; Luca, S.G.; Pastia, C.; Rosca, O.V. Analysis of practical application aspects for an active control strategy to civil engineering structures. In *Perspectives in Dynamical Systems II—Numerical and Analytical Approaches*; Awrejcewicz, J., Ed.; Springer Proceedings in Mathematics & Statistics; Springer: Cham, Switzerland, 2024; Volume 454. [CrossRef]
10. Zhou, X.Y.; Han, M.; Yang, L. Study on protection measures for seismic isolation rubber bearings. *ISET J. Earthq. Technol.* **2003**, *40*, 137–160.
11. Tonciu, O.; Bratu, P. Evaluation of the energetic parameters of viscous dissipation for dynamic systems excited in harmonic regime. *Acta Tech. Napoc.-Ser. Appl. Math. Mech. Eng.* **2023**, *66*, 367–374.
12. Lee, S.; Feng, M.Q.; Kwon, S.J.; Hong, S.H. Equivalent Modal Damping of Short-Span Bridges Subjected to Strong Motion. *J. Bridge Eng.* **2011**, *16*, 316–323. [CrossRef]
13. Bratu, P. Dynamic Rigidity of The Linear Voigt—Kelvin Viscoelastic Systems. *Rom. J. Acoust. Vib.* **2019**, *16*, 184–188.
14. Bratu, P. Modal Amortization Rate Equivalent to a Structural System with Elastomer Insulators. *Rom. J. Acoust. Vib.* **2020**, *17*, 27–31.
15. Potirniche, A. Assessments regarding effective configurations of vibration isolators based on displacements analysis. *Rom. J. Acoust. Vib.* **2015**, *12*, 126–131.
16. Sireteanu, T. Smart suspension systems. *Rom. J. Acoust. Vib.* **2016**, *13*, 2.
17. *SR EN 15129*; Anti-Seismic Devices. National Organization for Standardization—ASRO: Bucharest, Romania, 2018.
18. *SR EN 1337*; Support Devices for Structures. Part 3: Elastomer Bearing Devices. National Organization for Standardization—ASRO: Bucharest, Romania, 2005.
19. *Technical Solutions for Isolating the Viaducts for the "Transilvania" Highway in Romania*; Evaluation and Test Report; ICECON: Bucharest, Romania, 2021.
20. Chopra, A.K. *Dynamics of Structures: Theory and Applications to Earthquake Engineering*; Prentice Hall: Upper Saddle River, NJ, USA, 2012.
21. Tsai, H.C.; Kelly, J.M. Seismic response of heavily damped base isolation systems. *Earthq. Eng. Struct. Dyn.* **1993**, *22*, 633–645. [CrossRef]
22. Kelly, J.M.; Tsai, H.C. Seismic response of light internal equipment in base-isolated structures. *Earthq. Eng. Struct. Dyn.* **1985**, *13*, 711–712. [CrossRef]
23. Anajafi, H.; Medina, R.A.; Santini-Bell, E. Effects of the improper modeling of viscous amortization on the first-mode and higher-mode dominated responses of base-isolated buildings. *Earthq. Eng. Struct. Dyn.* **2020**, *49*, 51–73. [CrossRef]
24. Scutaru, M.L.; Vlase, S.; Marin, M.; Modrea, A. New analytical method based on dynamic response of planar mechanical elastic systems. *Bound. Value Probl.* **2020**, *2020*, 104. [CrossRef]
25. Hu, X.; Chen, Q.; Weng, D.; Zhang, R.; Ren, X. Estimation of Additional Equivalent Amortization Ratio of the Damped Structure Based on Energy Dissipation. *Adv. Civ. Eng.* **2019**, *2019*, 8052413. [CrossRef]
26. Suarez, L.E.; Gaviria, C.A. Equivalent frequencies and amortization ratios of buildings with viscous fluid dampers: A closed form formulation. In Proceedings of the 10 NCEE-Frontiers of Earthquake Engineering, Anchorage, AK, USA, 21–25 July 2014.
27. Bratu, P. Compound Amortization for Coupled Dynamic Systems. *Acta Tech. Napoc. Ser. Appl. Math. Mech. Eng.* **2020**, *63*, 239–242.
28. Bratu, P.; Mitu, A.M.; Vasile, O. Dissipation Capacity Evaluation for Neoprene Anti-Seismic Isolators Under Harmonic Dynamic Excitations. *Rom. J. Acoust. Vib.* **2011**, *8*, 67–71.
29. Murzea, P. Alternative Kinematic Assumptions for the Motions of Structures with Multiple Dynamic Degrees of Freedom Subjected to Seismic Actions. *Rom. J. Acoust. Vib.* **2014**, *11*, 53–58.
30. Radeș, M. Resonance Methods for the Dynamic Analysis of deformable structures. *St. Cerc. Mec. Apl.* **1993**, *32*, 607–633.

31. Meirovitch, L. *Principles and Techniques of Vibrations*, 1st ed.; Prentice Hall International Editions Series; Pearson: Piscataway, NJ, USA, 1996.
32. Integrative Concept of Digital Analysis of Data from Large-Scale Seismic Monitoring of the National Territory and the Built Environment, Intended for the Rapid Identification of the Destructive Potential Of Seismic Events Produced in Romania and in the Adjacent Regions Nucleus Programme PN23350101. Available online: https://incd.ro/proiecte-nationale/ (accessed on 30 July 2024).

buildings MDPI

Article

On-Site Measuring Robot Technology for Post-Construction Quality Assessment of Building Projects

Hangzi Wu [1,2,†], Minglei Ma [1,*,†], Yan Yang [1,2,*], Lifang Han [1] and Siyuan Wu [1]

[1] China Construction Eighth Engineering Division Co., Ltd., Shanghai 200112, China; wuhangzi@gmail.com (H.W.); hanlifang@cscec.com (L.H.); sy.wu92@gmail.com (S.W.)
[2] Key Lab of Structures Dynamic Behavior and Control of the Ministry of Education, Harbin Institute of Technology, Harbin 150090, China
* Correspondence: minglei.ma1983@gmail.com (M.M.); yangyan0615@cscec.com (Y.Y.)
† These authors contributed equally to this work.

Abstract: Post-construction quality assessment of building projects involves inspecting and verifying that completed construction works meet the specified standards. This process is traditionally conducted through manual methods, which can be inefficient and time-consuming. Existing measurement robots, typically integrating a robotic platform with 3D laser scanners, face challenges such as high storage demands, reliance on specialized post-processing software, and substantial costs. Additionally, robots with multiple sensors may face limitations in handling diverse measurement items. To address these issues, this article introduces a cost-effective and fully automated on-site measuring robot. A systematic approach was employed, including robot design, measurement algorithm development, validation experiments, and engineering applications. Firstly, a cost-effective hardware was designed, reducing expenses by 30% compared to commercial 3D laser scanners. Thereafter, the algorithm was developed by processing effective point cloud data to measure dimensions, wall evenness, alignments, floor heights, and corner angles, achieving a 90% reduction in data storage requirements. Subsequently, validation experiments were conducted, which verified the measurement accuracy of the developed robot. Furthermore, the robot was applied in two building projects, demonstrating a 40% improvement in efficiency over manual measurements and a minimum 50% reduction in labor costs. This investigation shows that the developed on-site measuring robot offers a practical and automated solution for post-construction quality assessment in building projects.

Keywords: post-construction quality assessment; on-site quality measurement; measuring robot

1. Introduction

Post-construction on-site quality assessment, conducted after the building structure is completed but before interior decoration begins, is a critical procedure for ensuring the quality of building projects, which is primarily carried out manually [1]. In the post-construction quality assessment of typical building projects, at least two inspectors are usually required: one to operate manual measurement tools (e.g., optical level, straight edge, measuring tape) and the other to write down the data [2]. Additionally, after the on-site measurements are completed, the engineer should input the data into electronic documents for storage in the office. This manual process has several shortcomings. Firstly, human errors may occur due to improper operation or the use of inaccurate inspection tools. Additionally, building projects typically involve tens of thousands of measurement items, such as column dimensions, wall evenness, wall alignments, and window sizes, rendering the process highly labor-intensive. Moreover, manually converting data into electronic documents for storage is time-consuming and carries the risk of data loss. In summary, the manual post-construction quality measurement process is time-consuming, labor-intensive, and dependent on manual efforts, highlighting the need for improvement.

Furthermore, as indicated by the United Nations (UN) population estimates and forecasts, the global population is expected to rise from 7.7 billion in 2019 to approximately 8.5 billion by 2030 and 9.7 billion by 2050 [3]. With the slowing down of global population growth, the construction sector is encountering serious global challenges, including sluggish productivity growth, rising costs, an aging workforce, and a shortage of skilled labor [4]. This highlights the necessity for the building construction industry to greatly enhance its productivity in order to address current and future demand challenges [5]. Since then, robotics, seen as a revolutionary innovation in construction technology, has gained growing interest from researchers, business leaders, and governments [6]. As the demand for construction robots surged after 2015, their designs have transitioned from large-scale machinery to more compact and mobile systems [7]. Meanwhile, the productivity improvements brought by industrial robots over the past decades have paved the way for new applications of robotics in on-site building construction [8–10], offering an alternative approach to updating the manual post-construction quality measurement process. With the development of vision-based sensors [11], researchers are increasingly focused on integrating machine vision techniques with robotics for autonomous measurement [12].

In recent years, researchers have developed automated on-site measuring robots to address the limitations of manual measurements, leveraging advancements in robotic technologies for civil engineering. One notable design solution involves combining 3D Laser Scanners with robot dogs. This method integrates the nimbleness of a robot dog with the extensive environmental data capture capabilities of a three-dimensional laser scanner. For instance, Trimble partnered with Boston Dynamics to equip Spot, the robot dog, with the ability to traverse hazardous construction sites for data collection in collaboration with Hensel Phelps [13,14]. Similarly, Foster + Partners utilized a 3D Laser Scanner along with Boston Dynamics' quadruped robot for data collection on a construction site. Additionally, Leica developed the measuring robot BLK ARC, which features a vision sensor mounted on a robot dog. A vision sensor was installed on a robotic dog to assist in performing frequent on-site inspections by repeatedly capturing the same data at set intervals, enabling efficient as-built verification [15]. These robots leverage the extensive measurement capabilities of 3D laser scanners and the adaptability of robot dog chassis to navigate complex construction site environments. However, the high volume of data they generate, reliance on advanced data processing software, and substantial costs—such as Spot's price of US $74,500 as listed on the Boston Dynamics online store—make them less feasible for widespread deployment.

Hence, robots that integrate multiple sensors (e.g., laser scanner, stereo vision) and motion platform have been designed to offer cost-effective and user-friendly solutions. For instance, Yan R et al. [16,17] developed a Quality Inspection and Assessment Robot known as QuicaBot. QuicaBot incorporates a thermal camera for detecting hollowness, a color camera for identifying cracks, a laser scanner for measuring evenness and alignments, and an inclinometer for assessing inclination. Different algorithms are applied for each of these five measurement parameters. QuicaBot is capable of comparing the inspection data with industry standards (CONQUAS® [18]) and conducting assessments to confirm the construction quality in different areas, thereby reducing labor requirements and enhancing efficiency to some extent. However, given that QuicaBot is primarily designed for use in completed buildings (post-interior decoration), its mobility platform is optimized for flat and well-prepared floors, which may limit its suitability for post-construction on-site quality measurement. Moreover, Nadeem Z et al. [19] suggested a method for gauging the distance to an adjacent wall utilizing an indoor mobile robot. Wang K et al. [20] developed a mobile measurement robot that includes the AVG trolley, an intelligent total station, an automatic leveling device, and a navigation system. Zahari M et al. [21] designed a Robot Measuring System by optical quadrature encoder to measure distances under various conditions, including both smooth and rough surfaces. Wang J et al. [22] introduced a unified mobile robotic measurement platform designed for precise and automated 3D measurements. This system consists of a mobile arm, a fringe projection scanner, and a stereo camera system, enabling it to capture point cloud data of objects, which is then processed to derive measurement parameters.

It can be found that current measurement robots, those relying on 3D laser scanners, face challenges related to the substantial storage requirements for point cloud data, reliance on specialized post-processing software, and elevated costs. Robots that utilize various sensors may also face limitations in the range of measurement items they can handle and often require specific working environments. To address these issues, this article developed an automated on-site measuring robot designed for post-construction quality assessment, aiming to provide a cost-effective and versatile solution for measuring a wide range of items.

The framework of the study is outlined as follows: Section 2 outlines the objectives of the robot. Section 3 details the robot's design, including the hardware system configuration and the development of key algorithms, and also discusses the validation experiments conducted to assess measurement accuracy. Section 4 describes the measurement process of the developed robot in building projects. Sections 5 and 6 elaborate on the robot's application and provide a summary of the findings and conclusions, respectively.

2. Objectives

Given the limitations of current on-site quality measurement practices in relation to evolving requirements and technological advancements, it is highly advisable to design and develop a robot capable of autonomously performing comprehensive on-site measurements for different measurement items. Specifically, this robot should be equipped for unmanned operations and precise non-contact measurements. For the current building projects, the robot must be able to measure dimensions (for columns, windows, doors), evenness, alignments, floor height, and corner angles. Furthermore, the robot should be engineered to automatically capture and store measurement data in accordance with the objects being measured. It should also integrate a construction quality assessment module capable of automatically comparing the measured data to predefined construction quality standards and generating an automatic evaluation of construction quality.

In summary, this investigation identifies three primary objectives for the designed robot:

(1) Unmanned oversight of the measurement process: the robot is capable of independently navigating and moving throughout the construction site, adapting to uneven and rugged terrain while executing its tasks.
(2) Automated measurement task planning: the robot can autonomously plan measurement tasks for site, including determining the measurement path, identifying measurement points and the corresponding parameters to be measured along the path.
(3) Non-contact acquisition of highly accurate measurement data: the robot should have the capability to obtain highly accurate measurement data using non-contact measurement methods.

3. Methodology

3.1. Design and Configuration of Hardware Systems

3.1.1. Overall Mechanical Design

Figure 1 presents the designed mechanisms for measuring robot. The robot's mechanisms primarily comprised the locomotion system, electrical system, 3D Scanner, pitch angle control module, and navigation system. The total hardware cost is approximately US $30,000, representing a reduction of approximately 30% compared to commercial 3D laser scanners (for example, the Trimble X7 Scanner 3D Scanning System is priced at around $45,000).

The locomotion system facilitates the robot's autonomous movement within the construction site, while the electrical system provides the necessary power for the robot to complete its measurement tasks. The 3D scanner is a critical component that captures point cloud data from the measured elements, providing essential data for measurement algorithms. It is connected to the pitch angle control module, which allows for adjustments of the scanner's angle, thereby enhancing the range of measurements. The 3D Scanner and pitch angle control module constitute the measurement system of the robot. Additionally, the navigation system facilitates unmanned movement. Detailed designs of each component will be presented in the following sections.

Figure 1. Designed mechanisms for measuring robot.

3.1.2. Locomotion System

As depicted in Figure 2, the locomotion system employs a four-wheel differential drive mechanism for precise control, enabling agile movement, excellent obstacle-climbing ability, and adaptability to complex environments. Each of the four wheels can be individually controlled in terms of speed.

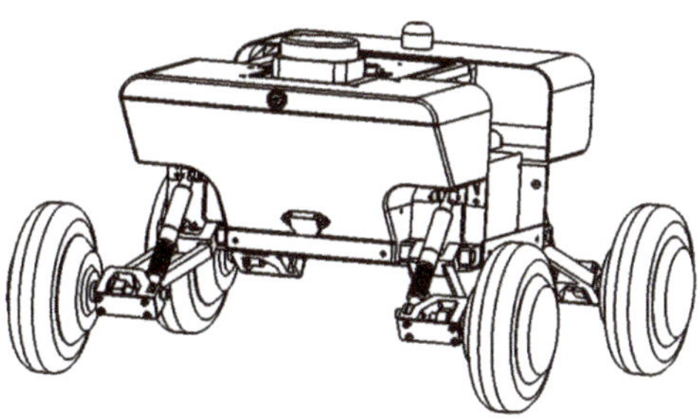

Figure 2. Locomotion system.

Considering the construction project sites and the storage requirements for the robot, the wheelbase width was set at 484 mm, with an axle distance of 430 mm, and the lowest point of the chassis was positioned 135 mm above the ground, which has been shown in Figures 3 and 4. Additionally, each wheel was fitted with independent suspension to reduce equipment vibrations, as illustrated in Figure 5.

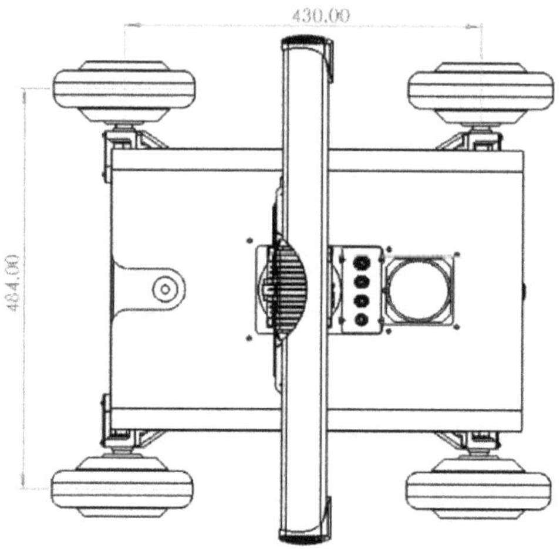

Figure 3. Vertical view of robot.

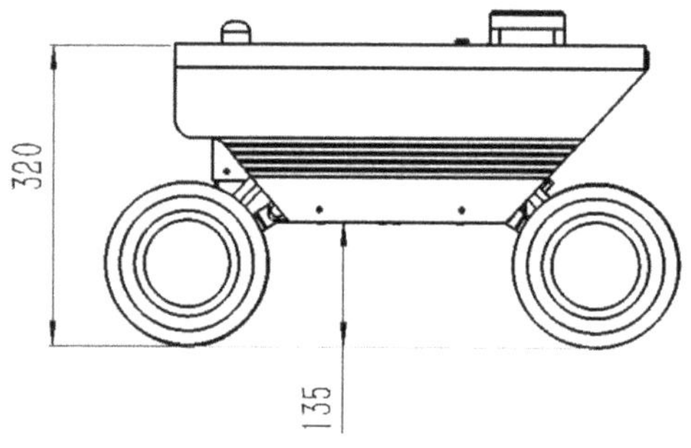

Figure 4. Lateral view of robot.

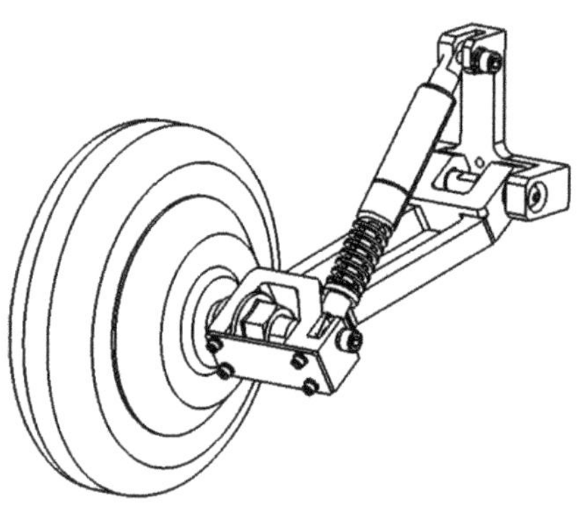

Figure 5. Suspension.

The locomotion system guarantees outstanding mobility and obstacle-crossing capability, achieving a maximum speed of up to 5 m/s. It can overcome ground obstacles with a height of up to 5 cm or a width of up to 3 cm. Videos demonstrating the robot navigating obstacles are available in Supplementery Files S1 and S2.

3.1.3. Electrical System

The electrical system primarily includes the drive motor, batteries, and their associated circuits.

To ensure the robot's capability to navigate slopes in the field, a force analysis was conducted to determine the appropriate drive motor model, as shown in Figure 6. Based on the calculations, the maximum power required for the robot to ascend a 15-degree slope while traveling at 0.3 m/s is 32.3 W.

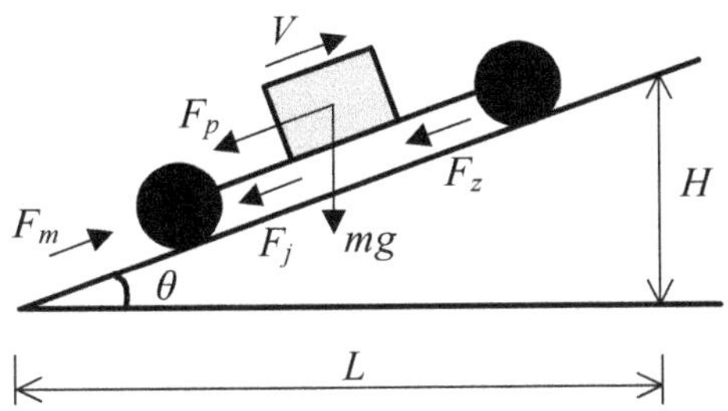

Figure 6. The force conditions of the robot while ascending a slope.

Based on these calculations, four 200 W drive motors have been selected for the robot. This motor configuration not only satisfies the current power requirements but also provides sufficient power for potential future upgrades.

Furthermore, a 24 V 17.5 Ah capacity battery is incorporated, and a Battery Management System (BMS) has been devised. The BMS discharge protection mechanism encompasses safeguards against short circuits, overloads, and over-temperature issues. It offers current and voltage relay protection in response to abrupt increases in current and sudden drops in voltage.

3.1.4. Measurement System

The 3D Scanner and pitch angle control constitute the measurement system of the robot. The 3D Scanner used is the PhoXi 3D Scanner L produced by Photoneo (Figure 7). This scanner is designed specifically for capturing detailed and precise measurements of stationary scenes. Its structured light projection method delivers accurate point cloud data, facilitating rapid localization of any specified component. The scanner's durability ensures high-quality scans, making it well-suited for the complex challenges encountered in building projects. The parameters of PhoXi 3D Scanner L have been listed in Table 1.

Table 1. D Scanner L.

No.	Parameters	Value
1	Resolution (3D points)	Up to 3.2 M
2	Scanning range	870~2150 mm
3	Optimal scanning distance	1239 mm
4	Scanning area at sweet spot	1080 × 772 mm
5	Point to point distance	0.524 mm
6	Calibration accuracy	0.2 mm
8	Scanning time	250~2750 ms
9	Dimensions	77 × 68 × 616 mm
10	Baseline	550 mm
11	3D points throughput	16 million points per second

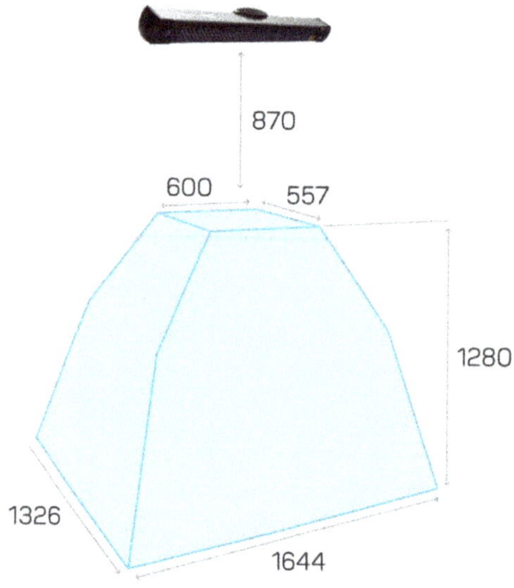

Figure 7. PhoXi 3D Scanner L.

Considering the extensive wall areas involved in on-site civil engineering measurements, a pitch angle control module was employed to extend the measurement range. This module allows the 3D Scanner to achieve various scanning angles, ensuring comprehensive coverage of the measurement object. The pitch control unit features a high-precision tilt sensor that provides real-time horizontal and tilt angles of the 3D scanner, facilitating accurate processing of measurement items based on point cloud data. The attached video in Supplementery File S3 demonstrates the robot adjusting the scanning angles to cover a broader range of measurement areas.

3.1.5. Navigation System

To attain autonomous navigation and positioning for the robot, the navigation system integrates LiDAR, an odometer, and a control center. The LiDAR utilized in the robot (RPLIDAR S2) employs the Time of Flight (TOF), offering a ranging radius of up to 30 m. It is positioned in the middle of the locomotion system, providing a 180° field of vision in front of the robot. This setup ensures comprehensive environmental scanning, enabling autonomous navigation and automatic obstacle avoidance, as demonstrated in Supplementery File S4.

The robot transmits odometer data at a frequency of 20 Hz, which encompasses its position and orientation. These data are sent to control the locomotion system through serial or network ports. In collaboration with autonomously developed navigation and positioning algorithms, precise information regarding the robot's position can be obtained. Subsequently, the control module generates speed and angular velocity data by integrating its position with the planned measurement task, facilitating more precise navigation to the designated measurement location.

3.2. Development of Key Algorithms

3.2.1. Method for Automated Measurement Task Planning

Given the robot's primary application in construction engineering, typical scenarios involve residential and office buildings characterized by standard structural and component design. In this context, an automated method for planning measurement tasks has been introduced by utilizing technology for converting and recognizing architectural drawing data.

A measurement points planning method (depicted in Figure 8) has been proposed to facilitate the automated planning of measurement tasks. Specifically, the method starts by identifying components such as doors, windows, and walls in the input drawings. Using a region segmentation algorithm, the processed map is divided into rooms. Following this, the five-point method (Figure 9) is employed to strategically plan measurement points for various parameters, including floor height, roof levelness, and ground flatness within the standard scene (positioning five points in an arrangement that includes up, down, left, right, and center within a room). If the initial five points fail to cover the entire measurement space, extra measurement points will be added based on the principles specified in Figure 10.

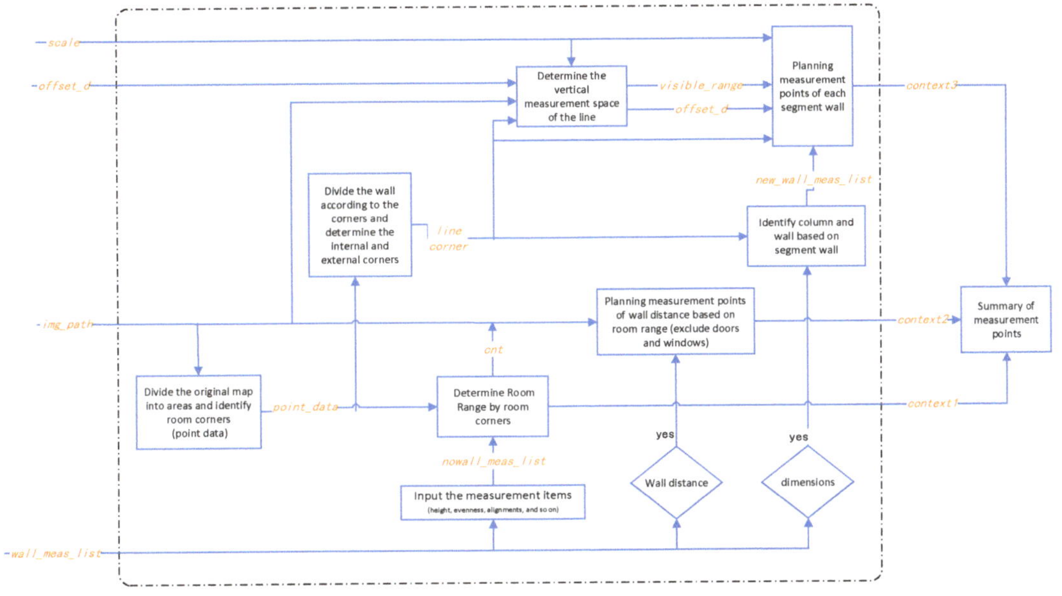

Figure 8. Measurement points planning method.

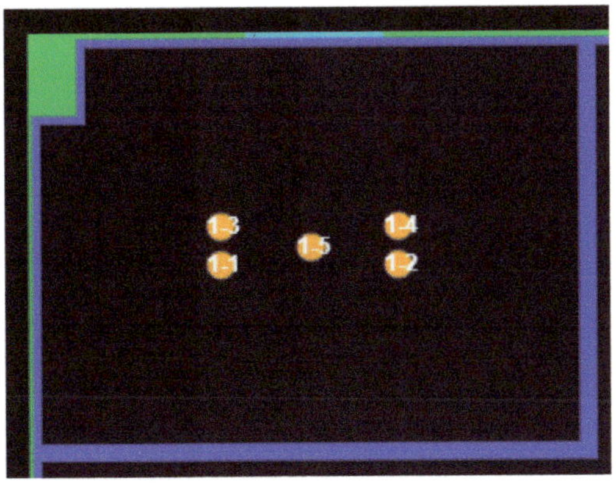

Figure 9. Five-point method.

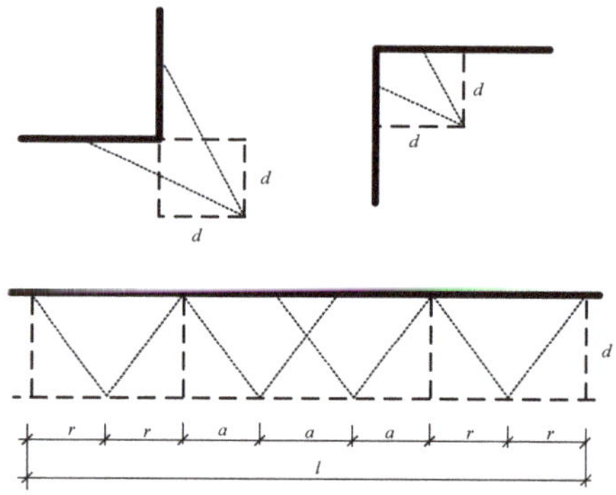

Figure 10. Arrangement of extra measurement points.

Once the measurement points have been determined, a measurement path optimization algorithm will be implemented to create an optimal measurement task that includes all points and parameters. Subsequently, the measurement task is devised, encompassing measurement points, measurement path, and measurement items.

The aforementioned method automatically generates measurement points and provides plans for measurement paths and content based on user-input drawing information, thereby significantly reducing the manual workload associated with planning measurement tasks.

3.2.2. Algorithms for Processing Measurement Data

The Point Cloud Library (PCL) was employed, and specific algorithms were developed for various measurement items based on data acquired from the PhoXi 3D Scanner L. These items include dimensions (such as column dimensions, window, and door sizes), wall evenness, alignments, floor height, and internal and external corner angles.

Moreover, the algorithm incorporates calibration and 3D data processing modules. For calibration, OpenCV is utilized to obtain high-precision spatial pose transformation parameters, enhancing the accuracy of the measurement function. The 3D data processing module utilizes various data processing methods, including 3D data downsampling, filtering, segmentation, principal component analysis, etc. The fundamental principles of these diverse algorithms are summarized in Table 2. Schematic diagrams processing evenness, alignments, and internal and external corner angles were primarily generated using the least squares method for plane fitting, as shown in Figure 11.

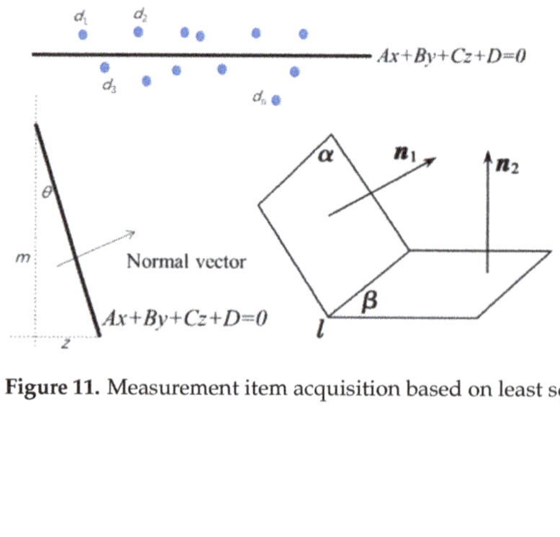

Figure 11. Measurement item acquisition based on least squares fitting plane.

Table 2. Fundamental principles of different algorithms.

Items	Fundamental Principles of Algorithms
Evenness	Based on the point cloud data, a plane is determined using the least squares method. The distances from all points in the calculation area to the plane are examined, and the extreme value is identified as the measurement output.
Alignment	Based on the point cloud data, a plane is fitted using the least squares method, and inclination sensor data is used to adjust the plane to align with the world coordinate system. The measurement output is calculated based on the normal vector of this plane.
Corner angles	Based on the point cloud data, two intersecting planes are determined using the least squares method. Following this, the angle between the normal vectors of these two planes is calculated and returned as the measurement output value.
Dimensions	Based on the point cloud data, a plane is determined using the least squares method. Subsequently, the boundary of the plane is traversed, and the dimensions of the boundary are returned as the measurement output value.
Height	Combining the height of the robot (1.2 m) and the point cloud depth.

In contrast to using all point cloud data, the proposed measurement algorithm utilizes key point cloud data for processing, as presented in Figures 12 and 13. Specifically, the blue points represent the obtained point cloud data, while the green points indicate the selected data, which accounts for about 10% of the total point cloud data. It significantly reduces storage costs while achieving the desired measurement functions. Taking the measurement of evenness as an example, the time from initiating the PhoXi 3D Scanner L to obtaining the measurement value is about 5 s.

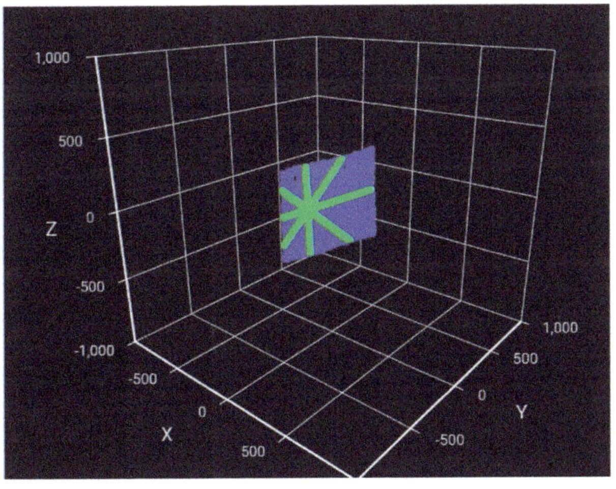

Figure 12. Key point of evenness.

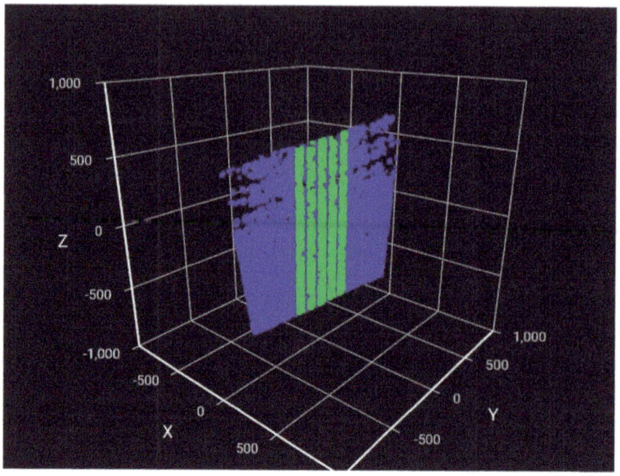

Figure 13. Key point of alignment.

3.3. Measuring Accuracy Evaluation

To evaluate the measurement accuracy of the developed robot, experiments were conducted to assess the alignment of a plumb glass plate (with a default alignment of 0 mm), the evenness of a tile floor (with default evenness set at 0 mm), and the external corner angle of a glass plate with a right angle. The measured results are listed in Table 3.

Table 3. Measured results of standard components (mm).

Items	1st Time	2nd Time	3rd Time	4th Time	5th Time	Average
Alignments	0.62	0.48	0.63	0.68	0.37	0.556
Evenness	0.65	0.87	0.56	0.82	0.77	0.734
External corner angle	0.71	0.83	0.79	0.7	0.69	0.744

The results indicate that the measurement accuracy of the developed robot is commendable. The average measurement accuracy for alignments, evenness, and external corner angle is within 1 mm.

Simultaneously, a further comparison was conducted between the robot and manual measurement methods. It involved using the developed robotic measurement system and conventional measuring tools, including straight edge rule, inside and outside right-angle measuring ruler, measuring tape, and total station. Varied construction quality control parameters for different measurement objects were acquired and presented in Table 4.

Table 4. Comparison between the robot and manual measurement methods (mm).

Items	Objects	Conventional Measuring Tools	Average Deviation	Maximum Deviation
Evenness	Wall	Straight edge rule	0.44	1.05
Alignments	Wall	Straight edge rule	0.99	1.95
Internal corner angle	Wall	Inside and outside right-angle measuring ruler	1.028	1.48
External corner angle	Wall	Inside and outside right-angle measuring ruler	1.12	1.46
Dimensions	Window	Measuring tape	1.99	2.47
Height	/	Total station	3.5	5

The data presented in the table indicate that the measured data obtained by the robot exhibit relatively minor discrepancies compared to manual measurement results for various items. Therefore, it can be concluded that the on-site measuring robot developed is well-suited for post-construction quality assessment of building projects.

4. Measuring Process

The hardware systems of the robots can be assembled and interconnected, enhancing the portability and transportability of the robot. This robot system includes a measurement system (PhoXi 3D Scanner L), a control terminal (for monitoring the robot's status, downloading measurement tasks, and receiving/uploading measurement data), the locomotion system, and connecting rods. As illustrated in Figure 14, each component can be packed into a portable box for easy transportation and later assembled to establish a functioning robot system.

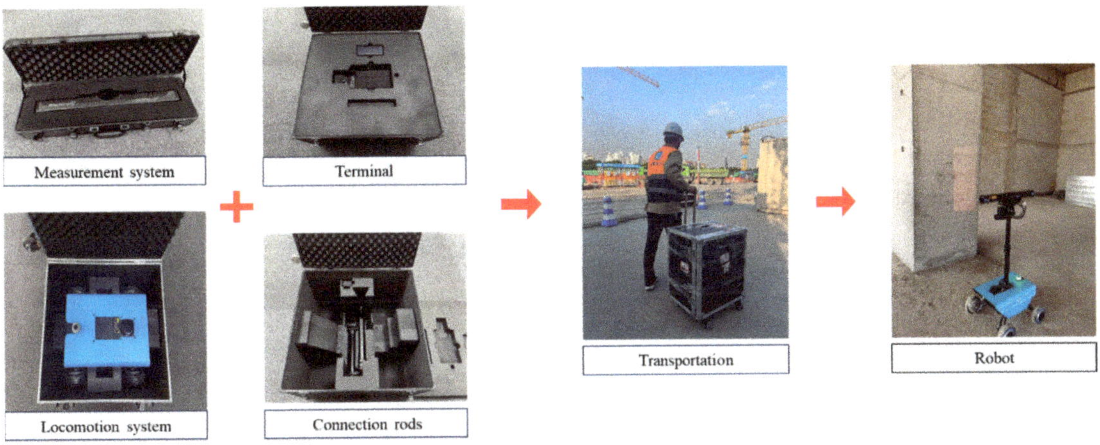

Figure 14. Assembly and transportation of the robot.

After completing the assembly of the robot, the measurement process follows the process outlined in Figure 15. The primary steps are as follows:

(1) Upload the architectural drawings to the web platform.
(2) Generate the measurement points on the web platform based on the architectural drawings.
(3) Create a measurement task that includes all planned measurement points.
(4) Upload the measurement task to the server.
(5) Download the measurement task onto the control terminal and send it to the robot.
(6) Specify the starting point of the robot's measurement task through the control terminal and issue the "Start Measurement Task" command.
(7) Robot executes the measurement task automatically.
(8) The control terminal receives the measurement data and uploads the data to the server once connected to the network.
(9) The web platform receives the data and automatically generates a measurement report with data evaluation.
(10) Print the report to complete the measurement process.

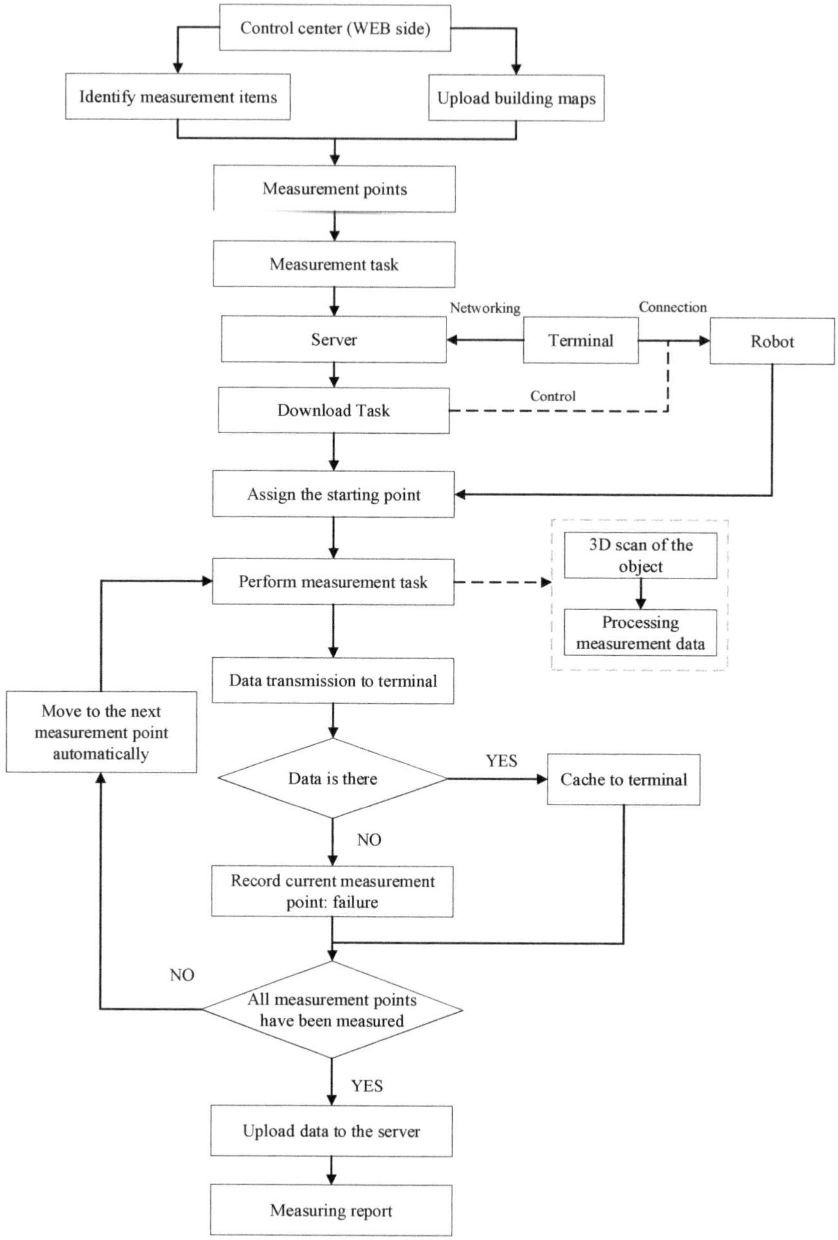

Figure 15. Measurement process.

5. Application

5.1. Shangxianfang Protective Renovation Project

The robot was applied in a building project in Shanghai, China, named Shangxianfang Protective Renovation Project (Figure 16). It is a structure comprising a steel pipe concrete frame and a concrete core tube. In this application, the developed robot was used to measure wall evenness, alignments, floor height, and the internal and external corner angles of the concrete core tube.

Figure 16. Shangxianfang Protective Renovation Project.

According to the measuring process presented in Section 4, the architectural drawings were first simplified to retain only the information about walls and openings (with walls represented by green lines and openings by yellow lines). The simplified drawing was then uploaded to the web platform to plan the measurement points and tasks. The original and processed drawings with designated measurement points have been presented in Figures 17 and 18. Subsequently, the measurement task was transmitted and assigned to the robot through the control terminal. The robot performing the measurement is illustrated in Figure 19.

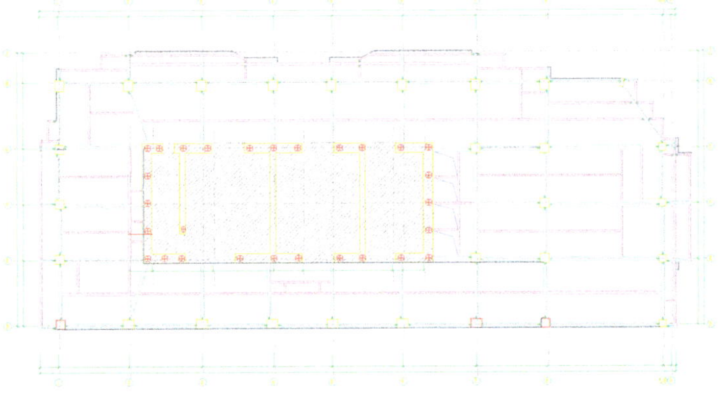

Figure 17. The original floor plan.

This measurement session involved a total of 24 measurement points, resulting in 131 measurement items. The data were automatically transmitted via the network, and a comprehensive measurement report was generated, as provided in Supplementery File S5.

The entire measurement process took approximately 30 min, with a rate of about 0.14 m^2/s. For comparison, manual measurement of a typical residence with a size of 85–125 m^2, including the same measurement items, would take at least 20 min (excluding data transcription) by skilled personnel, with a measurement speed of about 0.1 m^2/s.

Additionally, the manual measurement process typically requires collaboration between two or more inspectors, while the robot reduces the workforce to just one operator. Therefore, compared to manual measurement, the on-site measuring robot increases efficiency by approximately 40% and reduces labor costs by at least 50%.

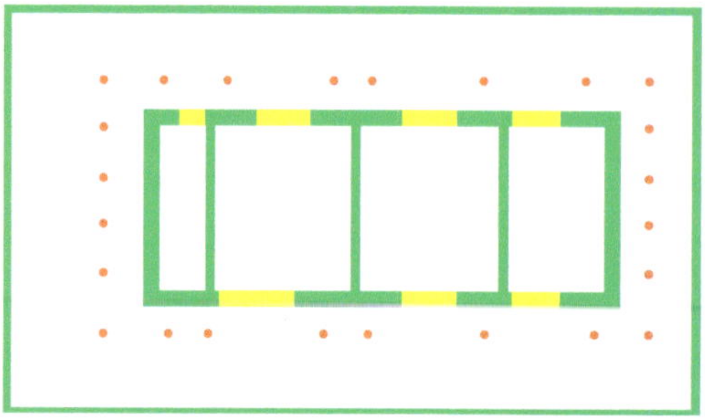

Figure 18. The processed floor plan.

(**a**) Measuring corner angle (**b**) Measuring wall evenness and alignments

Figure 19. The robot during the measurement.

5.2. Shenzhen International Convention Center

The robot was also applied in another building project in Shenzhen, China, named the Shenzhen international convention center. As shown in Figure 20, the building consists of steel–concrete columns, steel beam frames, and reinforced concrete shear walls.

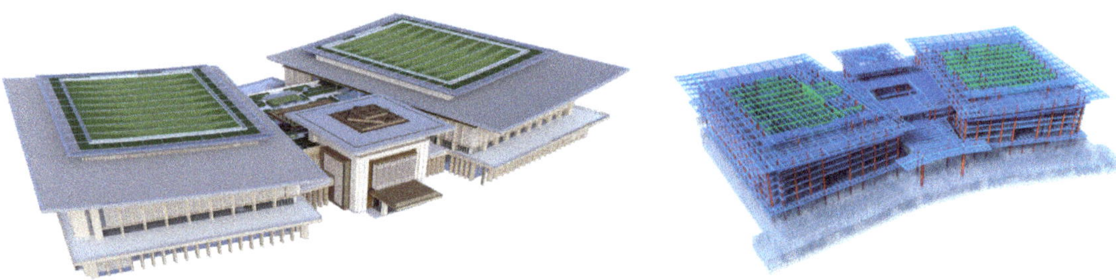

(**a**) Architecture (**b**) Structure

Figure 20. Shenzhen international convention center.

The robot was employed to assess the evenness and alignment of shear walls in the basement and brick walls in the above-ground structure. The measuring process follows the same steps as shown in Figure 15, with further details provided in Supplementery File S6. It can be seen that the robot autonomously navigated the site until it reached each designated measurement point, after which the task was executed, and the data were transmitted back to the terminal.

The measurement results are illustrated in Figures 21 and 22, with red arrows indicating areas that fail to meet construction quality standards. As presented, the 2D distribution of the wall measurement results, generated via the web interface, clearly illustrates the quality distribution of the walls, which allows on-site managers to easily identify areas with poor construction quality and take appropriate corrective actions.

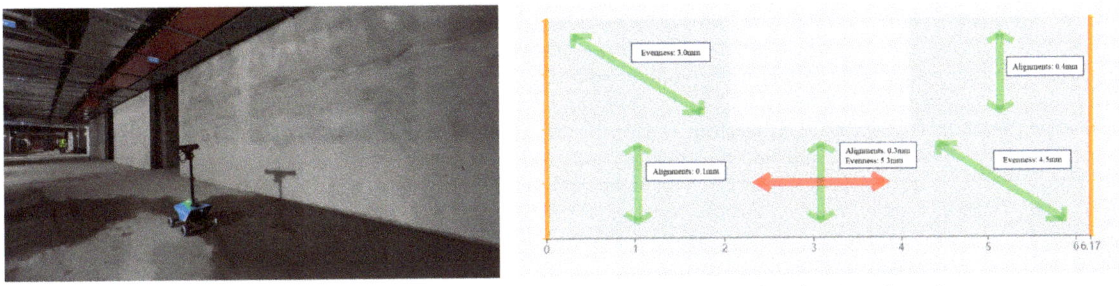

(**a**) Measurement (**b**) 2D distribution of results

Figure 21. Measurement of shear walls in the basement.

(**a**) Measurement (**b**) 2D distribution of results

Figure 22. Measurement of brick walls in the above-ground structure.

6. Limitations

Despite the significance of this investigation, there were still some limitations. For instance, the weight of the robot (33.2 kg) poses challenges in situations such as stair navigation or elevator use, where manual lifting is required. Reducing the robot's weight by using lightweight materials or optimizing its configuration will be necessary in the future. Additionally, the preprocessing of architectural drawings for task planning is time-consuming, and efforts will focus on refining and enhancing drawing recognition algorithms to streamline this process. Moreover, the robot cannot provide real-time output of measurement data during operation, as data are only uploaded upon task completion. This creates a risk of data loss from unexpected issues like power failures or obstacles. Developing more robust data-saving mechanisms and backup solutions would help ensure reliable and continuous performance.

7. Conclusions

Post-construction on-site quality measurement, traditionally performed manually, is essential for ensuring the quality of building projects. Recent developments in measurement robots, particularly those incorporating robot dogs and 3D laser scanners, have aimed to improve efficiency. However, these systems often face challenges, including substantial storage requirements for point cloud data, reliance on specialized post-processing software, and high costs. Although some cost-effective robots with multiple sensors have emerged, they still encounter limitations in handling diverse measurement tasks and often require specific environmental conditions. Therefore, this article addresses these challenges by developing an automated on-site measuring robot designed specifically for post-construction quality assessment and applied it in two building projects. The conclusions have been drawn as follows:

(1) The robot is equipped with a comprehensive hardware setup, including a locomotion system, electrical system, 3D scanner, pitch angle control module, and navigation system. This configuration enables unmanned, mobile, and continuous measurement operations. It also developed an automated method for planning measurement tasks and processing data using algorithms based on the Point Cloud Library (PCL). It is estimated that that the system provides approximately a 30% reduction in total investment compared to commercial 3D laser scanners.
(2) The robot can measure various parameters, including dimensions (e.g., column sizes and window and door dimensions), wall evenness, alignments, floor height, and internal and external corner angles. The proposed measurement algorithms focus on retaining only effective point cloud data, leading to a reduction of about 90% in data storage requirements. The accuracy and reliability of these measurements have been validated through experiments.
(3) In practical applications, the robot demonstrated a 40% improvement in efficiency over manual measurements for similar tasks and achieved a minimum of 50% reduction in labor costs, highlighting the robot's effectiveness in enhancing measurement productivity and reducing operational costs. It also shows that the developed on-site measuring robot offers a practical and automated solution for post-construction quality assessment in building projects.

Overall, this investigation demonstrates the potential of the developed robot to offer significant improvements in both efficiency and cost-effectiveness, advancing the field of post-construction quality measurement.

Supplementary Materials: The following supporting information can be downloaded at: https://www.mdpi.com/article/10.3390/buildings14103085/s1.

Author Contributions: Conceptualization, Y.Y.; software, S.W.; validation, H.W. and S.W.; investigation, H.W. and L.H.; resources, M.M. and Y.Y.; data curation, L.H.; writing—original draft preparation, H.W.; writing—review and editing, H.W.; project administration, M.M.; funding acquisition, Y.Y. All authors have read and agreed to the published version of the manuscript.

Funding: This research was funded by Shanghai Sailing Program (No. 23YF1452200), CCEED Technology R&D Program (CCEED-2022-3-14, CCEED-2022-2-30), National Key R&D Program of China (Grant No. 2022YFC3802203).

Data Availability Statement: The raw data supporting the conclusions of this article will be made available by the authors on request.

Conflicts of Interest: All authors are employed by China Construction Eighth Engineering Division Co., Ltd.

References

1. Liu, L.; Chen, I.M.; Kayacan, E.; Tiong, L.K.; Maruvanchery, V. Automated construction quality assessment: A review. In Proceedings of the 2015 10th International Symposium on Mechatronics and its Applications (ISMA), Sharjah, United Arab Emirates, 8–10 December 2015; IEEE: Piscataway, NJ, USA, 2015; pp. 1–6.
2. Lee, A.Y.; Seo, H.C.; Park, E.S. Development of a Manually Operated Mobile Robot That Prints Construction Site Layouts. *Machines* **2022**, *10*, 1192. [CrossRef]
3. Desa, U.N. World population prospects 2019: Highlights. *N. Y. (US) United Nations Dep. Econ. Soc. Aff.* **2019**, *11*, 125.

4. Bock, T. The future of construction automation: Technological disruption and the upcoming ubiquity of robotics. *Autom. Constr.* **2015**, *59*, 113–121. [CrossRef]
5. Gharbia, M.; Chang-Richards, A.; Lu, Y.; Zhong, R.Y.; Li, H. Robotic technologies for on-site building construction: A systematic review. *J. Build. Eng.* **2020**, *32*, 101584. [CrossRef]
6. You, S.; Kim, J.H.; Lee, S.H.; Kamat, V.; Robert, L.P., Jr. Enhancing perceived safety in human–robot collaborative construction using immersive virtual environments. *Autom. Constr.* **2018**, *96*, 161–170. [CrossRef]
7. Chang, S.; Siu, M.F.F.; Li, H.; Luo, X. Evolution pathways of robotic technologies and applications in construction. *Adv. Eng. Inform.* **2022**, *51*, 101529. [CrossRef]
8. De Santos, P.G.; Estremera, J.; Garcia, E.; Armada, M. A service robot for construction industry. In Proceedings of the Proceedings World Automation Congress, Seville, Spain, 28 June–1 July 2004; IEEE: Piscataway, NJ, USA, 2004; Volume 15, pp. 441–446.
9. King, N.; Bechthold, M.; Kane, A.; Michalatos, P. Robotic tile placement: Tools, techniques and feasibility. *Autom. Constr.* **2014**, *39*, 161–166. [CrossRef]
10. Nematollahi, B.; Xia, M.; Sanjayan, J. Current progress of 3D concrete printing technologies. In Proceedings of the ISARC, International Symposium on Automation and Robotics in Construction, Lille, France, 3–7 June 2024; IAARC Publications: Lyon, France, 2017; p. 34.
11. Koch, C.; Paal, S.G.; Rashidi, A.; Zhu, Z.; König, M.; Brilakis, I. Achievements and challenges in machine vision-based inspection of large concrete structures. *Adv. Struct. Eng.* **2014**, *17*, 303–318. [CrossRef]
12. Poorghasem, S.; Bao, Y. Review of robot-based automated measurement of vibration for civil engineering structures. *Measurement* **2023**, *207*, 112382. [CrossRef]
13. Construction Index. *Fetch, Spot! Robot Dog Gathers Construction Data*; Construction Index: March, UK, 22 October 2021. Available online: www.theconstructionindex.co.uk/news/view/fetch-spot-robot-dog-gathers-construction-data (accessed on 20 October 2021).
14. Frangoul, A. *Boston Dynamics' Four-Legged Robot Has Been Gathering Data at a Major Construction Site in London*; CNBC: Fort Lee, NJ, USA, 12 November 2020. Available online: www.cnbc.com/2020/11/12/boston-dynamics-robot-gathering-data-at-construction-site-in-london.html (accessed on 20 October 2021).
15. Şahan, N.H. *Get to Know the Leica BLK2FLY and Leica BLK ARC*; Surveying Group: Osborne Park, WA, USA, 4 August 2022. Available online: https://shop.leica-geosystems.com/leica-blk/blk2fly/blog/get-know-leica-blk2fly-and-leica-blk-arc (accessed on 20 March 2022).
16. Yan, R.J.; Kayacan, E.; Chen, I.M.; Tiong, L.K. A novel building post-construction quality assessment robot: Design and prototyping. In Proceedings of the 2017 IEEE/RSJ International Conference on Intelligent Robots and Systems (IROS), Vancouver, BC, Canada, 24–28 September 2017; IEEE: Piscataway, NJ, USA, 2017; pp. 6020–6023.
17. Yan, R.J.; Kayacan, E.; Chen, I.M.; Tiong, L.K.; Wu, J. QuicaBot: Quality inspection and assessment robot. *IEEE Trans. Autom. Sci. Eng.* **2018**, *16*, 506–517. [CrossRef]
18. Authority, E.M. Building and Construction Authority. In *Handbook for Solar Photovoltaic (PV) Systems*; Building and Construction Authority: Singapore, 2017.
19. Nadeem, Z.; Nadeem, H.; Khan, J.A.; Ullah, A. Wall-Distance Measurement for Indoor Mobile Robots. In *Intelligent Sustainable Systems: Selected Papers of WorldS4 2021, Volume 2*; Springer Nature: Singapore, 2021; pp. 1–9.
20. Wang, K.; Li, D.; Li, L.; Chi, S. The Research on the Mobile Measuring Robot Technology. *J. Phys. Conf. Ser.* **2022**, *2402*, 012037. [CrossRef]
21. Zahari, M.; Mutalib, N.A.; Affendi, N.N.; Hashim, N.; Hadi, D.A.; Johari, S.H.; Ab Wahab, N.; Ahmad, S. Development of mobile robot for measuring distance using optical quadrature encoder. *Indones. J. Electr. Eng. Comput. Sci.* **2020**, *18*, 745–749. [CrossRef]
22. Wang, J.; Tao, B.; Gong, Z.; Yu, S.; Yin, Z. A mobile robotic measurement system for large-scale complex components based on optical scanning and visual tracking. *Robot. Comput.-Integr. Manuf.* **2021**, *67*, 102010. [CrossRef]

buildings

MDPI

Article

Study on Cubic Stiffness Nonlinear Energy Sink Controlling Dynamic Responses of Multi-Degree-of-Freedom Structure by Shake Table Tests

Qinhua Wang [1,2,*], Xueshuang Yi [2], Dongxu Yang [2] and Yi Tang [1,3]

1 State Key Laboratory of Building Safety and Built Environment, Beijing 100013, China; tangyi@cabrtech.com
2 School of Civil Engineering and Architecture, Southwest University of Science and Technology, Mianyang 621010, China; yxs@mails.swust.edu.cn (X.Y.); ydx@mails.swust.edu.cn (D.Y.)
3 China Academy of Building Research, Beijing 100013, China
* Correspondence: gbqhwang@swust.edu.cn

Abstract: A nonlinear energy sink (NES) has such advantages as controlling broader band responses and better robustness than conventional control devices like tuned mass dampers (TMDs). In this research, a cubic stiffness NES mitigating the dynamic responses of a multi-degree-of-freedom structure under white noise, harmonic and seismic excitations was tested using a shake table, and the influences of the parameters of the NES on vibration mitigation effects were investigated. The test results indicate that the NES has the same vibration mitigation effects on the acceleration responses under the white noise and harmonic excitations as TMDs, even though the mass ratio of the NES is less than that of a TMD. The average control effects of the NES on the acceleration responses of the structure under the effect of 100 seismic waves are better than those of a TMD, which indicates that an NES has better robustness than a TMD.

Keywords: nonlinear energy sink; multi-degree-of-freedom structure; vibration control; shake table tests

1. Introduction

Various passive vibration control devices have been proposed to absorb vibration energy from a primary structure and mitigate the dynamic responses of the structure in the past decades [1]. In the civil engineering field, tuned mass dampers (TMDs) are widely applied for practical engineering around the world to control the dynamic responses of structures [2], such as the Hancock building (241 m) [3] in Boston, the Taibei 101 building (508 m high) [4], the Shanghai Center Tower (632 m high) [5] and the Ping-An Finance Center (600 m high) [6], due to their effectiveness in vibration mitigation and low maintenance costs. However, TMDs have a disadvantage because of their narrow tuning band; hence, their control robustness decreases as the primary structure is subjected to broadband excitations or the dynamic characteristics of the primary characteristics change. The nonlinear energy sink (NES) [7], composed of a nonlinear spring, damping element and auxiliary mass, may overcome the disadvantages of TMDs. The key difference between NESs and TMDs is the nonlinear stiffness aspect of NESs, which means their response varies in a complex manner. This nonlinear characteristic allows NESs to effectively engage with a wide range of frequencies, enabling them to resonate across a broad spectrum [8]. NESs require little mass to achieve excellent vibration mitigation, but they need a limited initial energy range to be effective [9,10]. A lot of research on the various types of NES, controlling the dynamic responses of structures, has been carried out using theoretical, numerical and test methods. Gourdon et al. [11] studied the vibration response of an eight-story structure with a fundamental frequency of 0.83 Hz controlled by an NES under seismic and stochastic excitations through numerical analysis. The numerical results show that the NES had excellent robustness. Nucera et al. [12,13] used an attached vibro-impact NES (VI NES) to control the vibration response of a three-story structure with the first natural frequency of 4.82 Hz. The results of the shaking table test and numerical simulation consistently verified that the VI NES showed good vibration control effects and excited higher structure modes under seismic excitations. Vaurigaud et al. [14] proposed a method to optimize the NES parameters. The feasibility of this method was verified experimentally and by the numerical simulation of a four-story structure controlled by two parallel NES structures under harmonic excitation. Wierschem et al. [15,16] used NES to control the vibration response of a nine-degree-of-freedom structure with a fundamental frequency of 1.74 Hz under impact loads. These studies show that NESs have excellent vibration mitigation effects and can quickly reduce the maximum response of a structure. Luo et al. [17] studied two types of NES mitigating the dynamic responses of a structure through numerical simulation and shaking table tests. Their investigation demonstrates that the combined use of two different NES systems has certain vibration mitigation effects on different earthquake excitations. Wierschem et al. [18] installed a type III NES on a six-degree-of-freedom structure subjected to impact forces and carried out tests by shaking table and numerical simulation. Their results show that the type III NES can reduce most of the vibration responses of the structure. Wang et al. [19] employed a track nonlinear energy sink (track NES) and a single-sided vibro-impact track nonlinear energy sink (SSVI track NES) to mitigate the vibration responses of a 32-story steel structure with a fundamental frequency of 0.24 Hz under the action of an earthquake. The numerical results from their study show that track NES and SSVI track NES show excellent seismic reduction effects and are suitable as control devices for high-rise buildings. Through experimental testing and numerical analysis, Wang et al. [20] evaluated the vibration mitigation performance of a two-story structure with a fundamental frequency of 1.63 Hz controlled by track nonlinear energy sink under seismic and pulse excitations. Analytical and numerical studies indicate that the vibration responses of the structure were reduced faster compared to when a conventional NES was used. Later, Wang et al. [21] proposed an asymmetric nonlinear energy sink (Asym NES) and applied it to control the vibration responses of a three-degree-of-freedom structure and a six-degree-of-freedom structure under seismic excitations. Their research shows that Asym NES has strong robustness to structural frequency changes. Following on their research, Wang et al. [22] used track bistable nonlinear energy sink (track BNES) to control the vibration response of this structure under impulse and earthquake excitations; the results show that track BNES shows better robustness than Asym NES when the frequency of the main structure changes. Yang et al. [23] carried out a shaking table test in order to study the vibration mitigation effects of an NES on the dynamic responses of a five-story concrete structure with a fundamental frequency of 5.0 Hz in the X direction and 4.75 Hz in the Y direction under earthquake action. The NES played a good role in reducing vibration responses and showed its robustness. Ndemanou et al. [24] studied the vibration response of super-high-rise buildings controlled by an NES under wind loads through numerical simulation. Their results demonstrate that the control rate of the NES on the top floor vibration response of the building was over 79%, which is better than that of a TMD. Wang et al. [25] analyzed the vibration responses of a 300-meter-high building controlled by a nonlinear energy sink inerter (NESI) under wind loads. Their numerical simulation results show that the vibration control effect and robustness of an NESI are better than that of a tuned mass damper inerter (TMDI).

However, most of the previous work focused on the NES controlling dynamic responses of structures under only one type of excitation. Some structures, like high-rise buildings, may be subjected to different types of loads, for instance, earthquake excitations, turbulent buffeting (a broadband stochastic process) and vortex-shedding aerodynamic forces (similarly, harmonic excitations) during their designed lifespan. In this research, an NES controlling a multi-degree-of-freedom structure under white noise, harmonic excitations and earthquake excitations was designed and tested by the shake table method.

The effects of parameters of the NES on vibration mitigation effects of the structure under different types of excitations were investigated and compared with that of a TMD.

2. Test Introduction

2.1. Introduction of Multi-Degree-of-Freedom Structure Model

As shown in Figure 1a, the multi-degree-of-freedom structure model is composed of eight rigid plates, four elastic columns with a rectangle cross-section, which simulated the stiffness of the structure, and a base foundation. The plates, columns and base foundation are joined by rigid connection. The model is 125 mm in width and 716 mm in height, as shown in Figure 1b. The total mass of the model is 3.29 kg. The stiffness of the model is larger at the y-axis direction than that at the x-axis, and the dynamic responses at the x-axis may be greater than the responses at the y-axis; hence, the NES is designed to control the dynamic responses of the model at the x-axis. The five sensors from $1^{\#}$ to $5^{\#}$ were distributed along the height of the model, as shown in Figure 1b, to measure acceleration responses at the x-axis of the model according to the standard [26]. The free vibration tests in Figure 2 indicate that the first natural frequency of the model is 5.88 Hz at the x-axis and the corresponding damping ratio is 0.4%.

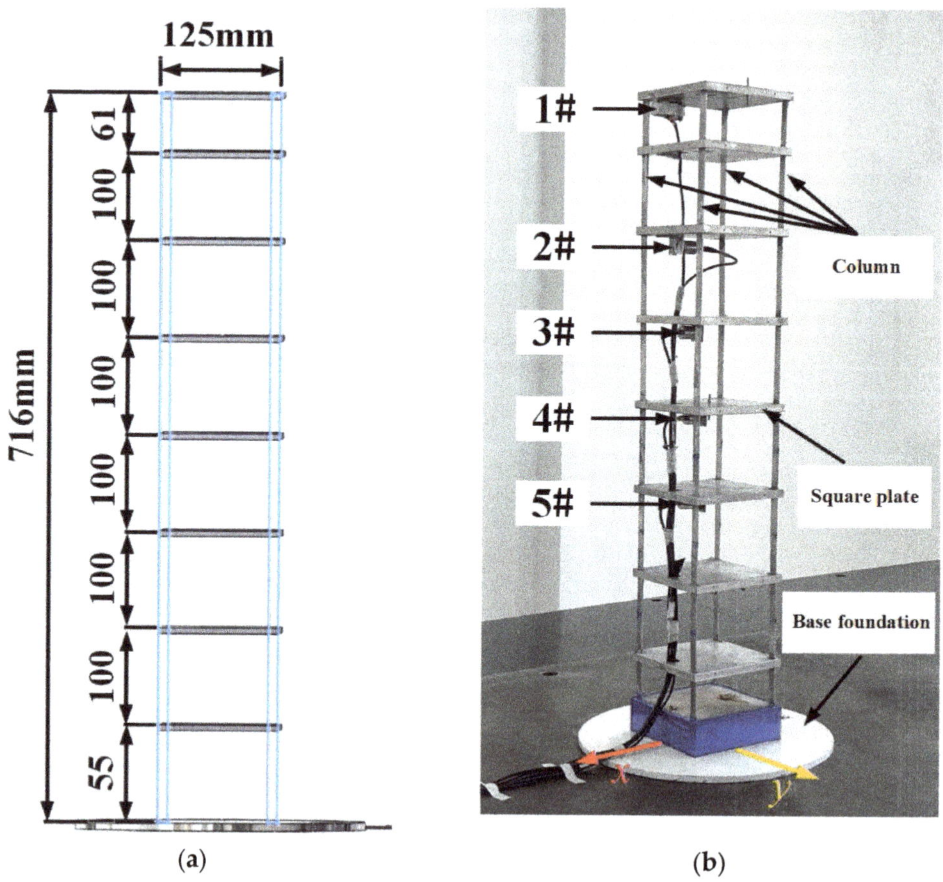

Figure 1. Multi-degree-of-freedom structure model. (**a**) Elevation of the model. (**b**) Physical model.

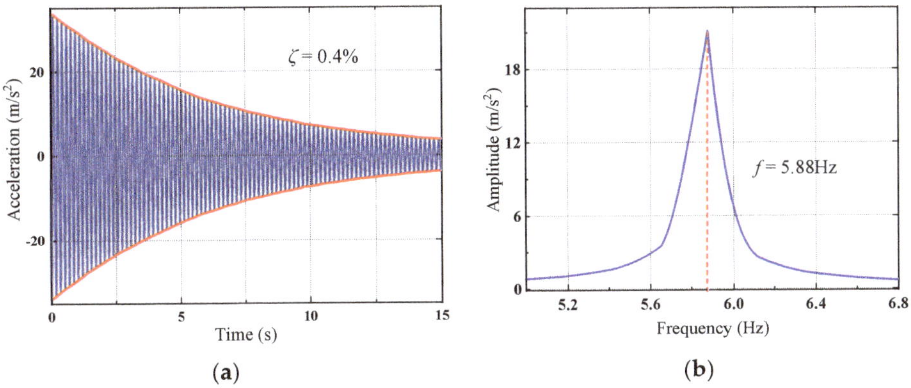

Figure 2. Free vibration tests. (**a**) Time history of free vibration. (**b**) Acceleration response spectrum of the free vibration.

2.2. Model of NES Controlling Multi-Degree-of-Freedom Structure

As shown in Figure 3a, the NES comprises a sliding rail, a pair of springs and an auxiliary mass, which can freely move along the x-axis. The NES was installed on the top floor of the model by rigid connection. The stiffness of the spring and auxiliary mass in the NES can be changed to investigate the vibration mitigation effects of the different parameters of the NES on the dynamic responses of the model under different excitations. Three sets of auxiliary mass, e.g., 100 g, 150 g and 200 g, were adopted in this research. The corresponding mass ratio μ, which is the ratio of the auxiliary mass of the NES to the mass of the structure, was $\mu_1 = 3.0\%$, $\mu_2 = 4.5\%$ and $\mu_3 = 6.0\%$, respectively. The stiffness of the NES was set as $k_1 = 105.0$ N/m and $k_2 = 330.0$ N/m, respectively. The variation in auxiliary mass of the NES changes the natural frequency of the structure–NES system, when the NES is locked. Figure 4 shows the frequency response function of the structure–NES system when the NES is locked. It indicates that the natural frequencies are 5.71 Hz, 5.64 Hz and 5.57 Hz when μ_1, μ_2 and μ_3 are equal to 3.0%, 4.5% and 6.0%, respectively. Table 1 gives the first three natural frequencies of the system when the NES locked. The first three natural frequencies are decreasing as the masses of the system are increasing.

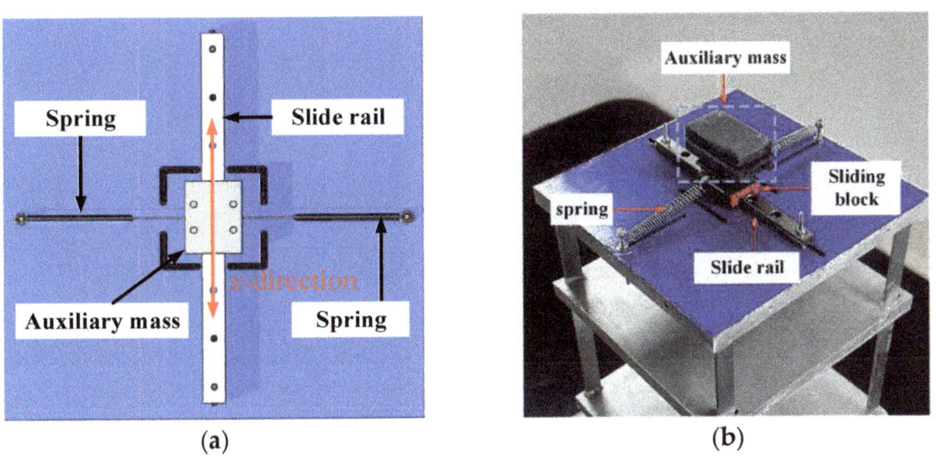

Figure 3. NES model diagram. (**a**) NES top view. (**b**) Diagram of the NES installed on the model.

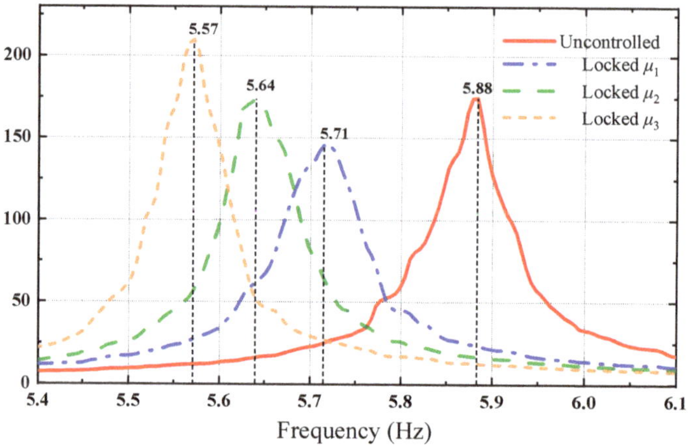

Figure 4. The frequency response function when the NES is locked.

Table 1. The first three natural frequencies of the system when the NES is locked.

Mass Ratio	f_1 (Hz)	f_2 (Hz)	f_3 (Hz)
0.0 (no NES)	5.88	18.75	31.50
μ_1 = 3.0%	5.71	18.40	31.17
μ_2 = 4.5%	5.64	18.26	30.90
μ_3 = 6.0%	5.57	18.14	30.65

2.3. Shake Table Tests of NES Controlling MDOF Structure

The MDOF of the structural model was installed on a shake table, which was controlled by a computer terminal, and accelerometers on the model and data acquisition device were used to measure and collect the dynamic responses of the model, as shown in Figure 5. The shake table has six degrees of freedom, and the platform is 2.5 m in length and 1.5 m in width. The range of working frequencies of the shake table is from 0.0 Hz to 30.0 Hz, and the parameters of the shake table are shown in Table 2. The model was installed on the shake table as shown in Figure 6. The time histories of the dynamic responses were measured by the accelerometers at 1000.0 samples per second.

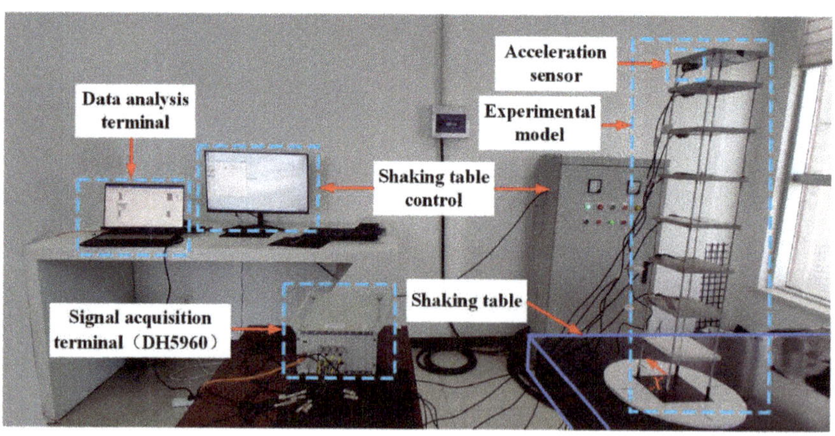

Figure 5. Apparatus for the tests.

Table 2. Parameters of the shaking table.

Direction	Range of Displacement (mm)	Direction	Range of Acceleration (g)
X	(−185, 260)	X	(2, 2)
Y	(−200, 200)	Y	(−2, 2)
Z	(−111, 111)	Z	(−2, 2)

Figure 6. The model installed on the shake table.

Tests of the NES with six sets of parameters controlling the dynamic responses of the structure were carried out to explore its optimal vibration mitigation effects. The corresponding tests of a TMD with optimal parameters were performed to compare its vibration mitigation effects with those of the NES. These test conditions are shown in Table 3. In Table 3, the mass ratios μ_1, μ_2 and μ_3 are 3.0%, 4.5% and 6.0%, respectively. The stiffness values k_1 and k_2 are 105.0 N/m and 330.0 N/m. The frequency ratio f and the mass ratio of the TMD are 0.98 and 4.5%, respectively.

Table 3. Test conditions.

Number	Test Conditions	Mass Ratio	Stiffness	Frequency Ratios	Conditions of Control Device
1	Uncontrolled	--	--	--	--
2	NES	μ_1	--	--	Locked
3	NES–1	μ_1	k_1	--	Unlocked
4	NES–2	μ_1	k_2	--	Unlocked
5	NES	μ_2	--	--	Locked
6	NES–3	μ_2	k_1	--	Unlocked
7	NES–4	μ_2	k_2	--	Unlocked
8	NES	μ_3	--	--	Locked
9	NES–5	μ_3	k_1	--	Unlocked
10	NES–6	μ_3	k_2	--	Unlocked
11	TMD	μ_2	--	--	Locked
12	TMD	μ_2	--	f	Unlocked

3. Results and Discussion

3.1. NES Mitigation Effects on Dynamic Responses of MDOF Structure Under White Noise Excitations

The NES and TMD controlling the dynamic responses of the MODF structure under the white noise excitations, with peak acceleration being 7.5 m/s^2, as shown in Figure 7, were investigated in this section. Figure 8 presents the time history of acceleration responses on the top of the structure with and without NES or TMD installation under white noise excitations. Figure 8 indicates that the NES and TMD can dramatically decrease the acceleration responses of the structure compared with the uncontrolled structure.

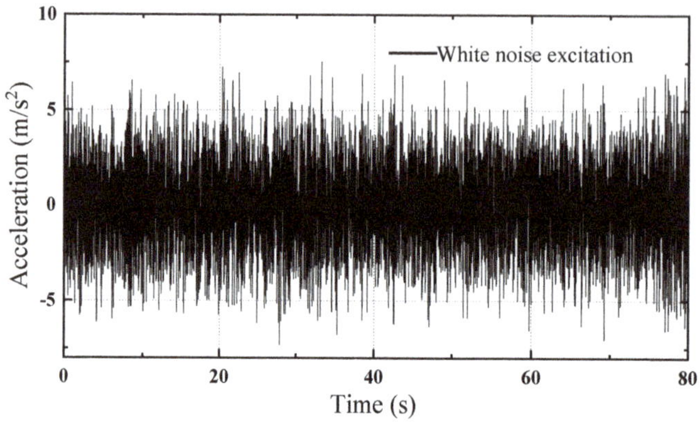

Figure 7. Time history of white noise excitation.

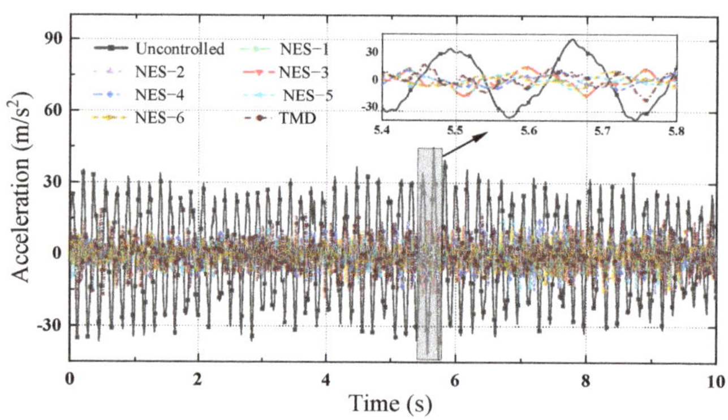

Figure 8. Acceleration time history curve of the top of the structure.

The time histories of these acceleration responses of the structure with locked and unlocked TMD were analyzed in the frequency domain by power spectrum density (PSD), as shown in Figure 9. Figure 9 indicates that the TMD can control the dynamic responses corresponding to the first vibration mode and cannot decrease the responses of the higher vibration modes. Figure 10 presents the results of the NES controlling acceleration responses on the top of the structure in the frequency and time domain by wavelet transform. Figure 10 demonstrates that the energy of the dynamic responses of the structure comes entirely from the first vibration mode when the NES is locked. The energy of the dynamic responses corresponding to the first vibration mode were reduced and the dynamic responses corresponding to higher vibration modes appeared as the NES was working. Compared with the results in Figure 10b,c,e,f,h,i, the NES with parameters μ_2 and k_2 in Figure 10f has the best mitigation effects on the acceleration responses on the top of the structure. The maximum values of the acceleration responses were investigated based on the time histories of the responses measured by the acceleration sensors 1# to 5# shown in Figure 1b. Figure 11 shows the variation in the maximum values along the height of the structure, and that the NES and TMD can effectively mitigate the maximum values of the acceleration responses on different heights of the structure. With regard to the responses on the top in Figure 11, the vibration mitigation ratio of these control devices reaches about 50.0%; NES–4 has the best mitigation effects among these control devices, which is consistent with the result in Figure 10, and its vibration mitigation ratio is 56.1%.

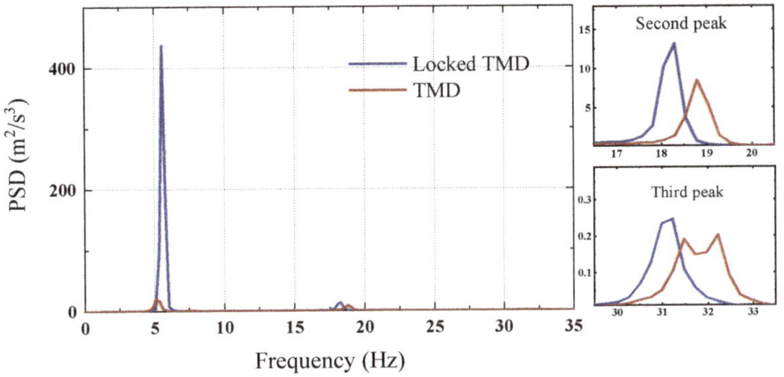

Figure 9. Acceleration response spectrum of the TMD.

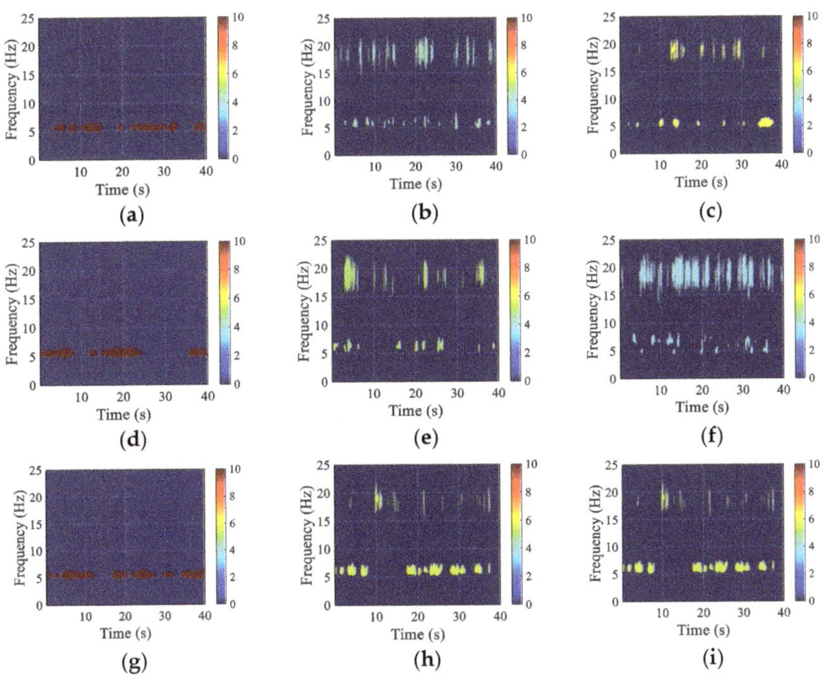

Figure 10. Wavelet analysis of NES controlling dynamic responses under white noise excitation. (**a**) Locked NES–1 or –2, (**b**) NES–1, (**c**) NES–2, (**d**) locked NES–3 or –4, (**e**) NES–3, (**f**) NES–4, (**g**) locked NES–5 or –6, (**h**) NES–5, (**i**) NES–6.

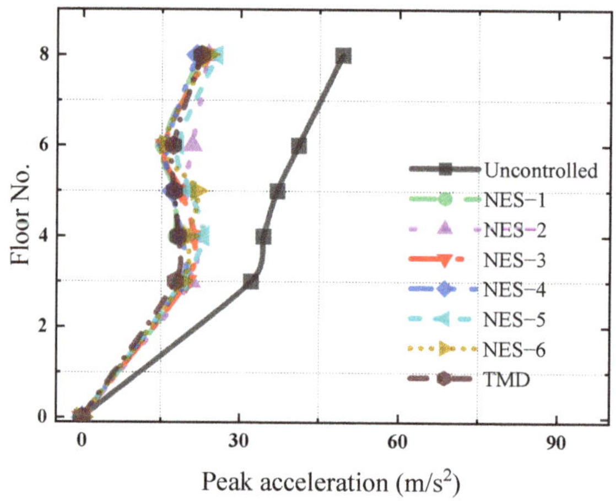

Figure 11. Variation in the maximum responses along the height of the structure.

3.2. NES Mitigation Effects on the Dynamic Responses of the MDOF Structure Under Harmonic Excitations

The vibration mitigation effects of the NES on the dynamic responses of the structure under harmonic excitation, the frequency of which is equal to that of the structure with the NES installation, were investigated in this section, as shown in Figure 12. It is noted that the frequency of harmonic excitation varies along the structure with the NES having a different mass ratio μ, which can change the natural frequency of the structure–NES system. Figure 13a–d present the time histories of the acceleration responses on the top of the structure when the NES or TMD is locked or unlocked. Both NES and TMD can decrease the amplitude of the dynamic responses from 50.0 m/s^2 to 5.0 m/s^2. Figure 13a–d indicate that both the NES with different parameters such as mass ratio μ and stiffness k and the TMD have the same vibration mitigation effects on the acceleration responses, except for NES–2, even though the mass ratio of the NES is smaller than that of the TMD. Figure 13 indicates that NES–2 has the worst vibration mitigation effects on the acceleration responses of the structure under harmonic excitations among these control devices. Its vibration mitigation ratio is 81.84%, as shown in Figure 13a.

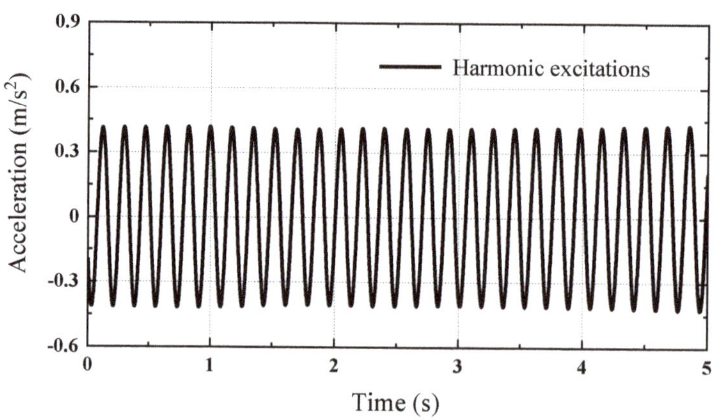

Figure 12. Time history of harmonic excitations.

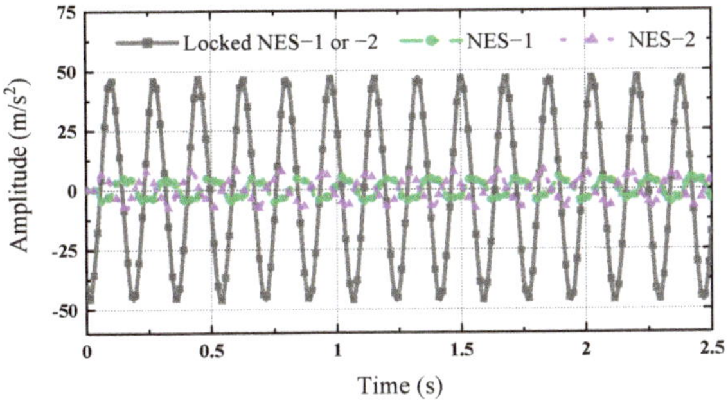

(a)

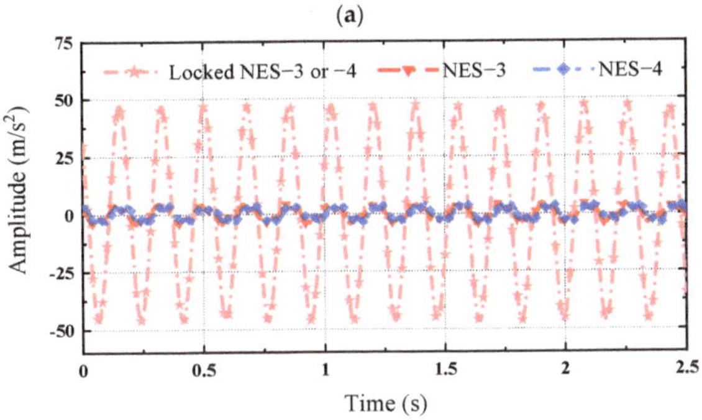

(b)

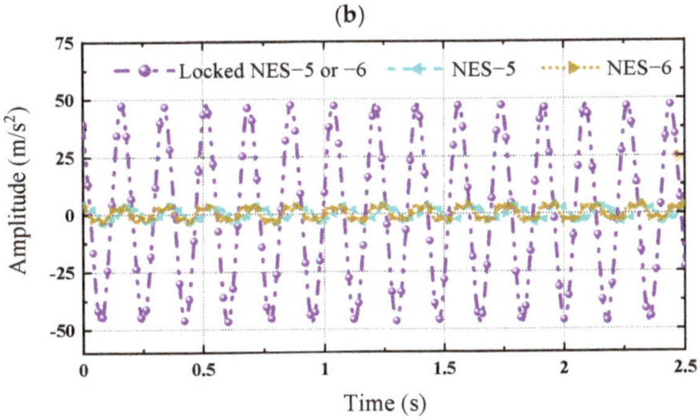

(c)

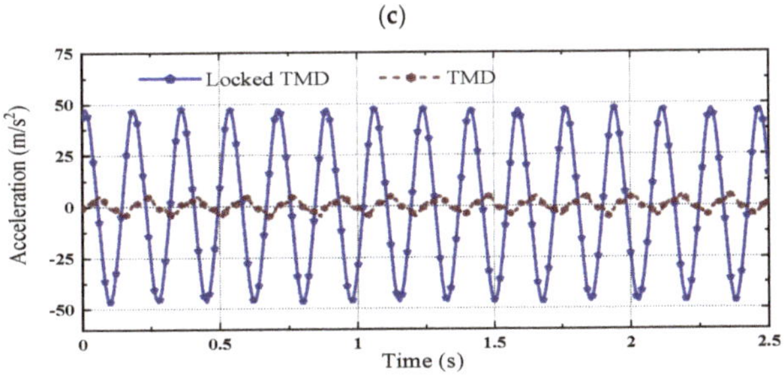

(d)

Figure 13. Time history of acceleration responses on the top of the structure with control devices. (**a**) Vibration mitigation effects of NES–1 or NES–2 under 5.70 Hz harmonic excitation. (**b**) Vibration mitigation effects of NES–3 or NES–4 under 5.65 Hz harmonic excitation. (**c**) Vibration mitigation effects of NES–5 or NES–6 under 5.60 Hz harmonic excitation. (**d**) Vibration mitigation effects of TMD under 5.65 Hz harmonic excitation.

The time histories in Figure 13 can be transformed into the frequency and time domains by wavelet analysis to investigate the vibration mitigation effects, as shown in Figure 14. Figure 14a,d,g,j show that the energy of the structural dynamic responses entirely comes from the first vibration mode and does not vary with time, when the NES and TMD are locked. As shown in Figure 14b,c,e,f,h,i,k, when the NES and TMD work, the energy of the dynamic responses corresponding to the first vibration mode significantly decreases, while the higher vibration mode around 18.0 Hz begins to be involved in the vibration. Figure 14 demonstrates that the different NESs have the same vibration mitigation effects as that of the TMD except for NES–2, and the vibration mitigation effects of NES–2 are worse than those of the others. The variation in peak values of the time history in Figure 13 along the height of the structure can be calculated as shown in Figure 15. For the responses on the top of the structure in Figure 15, all control devices except for NES–2 achieved a vibration mitigation ratio of over 88%, and NES–3 and NES–4 exhibited the best mitigation effects at 90.1% and 90.0%, respectively, which is consistent with the results from Figure 13.

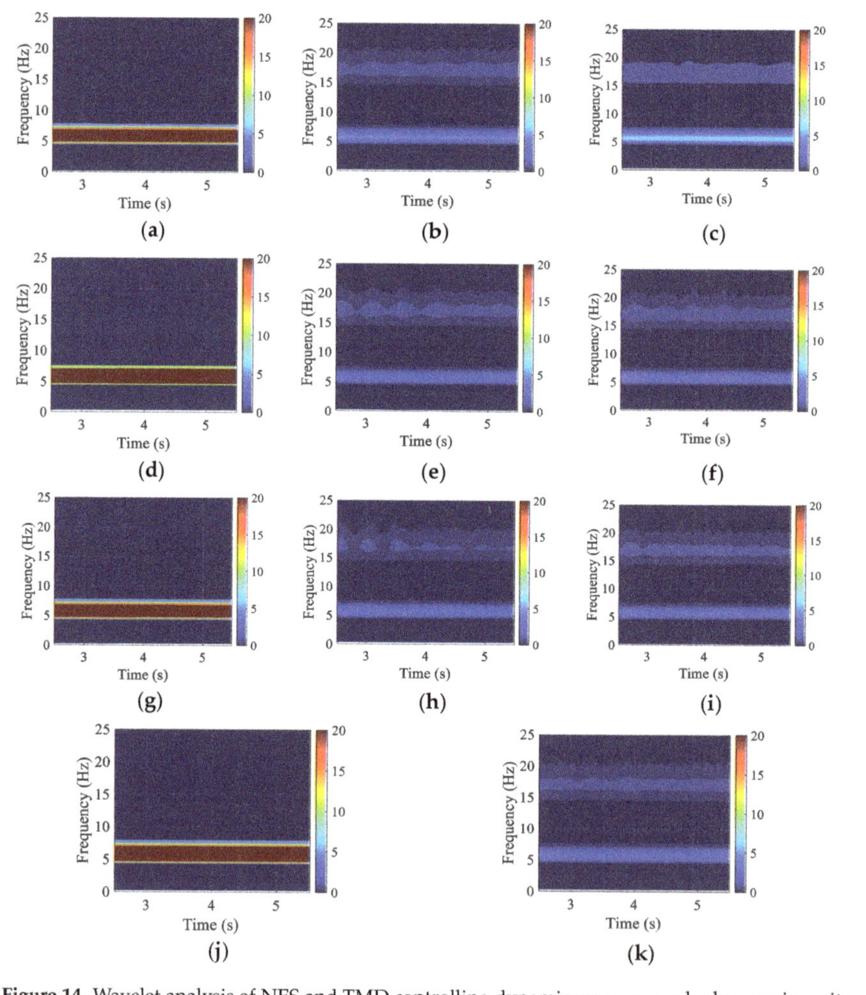

Figure 14. Wavelet analysis of NES and TMD controlling dynamic response under harmonic excitation. (**a**) Locked NES–1 or –2, (**b**) NES–1, (**c**) NES–2, (**d**) locked NES–3 or –4, (**e**) NES–3, (**f**) NES–4, (**g**) locked NES–5 or –6, (**h**) NES–5, (**i**) NES–6, (**j**) locked TMD, (**k**) TMD.

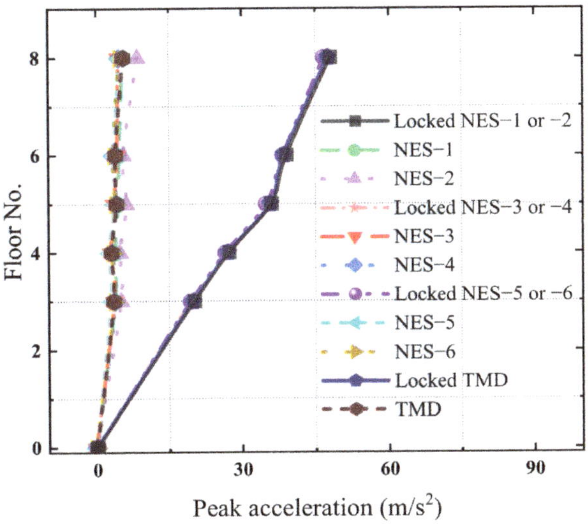

Figure 15. Variation in the maximum responses along the height of the structure.

3.3. NES Mitigation Effects on the Dynamic Responses of the MDOF Structure Under Seismic Excitations

The NES controlling the dynamic responses of the MODF structure under stochastic excitations was examined in this section. Figure 16 shows the spectral analysis of 100 stochastic excitations.

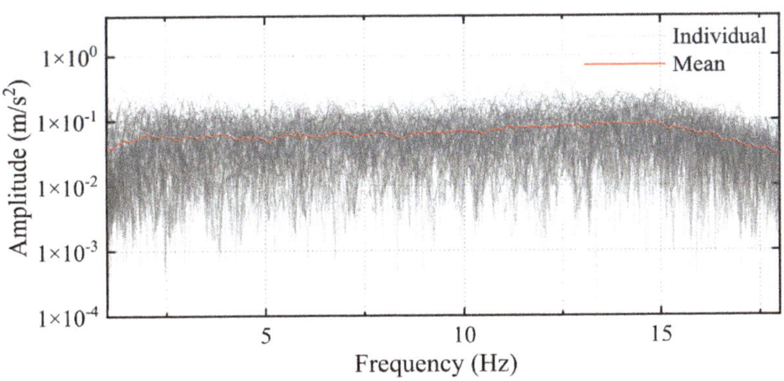

Figure 16. Spectral analysis of 100 stochastic excitations.

Root mean square (RMS) values of the time history of the acceleration responses measured by the acceleration sensors, which were distributed on the structure, under 100 seismic waves were estimated by a statistic method, as shown in Figure 17. In Figure 17, the average value of the uncontrolled structure denotes the mean RMS value of these responses under 100 seismic excitations. Figure 17 shows that the average mitigation effects of the NES on the RMS values of the acceleration responses of the structure under 100 seismic waves are better than that of the TMD. This indicates that the robustness of the NES is better than that of the TMD.

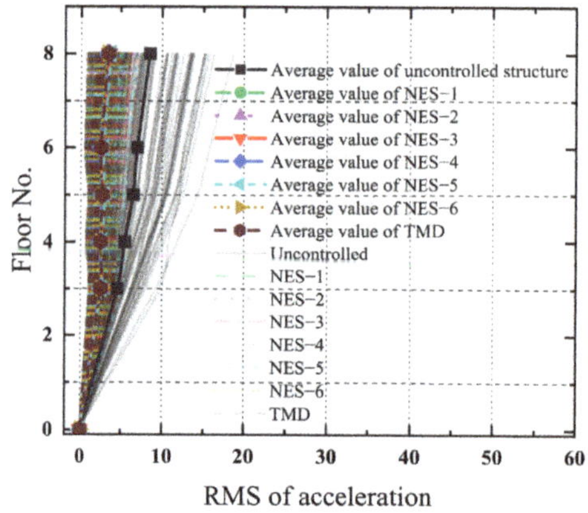

Figure 17. RMS values of acceleration responses.

One of the one hundred seismic waves was selected to show NES vibration mitigation effects on the dynamic responses of the structure under its excitation. Figure 18a displays the time history of the selected seismic wave. Figure 18b presents the PSD of the seismic waves, and indicates that its dominant frequency concentrates at the range of 1.3 Hz to 1.9 Hz.

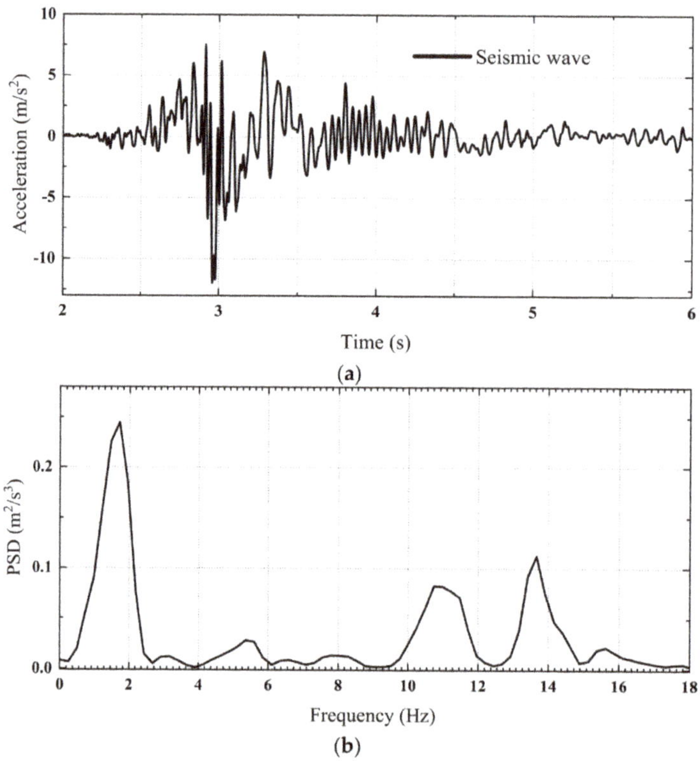

Figure 18. Seismic waves: (**a**) time history of one seismic wave and (**b**) PSD of one seismic wave.

The time histories of the acceleration responses on the top of the structure with and without NES or TMD installation under the seismic wave are shown in Figure 19. Figure 19 demonstrates that the NES and TMD have no considerable vibration mitigation effects on the acceleration responses of the structure under the seismic wave press before 3.5 s, and the vibration mitigation effects begin to become significant after this time. The acceleration response of the uncontrolled structure reaches 12.5 m/s^2 at 7.0 s. At this time, the responses decrease to 1.8 m/s^2, 0.54 m/s^2, 1.2 m/s^2, 0.3 m/s^2, 1.13 m/s^2, 0.15 m/s^2 and 0.66 m/s^2 corresponding to the control conditions of NES–1~NES–6 and TMD, respectively. Among them, NES–6 has the best vibration mitigation effects, and its mitigation ratio is 98.8%.

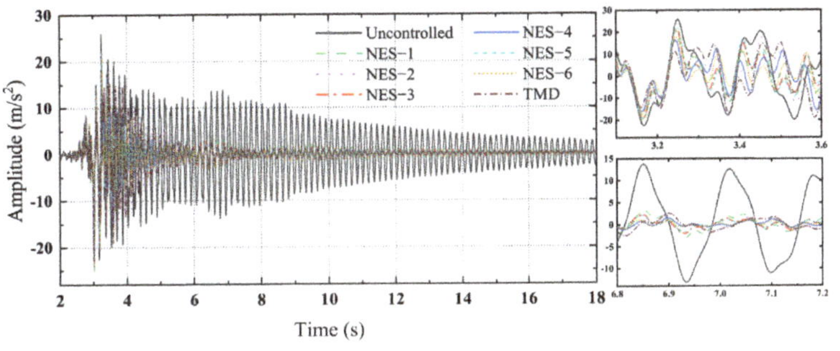

Figure 19. Time history of acceleration responses on the top of the structure under the seismic wave.

Through wavelet analysis, the time history in Figure 19 can be transformed into the frequency and time domain to investigate the vibration mitigation effects, as shown in Figure 20. Figure 20 shows that the dynamic responses mainly come from the first vibration mode, and the amplitude of dynamic responses corresponding to the first natural frequency around 5.88 Hz (the black line) decreases with time for the uncontrolled structure due to the values of stochastic excitations become smaller and smaller after 4.0 s, as shown in Figure 18a. It is clear that both NES and TMD can considerably decrease the structural dynamic responses corresponding to the first vibration mode, and do not affect the responses of the higher vibration modes, as shown in Figure 20.

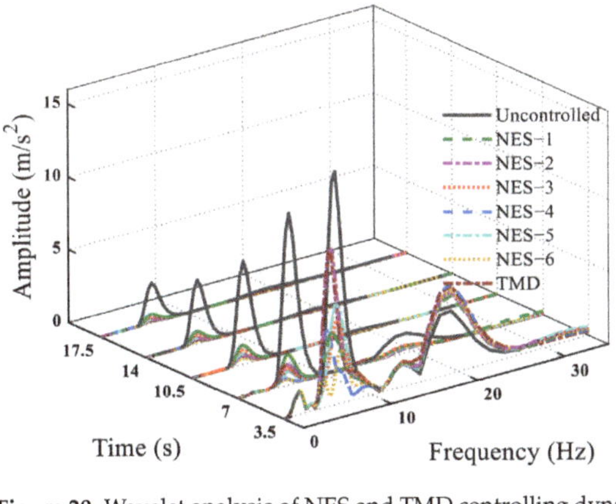

Figure 20. Wavelet analysis of NES and TMD controlling dynamic responses under one seismic wave.

The RMS values of the time history of the acceleration responses along the height of the structure under the seismic wave were estimated to evaluate the vibration mitigation effects of these control devices, as shown in Figure 21. Figure 21 manifests that these control devices can effectively alleviate the RMS values of the dynamic responses along the height of the structure. Among them, NES–3 has the best vibration reduction effects. The RMS value is 4.99 m/s^2 at the top of the uncontrolled structure. NES–1~NES–6 and TMD reduce the values to 2.3 m/s^2, 1.6 m/s^2, 1.52 m/s^2, 1.82 m/s^2, 1.73 m/s^2, 1.74 m/s^2 and 2.18 m/s^2, respectively, and the corresponding vibration mitigation ratios are 54.1%, 67.8%, 69.4%, 63.5%, 65.3%, 65.0% and 56.3%.

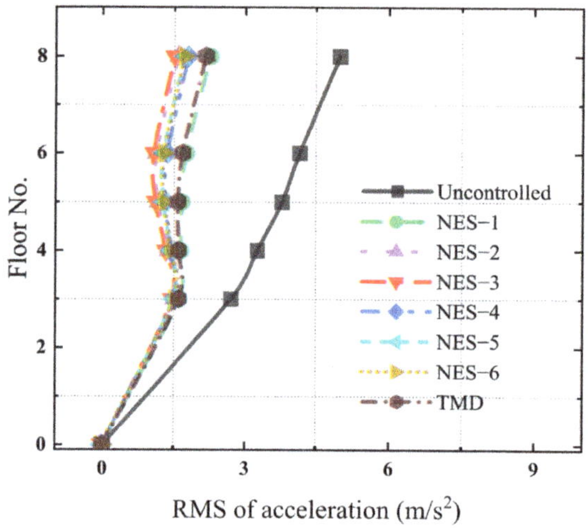

Figure 21. Variation in RMS values along the height of the structure under the seismic wave.

4. Conclusions

In this research, tests of the cubic stiffness NES controlling the dynamic responses of the MDOF structure under white noise, harmonic and seismic excitations were carried out using a shake table. The effects of the parameters of the NES on the vibration reduction ratios were explored. The main results and conclusions are summarized as follows.

(1) Under white noise excitations, these six NESs with different parameters and TMD have effective vibration mitigation effects, and their mitigation ratio of the acceleration responses at the top of the structure can reach about 50.0%. NES–4 has the best mitigation effects among these control devices.

(2) With regard to harmonic excitation, the NESs with different parameters and TMD almost have the same vibration mitigation effects on the acceleration responses except for NES–2, even though the mass ratio of the NES is less than that of the TMD. The vibration mitigation effects of NES–2 are worse than those of the other control devices under harmonic excitations. The control devices can achieve a vibration mitigation ratio of over 88.0% on the acceleration responses of the structure.

(3) One hundred seismic waves were inputted in the MODF structure using a shake table to investigate the vibration mitigation effects of the NES on the dynamic responses. The NES can alleviate the dynamic responses of the structure under the 100 seismic waves. The average mitigation effects of the NES on the RMS values of the acceleration responses of the structure under 100 seismic waves are better than that of the TMD. The robustness of the NES is better than that of the TMD.

To investigate the energy transmission from the structure to the NES under dynamic excitations, further shake table tests will be carried out to measure the responses of the structure as well as the dynamic responses of the NES.

Author Contributions: Conceptualization, Q.W. and Y.T.; methodology, Q.W. and Y.T.; software, X.Y. and D.Y.; validation, Y.T., X.Y. and Q.W.; formal analysis, X.Y. and D.Y.; investigation, Q.W., X.Y. and D.Y.; writing—original draft preparation, Q.W. and X.Y.; writing—review and editing, Q.W. and X.Y.; supervision, Q.W.; funding acquisition, Q.W. and Y.T. All authors have read and agreed to the published version of the manuscript.

Funding: This research was funded by Opening Funds of State Key Laboratory of Building Safety and Built Environment and the National Engineering Research Center of Building Technology (BSBE2021-11). This support is gratefully acknowledged.

Data Availability Statement: Data availability as required.

Conflicts of Interest: The authors declare no conflicts of interest.

References

1. Housner, G.W.; Bergman, L.A.; Caughey, T.K.; Chassiakos, A.G.; Claus, R.O.; Masri, S.F.; Skelton, R.E.; Soong, T.T.; Spencer, B.F.; Yao, J.T.P. Structural Control: Past, Present, and Future. *J. Eng. Mech.* **1997**, *123*, 897–971. [CrossRef]
2. Gutierrez Soto, M.; Adeli, H. Tuned Mass Dampers. *Arch. Comput. Methods Eng.* **2013**, *20*, 419–431. [CrossRef]
3. Mahmoodi, P. Structural Dampers. *J. Struct. Div.* **1969**, *95*, 1661–1672. [CrossRef]
4. Li, Q.S.; Zhi, L.-H.; Tuan, A.Y.; Kao, C.-S.; Su, S.-C.; Wu, C.-F. Dynamic Behavior of Taipei 101 Tower: Field Measurement and Numerical Analysis. *J. Struct. Eng.* **2011**, *137*, 143–155. [CrossRef]
5. Lu, X.; Chen, J. Mitigation of wind-induced response of Shanghai Center Tower by tuned mass damper. *Struct. Des. Tall Spec. Build.* **2011**, *20*, 435–452. [CrossRef]
6. Li, Q.; He, Y.; Zhou, K.; Han, X.; He, Y.; Shu, Z. Structural health monitoring for a 600 m high skyscraper. *Struct. Des. Tall Spec. Build.* **2018**, *27*, e1490. [CrossRef]
7. Vakakis, A.F. Inducing Passive Nonlinear Energy Sinks in Vibrating Systems. *J. Vib. Acoust.* **2001**, *123*, 324–332. [CrossRef]
8. Lee, Y.S.; Vakakis, A.F.; Bergman, L.A.; McFarland, D.M.; Kerschen, G.; Nucera, F.; Tsakirtzis, S.; Panagopoulos, P.N. Passive non-linear targeted energy transfer and its applications to vibration absorption: A review. *Proc. Inst. Mech. Eng. Part K J. Multi-Body Dyn.* **2008**, *222*, 77–134. [CrossRef]
9. McFarland, D.M.; Kerschen, G.; Kowtko, J.J.; Lee, Y.S.; Bergman, L.A.; Vakakis, A.F. Experimental investigation of targeted energy transfers in strongly and nonlinearly coupled oscillators. *J. Acoust.* **2005**, *118*, 791–799. [CrossRef]
10. Vakakis, A.F.; Bergman, L.A.; McFarland, D.M.; Gendelman, O.V.; Lee, G.K.Y.S. *Nonlinear Targeted Energy Transfer in Mechanical and Structural Systems*; Springer Science & Business Media: Berlin/Heidelberg, Germany, 2009.
11. Gourdon, E.; Lamarque, C.H. Energy Pumping with Various Nonlinear Structures: Numerical Evidences. *Nonlinear Dyn.* **2005**, *40*, 281–307. [CrossRef]
12. Nucera, F.; Lo Iacono, F.; McFarland, D.M.; Bergman, L.A.; Vakakis, A.F. Application of broadband nonlinear targeted energy transfers for seismic mitigation of a shear frame: Experimental results. *J. Sound Vib.* **2008**, *313*, 57–76. [CrossRef]
13. Nucera, F.; McFarland, D.M.; Bergman, L.A.; Vakakis, A.F. Application of broadband nonlinear targeted energy transfers for seismic mitigation of a shear frame: Computational results. *J. Sound Vib.* **2010**, *329*, 2973–2994. [CrossRef]
14. Vaurigaud, B.; Savadkoohi, A.T.; Lamarque, C.H. Efficient Targeted Energy Transfer with Parallel Nonlinear Energy Sinks: Theory and Experiments. *J. Comput. Nonlinear Dyn.* **2011**, *6*, 041005. [CrossRef]
15. Wierschem, N.E.; Hubbard, S.A.; Luo, J.; Fahnestock, L.A.; Spencer, B.F.; Quinn, D.D.; McFarland, D.M.; Vakakis, A.F.; Bergman, L.A. Experimental blast testing of a large 9-story structure equipped with a system of nonlinear energy sinks. In Proceedings of the ASME 2013 International Design Engineering Technical Conferences and Computers and Information in Engineering Conference, Portland, OR, USA, 4–7 August 2013.
16. Wierschem, N.E.; Hubbard, S.A.; Luo, J.; Fahnestock, L.A.; Spencer, B.F., Jr.; McFarland, D.M.; Quinn, D.D.; Vakakis, A.F.; Bergman, L.A. Response attenuation in a large-scale structure subjected to blast excitation utilizing a system of essentially nonlinear vibration absorbers. *J. Sound Vib.* **2017**, *389*, 52–72. [CrossRef]
17. Luo, J.; Wierschem, N.E.; Hubbard, S.A.; Fahnestock, L.A.; Quinn, D.D.; McFarland, D.M.; Spencer, B.F.; Vakakis, A.F.; Bergman, L.A. Large-scale experimental evaluation and numerical simulation of a system of nonlinear energy sinks for seismic mitigation. *J. Struct. Eng.* **2014**, *77*, 34–48. [CrossRef]
18. Wierschem, N.E.; Luo, J.; Al-Shudeifat, M.; Hubbard, S.; Ott, R.; Fahnestock, L.A.; Quinn, D.D.; McFarland, D.M.; Spencer, B.F.; Vakakis, A.; et al. Experimental Testing and Numerical Simulation of a Six-Story Structure Incorporating Two-Degree-of-Freedom Nonlinear Energy Sink. *J. Struct. Eng.* **2014**, *140*, 04014027. [CrossRef]
19. Wang, J.J.; Wierschem, N.E.; Wang, B.; Spencer Jr, B.F. Multi-objective design and performance investigation of a high-rise building with track nonlinear energy sinks. *Struct. Des. Tall Spec. Build.* **2020**, *29*, e1692. [CrossRef]
20. Wang, J.J.; Wierschem, N.; Spencer, B.F.; Lu, X. Experimental study of track nonlinear energy sinks for dynamic response reduction. *Eng. Struct.* **2015**, *94*, 9–15. [CrossRef]
21. Wang, J.J.; Wang, B.; Liu, Z.B.; Zhang, C.; Li, H.B. Experimental and numerical studies of a novel asymmetric nonlinear mass damper for seismic response mitigation. *Struct. Control Health Monit.* **2020**, *27*, e2513. [CrossRef]
22. Wang, J.J.; Zhang, C.; Li, H.; Liu, Z. Experimental and numerical studies of a novel track bistable nonlinear energy sink with improved energy robustness for structural response mitigation. *Eng. Struct.* **2021**, *237*, 112184. [CrossRef]
23. Yang, H.X.; Li, Z.; Wang, H.B.; Zhang, M.H.; Ni, S.Y. Shaking table test study on seismic performance of RC frame structure with NES. *Structures* **2023**, *47*, 153–164. [CrossRef]
24. Ndemanou, B.P.; Oumbé Tékam, G.T.; Kingni, S.T.; Nana Nbendjo, B.R. Energy sink outrigger control of a tall building under wind loads. *Eng. Struct.* **2023**, *288*, 116158. [CrossRef]

25. Wang, Q.; Wu, H.; Qiao, H.; Yu, X.; Huang, P. Asymmetric and Cubic Nonlinear Energy Sink Inerters for Mitigating Wind-Induced Responses of High-Rise Buildings. *Struct. Control Health Monit.* **2023**, *2023*, 1150525. [CrossRef]
26. *ISO 5348:1998-05*; Mechanical Vibration and Shock—Mechanical Mounting of Accelerometers. ISO: Geneva, Switzerland, 1998.

buildings

MDPI

Article

Accurate Dynamic Analysis Method of Cable-Damper System Based on Dynamic Stiffness Method

Hui Jiao [1], Bin Xu [2], Zhengkai Jiang [1], Can Cui [3] and Haoxiang Yang [4,*]

1 Investment Company of China Railway Seventh Group Co., Ltd., Zhengzhou 450016, China; jiaohui@crecg.com (H.J.); jiangzhengkai@crecg.com (Z.J.)
2 School of Civil Engineering, Zhengzhou University, Zhengzhou 450001, China; xubin_tmxy@zzu.edu.cn
3 School of Water Conservancy and Transportation, Zhengzhou University, Zhengzhou 450001, China; cuicanoffice@zzu.edu.cn
4 School of Civil Engineering, Southeast University, Nanjing 211189, China
* Correspondence: 230249658@seu.edu.cn

Abstract: To suppress large vibrations of the cable in cable-stayed bridges, it is common to install transverse dampers near the end of the cable. This paper focuses on the cable-damper system; based on the dynamic stiffness method, an accurate dynamic analysis method considering cable parameters, damper parameters, and cable forces is proposed. First, a mechanical analysis model is established which is closer to the cable with a transverse damper installed in the bridge. The model considers the cable bending stiffness, sag, inclination angle, cable force, damper stiffness, damping coefficient, and damper installation height. Then, the characteristic frequency equation of the cable-damper system is established, and a solution method that combines the initial value method and Newton–Raphson method is proposed. This method is confirmed to provide more accurate frequency analysis for the cable-damper system. Finally, using this method, the effect of the damper parameters on the dynamic characteristics of the system is investigated.

Keywords: cable; damper; dynamic analysis; dynamic stiffness method; frequency equation

1. Introduction

Cable-stayed bridges are a cornerstone of modern civil engineering, widely celebrated for their ability to span vast distances with both structural efficiency and aesthetic appeal. These bridges are increasingly favored for their versatility in accommodating various load demands, as well as their ability to integrate into complex urban infrastructure. With the increase in cable length, their vibration problem under the excitation of environmental and traffic loads becomes more and more prominent [1,2]. To restrain large vibrations of the cable, installing transverse dampers near the anchor position of the cable near the main beam has become a very common measure to restrain the vibration [3,4]. The cable-damper system not only enhances the vibration response of individual cables, but also provides supplementary damping for the entire bridge [5]. This makes the cable system complicated and leads to inaccuracy when analyzing the dynamic characteristics of the cable using pure cable theory. Therefore, it is very important to analyze the system composed of cable and transverse damper as a whole to accurately analyze the dynamic characteristics of the cable and maintenance of a cable-supported bridge. In this paper, a cable with a transverse damper is called a cable-damper system.

At present, the research on the cable-damper system focuses on the optimization of the damper. Therefore, in the modeling of a cable-damper system, the cable is often simplified to a tensioned string model [6,7]. Neglecting sag and bending stiffness in existing models leads to inaccurate dynamic predictions, underestimating vibration amplitudes and damping effects [8]. Generally speaking, the sag and bending stiffness of the cable in a cable-stayed bridge are relatively small, but even a small sag will have a significant negative impact on the performance of the linear damper [9,10]. Sun et al. [11] incorporated the effects of cable sag and stiffness, enhancing the cable-reduced order model by including a viscous damper. The modified model employs one or two viscous dampers with varying nonlinearities to numerically analyze the cable's behavior under different sag extensibility parameters. Mehrabi et al. [12] proposed a motion formula for a cable-damper system using finite difference methods, incorporating the effects of cable bending stiffness, sag, variable cross-section, and boundary conditions. The accuracy and computational cost of this finite difference (FD) model are strongly influenced by spatial and temporal discretization. However, that study is limited to the first vibration mode of the cable. Fujino and Hoang [13] developed a free vibration analysis model for an inclined uniform cable with transverse dampers, which considered the influences of cable bending stiffness, sag, and damper support, each of which was reflected by the corresponding reduction factor. Pacheco et al. [6] proposed a straightforward design methodology for external viscous dampers in cable systems, utilizing a general estimation curve. This approach integrates the cable's modal damping ratio with key parameters such as length, mass per unit length, and fundamental frequency, alongside factors like additional dampers, their size, and placement. Javanbakht [14] proposed a control-oriented numerical model for evaluating the dynamic response of a stay cable equipped with a damper/actuator, proposed based on the mode superposition method (MSM) with enhanced shape functions. This enhancement accounts for the effect of the truncated higher modes by introducing a static correction term. Zhou et al. [15] performed a stochastic analysis of the coupled in-plane and out-of-plane vibrations in the cable-damper system. By selecting the static deflection shape at the damper location and the first sine term as shape functions, they derived a reduced four-degree-of-freedom system of nonlinear stochastic ordinary differential equations to characterize the cable's dynamic response. These two methods primarily address the dynamic force response of the cable.

In the research mentioned above, the damper and cable are simplified to varying extents, and the system description lacks sufficient comprehensiveness. Unlike the cable-damper systems used in practical engineering, the accuracy of the cable modal parameters obtained through this method is also limited. Furthermore, existing approaches that employ reduction factors or fitting techniques to account for the damper's effect fail to distinguish between the impacts of the cable and damper parameters. As a result, these methods are unsuitable for solving inverse problems with high accuracy, especially for cables in service. Therefore, it is essential to develop a high-precision analytical frequency characteristic equation that can be applied to both forward and inverse analyses of the cable-damper system. The dynamic stiffness method (DSM) is an efficient and accurate dynamic analysis method that can solve the problems of accuracy and consideration of all relevant design parameters in the dynamic response [16,17].

In this paper, an accurate dynamic analysis method of the cable-damper system is proposed based on the dynamic stiffness theory with minimal simplification of the cable. In this method, the cable force, sag, bending stiffness, inclination angle, and mass per unit length of the cable are considered, and the damper stiffness, damping coefficient, and damper position, to more accurately represent the real-world behavior of the cable system. In the second part, the dynamic analysis theory of the cable-damper system is established based on the dynamic stiffness theory, and the analytic frequency characteristic equation of the system is presented. In the third part, by studying the characteristics of the frequency characteristic function, a numerical method for solving the frequency characteristic equation is proposed. In the fourth part, the accuracy of the proposed frequency equation is verified by the cable experiments in the literature and the real cable monitoring of Shaozhou Bridge. In the fifth part, the influence of damper parameters on the dynamic characteristics of the cable is analyzed. The conclusion of this paper is given in the sixth part.

2. Dynamic Analysis Theory of Cable-Damper System

For cables installed with transverse dampers in a cable-stayed bridge, the mechanical model can be simplified as in Figure 1. The model considers the cable force, sag, bending stiffness, inclination angle boundary condition, and mass per unit length, along with the damper stiffness, damping coefficient, and position, providing a more accurate representation of the actual cable system. In the model, the transverse damper is approximated as a linear damper, represented by the combination of a linear spring and a viscous damping element [18]. In Figure 1, the stiffness of the linear spring is k_d, and the damping coefficient of the viscous damper is c_d. Under static action, the transverse damper has no impact on the cable. $y(x_0)$ represents the static configuration of the cable, accounting for sag, angle, and cable force. For shallow cables, this configuration can be approximated as a quadratic parabola. When the cable vibrates, the external damper will affect the dynamic configuration of the cable, as shown in $v(x_j, t)$. When analyzing the cable-damper system in Figure 1, the cable is divided into two cable segments at the position C of the damper, and coordinate systems (x_j, y_j) are established for the original cable and the two cable segments. $j = 0$ indicates the original cable AB with l_0 length; $j = 1$ indicates the lower cable segment AC with l_1 length; $j = 2$ indicates the upper cable segment CB with l_2 length. θ is the angle between the cable and the horizontal plane, and d is the sag of the cable in the middle span. The cable force H is the average cable force, regardless of the change along its length.

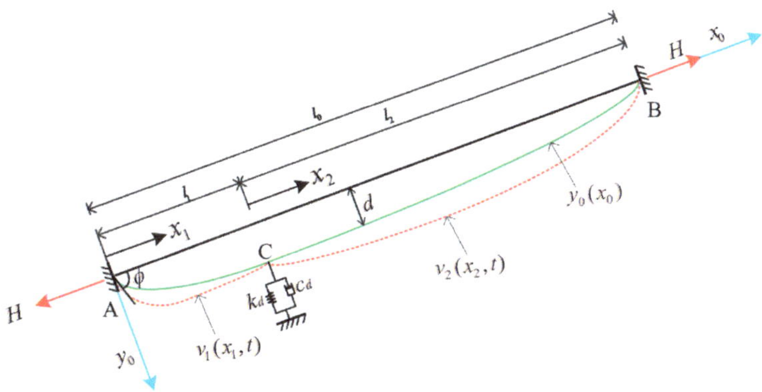

Figure 1. Cable-damper system mechanical model.

In the dynamic analysis of the cable segment, the effects of internal damping and shear force are neglected. Using Hamilton's principle, the control differential equation for in-plane vibration under free vibration is derived, accounting for the bending stiffness and sag of the cable segment.

$$EI\frac{\partial^4 v(x_j,t)}{\partial x_j{}^4} - H\frac{\partial^2 v(x_j,t)}{\partial x_j{}^2} - h_j(t)\frac{\mathrm{d}^2 y(x_0)}{\mathrm{d}x_0{}^2} + m\frac{\partial^2 v(x_j,t)}{\partial t^2} = 0 \quad (1)$$

where EI is the bending stiffness of the cable; m is the cable mass; $h_j(t)$ is the vibration-induced additional cable force [19]; and t is the time.

For Equation (1), by separating variables, we assume the solution takes the form of

$$v(x_j,t) = \varphi(x_j)\mathrm{e}^{\mathrm{i}\omega_d t} \quad (2)$$

where $\varphi(x_j)$ is the displacement function of the cable segment; $\mathrm{i} = \sqrt{-1}$, ω_d is the natural circular frequency of the cable-damper system.

Substitute Equation (2) into Equation (1), and let $\xi_j = x_j/l_0$; the dimensionless cable vibration equation is

$$\hat{\varphi}^{IV}(\xi_j) - \gamma^2\hat{\varphi}''(\xi_j) - \widetilde{\omega}^2\hat{\varphi}(\xi_j) = -\hat{h}_j \tag{3}$$

where $\hat{\varphi}(\xi_j) = \varphi(x_j) \cdot EI/\left(mgl_0{}^4\right), \gamma^2 = Hl_0{}^2/EI, \widetilde{\omega} = \omega_d l_0{}^2/\sqrt{EI/m}, \hat{h}_j = h_j(t)\cos\theta/(He^{i\omega_d t})$.

The general solution of Equation (3) is

$$\hat{\varphi}(\xi_j) = \left(\mathbf{\Phi}(\xi_j) + \frac{\eta_j}{\widetilde{\omega}^2 - \eta_j}\int_0^1 \mathbf{\Phi}(\xi_j)\mathrm{d}\xi_j\right) \cdot \left\{A_1{}^{(j)} A_2{}^{(j)} A_3{}^{(j)} A_4{}^{(j)}\right\}^T \tag{4}$$

where $\mathbf{\Phi}(\xi_j) = \left[\mathrm{e}^{-p\xi_j} \quad \mathrm{e}^{-p(1-\xi_j)} \quad \cos(q\xi_j) \quad \sin(q\xi_j)\right]$, $\left.\begin{matrix} p \\ q \end{matrix}\right\} = \sqrt{\sqrt{\frac{\gamma^4}{4} + \widetilde{\omega}^2} \pm \frac{\gamma^2}{2}}$, $\eta_j = \frac{64Al_0d^2}{Il_j^s}$, $l_j{}^s = \int_0^{l_j}\left(\frac{\mathrm{d}s}{\mathrm{d}x_j}\right)^3 dx_j$ is the effective length of the cable; $A_1^{(j)} \sim A_4^{(j)}$ are the undetermined parameters.

In order to determine the undetermined parameters in Equation (4), the cable dynamic displacement boundary conditions and force edge boundary conditions as shown in Figure 2 are introduced.

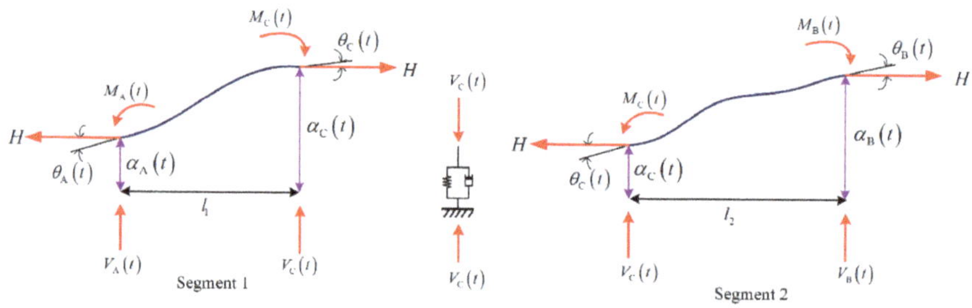

Figure 2. Force boundary condition and dynamic displacement boundary conditions of the cable.

Similarly, the dynamic displacement and internal force of the cable in the time domain are expressed by separating variables. Using the dimensionless method described earlier, it follows that the force and dynamic displacement boundary conditions for each cable segment satisfy the following relationship.

$$\begin{cases} V_*(t) = \widetilde{V}_*\mathrm{e}^{i\omega_d t} = mgl_0\left(\hat{\varphi}'''(\xi_j) - \gamma^2\hat{\varphi}'(\xi_j)\right) \cdot \mathrm{e}^{i\omega_d t} \\ M_*(t) = \widetilde{M}_*\mathrm{e}^{i\omega_d t} = mgl_0{}^2\hat{\varphi}''(\xi_j)\mathrm{e}^{i\omega_d t} \\ \alpha_*(t) = \widetilde{\alpha}_*\mathrm{e}^{i\omega_d t} = \frac{mgl_0^4}{EI}\hat{\varphi}(\xi_j)\mathrm{e}^{i\omega_d t} \\ \theta_*(t) = \widetilde{\theta}_*\mathrm{e}^{i\omega_d t} = \frac{mgl_0^3}{EI}\hat{\varphi}'(\xi_j)\mathrm{e}^{i\omega_d t} \end{cases} \tag{5}$$

where $\widetilde{M}_*$, $\widetilde{V}_*$, $\widetilde{\alpha}_*$, $\widetilde{\theta}_*$ represent the bending moment amplitude, shear amplitude, lateral dynamic displacement amplitude, and angular dynamic displacement amplitude at nodes $*$; $*$ represents nodes A, B, and C in the cable segment.

By substituting Equation (4) into Equation (5), the dynamic equilibrium equation for the cable section is obtained as follows.

$$\begin{cases} \left[\widetilde{V}_A \quad \widetilde{M}_A/l_0 \quad \widetilde{V}_C \quad \widetilde{M}_C/l_0\right]^T = \mathbf{K}^{(1)} \cdot \left[\widetilde{\alpha}_A \quad \widetilde{\theta}_A l_0 \quad \widetilde{\alpha}_C \quad \widetilde{\theta}_C l_0\right]^T \\ \left[\widetilde{V}_C \quad \widetilde{M}_C/l_0 \quad \widetilde{V}_B \quad \widetilde{M}_B/l_0\right]^T = \mathbf{K}^{(2)} \cdot \left[\widetilde{\alpha}_C \quad \widetilde{\theta}_C l_0 \quad \widetilde{\alpha}_B \quad \widetilde{\theta}_B l_0\right]^T \end{cases} \tag{6}$$

where $\mathbf{K}^{(j)}$ is the dynamic stiffness matrix of the cable segment, which is a function of frequency, force, and cable parameters. Its expression is shown in Equation (7).

$$\mathbf{K}^{(j)} = \frac{EI}{l_0^3}\begin{bmatrix} \mathbf{\Phi}'''\left(\xi_j|_{=0}\right) - \gamma^2 \cdot \mathbf{\Phi}'\left(\xi_j|_{=0}\right) \\ -\mathbf{\Phi}''\left(\xi_j|_{=0}\right) \\ -\mathbf{\Phi}'''\left(\xi_j|_{=\mu_j}\right) + \gamma^2 \cdot \mathbf{\Phi}'\left(\xi_j|_{=\mu_j}\right) \\ \mathbf{\Phi}''\left(\xi_j|_{=\mu_j}\right) \end{bmatrix} \cdot \left(\begin{bmatrix} \mathbf{\Phi}\left(\xi_j|_{=0}\right) \\ \mathbf{\Phi}'\left(\xi_j|_{=0}\right) \\ \mathbf{\Phi}\left(\xi_j|_{=\mu_j}\right) \\ \mathbf{\Phi}'\left(\xi_j|_{=\mu_j}\right) \end{bmatrix} + \begin{bmatrix} 1 \\ 0 \\ 1 \\ 0 \end{bmatrix} \cdot \frac{\eta_j}{\widetilde{\omega}^2 - \eta_j}\int_0^1 \mathbf{\Phi}(\xi_j)\mathrm{d}\xi_j \right)^{-1} \tag{7}$$

where $\mu_j = l_j/l_0$ represents the installation position of the damper.

In engineering application, the damper is usually installed on the long cable and perpendicular to the cable. For the medium and long cable, the boundary conditions have little effect on its dynamic characteristics, and both ends of the cable are assumed to be fixed. In the analysis, it is assumed that the damper only provides transverse force and ignores the bending moment. Based on the above assumptions, the dynamic displacement amplitude vector and dynamic amplitude vector of the cable can be expressed as Equation (8).

$$\begin{cases} \begin{bmatrix} \widetilde{\alpha}_A & \widetilde{\theta}_A & \widetilde{\alpha}_C & \widetilde{\theta}_C & \widetilde{\alpha}_B & \widetilde{\theta}_B \end{bmatrix} = \begin{bmatrix} 0 & 0 & \widetilde{\alpha}_C & \widetilde{\theta}_C & 0 & 0 \end{bmatrix} \\ \begin{bmatrix} \widetilde{V}_A & \widetilde{M}_A & \widetilde{V}_C & \widetilde{M}_C & \widetilde{V}_B & \widetilde{M}_B \end{bmatrix} = \begin{bmatrix} 0 & 0 & \widetilde{V}_C & 0 & 0 & 0 \end{bmatrix} \end{cases} \tag{8}$$

For a general cable external damper, the mechanical model can be simplified to a linear spring and a linear viscous damper, whose transverse force amplitude in the frequency domain can be expressed as follows.

$$\widetilde{V}_C = -k_c\widetilde{\alpha}_C = -(k_d + ic_d\omega_d)\widetilde{\alpha}_C \tag{9}$$

According to the assembly method of the stiffness matrix of finite element method, grouping Equation (6), and introducing the boundary conditions of Equations (9) and (10), the mechanical equilibrium equation of the cable which provides the transverse force damper can be obtained as follows.

$$\begin{bmatrix} k_{33}^{(1)} + k_{11}^{(2)} + k_c & k_{34}^{(1)} + k_{12}^{(2)} \\ k_{43}^{(1)} + k_{21}^{(2)} & k_{44}^{(1)} + k_{22}^{(2)} \end{bmatrix} \begin{bmatrix} \widetilde{\alpha}_C \\ \widetilde{\theta}_C l_0 \end{bmatrix} = \begin{bmatrix} 0 \\ 0 \end{bmatrix} \tag{10}$$

where $k_{n_1n_2}^{(j)}$ represents the element corresponding to the n_1th row and the n_2th column in the dynamic stiffness matrix $\mathbf{K}^{(j)}$ of cable segment j, that is, the stiffness component corresponding to the lateral displacement and angular displacement at the installation position C of the damper. In particular, the dynamic stiffness matrix of the cable-damper system is defined as

$$\mathbf{K} = \begin{bmatrix} k_{33}^{(1)} + k_{11}^{(2)} + k_c & k_{34}^{(1)} + k_{12}^{(2)} \\ k_{43}^{(1)} + k_{21}^{(2)} & k_{44}^{(1)} + k_{22}^{(2)} \end{bmatrix} \tag{11}$$

According to Equation (11) for a cable-damper system with vibration, the condition that the equation has a trivial solution is that the value of the determinant of the dynamic stiffness matrix is 0. Therefore, the frequency characteristic equation of the cable-damper system can be defined as

$$|\mathbf{K}| = 0 \tag{12}$$

In Equation (12), the dynamic stiffness matrix K of the cable-damper system is a function of the cable frequency, cable force, cable parameters, and damper parameters. Once the cable force, cable parameters, and damper parameters are known, the system's frequency can be determined by solving Equation (12), enabling the analysis of its dynamic characteristics. Since Equation (12) is a complex transcendental equation, an analytical solution for the frequency cannot be directly obtained and must be solved numerically.

3. Solution of the Frequency Characteristic Equation for the Cable-Damper System

When considering the damper effect, $|\mathbf{K}|$ is a complex number, that is,

$$|\mathbf{K}| = \mathrm{Re}(|\mathbf{K}|) + \mathrm{Im}(|\mathbf{K}|)\mathrm{i} \tag{13}$$

For Equation (13) to be true, both $\mathrm{Re}(|\mathbf{K}|)$ and $\mathrm{Im}(|\mathbf{K}|)$ in Equation (14) must be 0. Considering the sparse characteristic of the cable frequency, this paper defines the characteristic frequency function of the cable-damper system as

$$F\left(f \middle| H, P_{\text{cable}}, P_{\text{damper}}\right) = \frac{1}{|\mathrm{Re}(|\mathbf{K}|)| + |\mathrm{Im}(|\mathbf{K}|)|} \tag{14}$$

where $f = \frac{\omega_d}{2\pi}$, $P_{\text{cable}} = (l_0, m, EI, D)$, $P_{\text{damper}} = (k_d, c_d, \mu_1)$.

Thus, the problem of determining the frequency of the cable-damper system is transformed into finding the maximum value of Equation (14). The parameters of the cable-damper system in Table 1, which correspond to an actual existing cable, are used to analyze the characteristics of the frequency function of the system. During the analysis, for C1, the variation range of frequency f is 0.01 Hz–20 Hz; for C2, the variation range of frequency f is 0.01 Hz–10 Hz, and the frequency change intervals are all 0.01 Hz. The variation curve of the characteristic frequency function with frequency is obtained by Equation (14), as shown in Figure 3.

Table 1. Cable-damper system parameters.

Cable	l_0(m)	m(kg/m)	D(m)	θ(°)	EI(kN·m^2)	H(kN)	μ_1	k_d(kg/m)	c_d(kg/m)
C1	30	45	0.160	60	2.5	1500	0.1	10	20
C2	80	65	0.250	36	3.5	2500	0.1	10	20

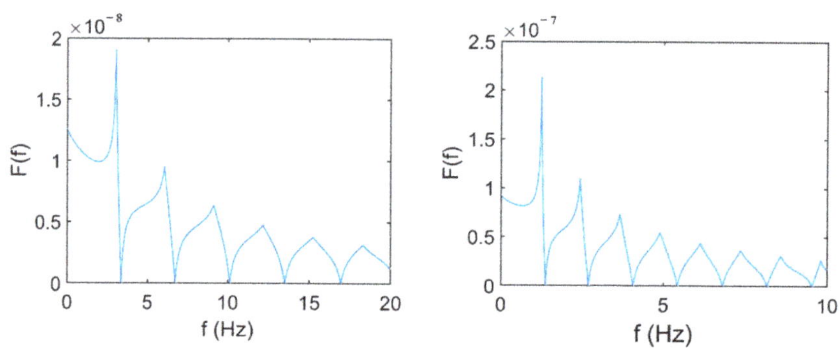

(**a**) C1 (**b**) C2

Figure 3. Variation curve of characteristic frequency function of cable-damper system with frequency.

As shown in Figure 3, the characteristic frequency function of the cable-damper system exhibits multiple peaks as the frequency varies. The frequencies at these peaks correspond to the potential resonant frequencies of the system. To obtain a high precision cable frequency from the function image usually requires a lot of computation and human intervention. The numerical method can usually determine the solution of the equation according to the set calculation accuracy, but, for the functions with multiple peaks as shown in Figure 3, there is a case of missing roots. To accurately obtain the system frequencies of all orders, this paper proposes a two-step calculation method, tailored to the characteristics of the frequency function. First, the characteristic frequency function image is drawn through large frequency changes, and the interval where the frequency is located is preliminarily determined. Then, the Newton–Raphson Method is applied in each interval to obtain the frequency of cable-damper system that meets the accuracy requirements [20].

4. Accuracy Verification of the Characteristic Frequency Equation of the Cable-Damper System

In order to verify the accuracy of the characteristic frequency equation of the cable-damper system and the solution method proposed in this paper, the experimental method in the literature [21] and the cable-damper system of a real bridge were used for verification.

4.1. Literature Experimental Verification

The experimental field photos of the cable experiment with damper installed as in the literature [21] are shown in Figure 4. The cable experiment adopts a horizontal cable with a fixed length of 61.8 m at both ends, and the damper is installed at a position l_1 away from the end. Install a load cell on the right end of the cable. During the experiment, the cable between the damper position and the right end is struck with a hammer, and the resulting vibration is measured using an accelerometer installed on the cable. For the validation of this paper, three working conditions were selected for validation, and the tension cable parameters and damper parameters are shown in Table 2.

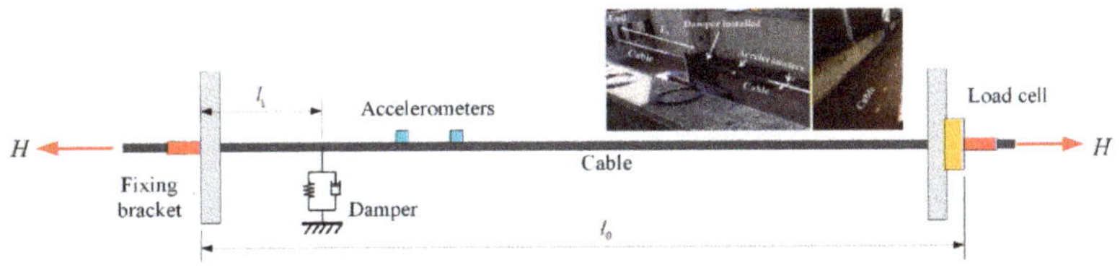

Figure 4. Schematic diagram of experimental setting and photograph of experimental setting [21].

Table 2. Parameters of experimental cable-damper system in the literature [21].

Cases	l_0(m)	m(kg/m)	D(m)	EI(kN·m^2)	H(kN)	$\mu_1=\frac{l_1}{l_0}$	k_d(kg/m)	c_d(kg/m)
1	377.92	61.8	4.26	3.219	0.0286	0.15	29.3	17.9
2	376.75	61.8	4.26	3.219	0.0286	0.15	57.1	31.4
3	377.75	61.8	4.26	3.219	0.0286	0.15	84.7	52.5

The measurement frequency, the frequency calculated by the method in this paper and the frequency calculated in the literature are listed in Table 3. The 5th order frequency of the cable-damper system is not measured in the literature, so the corresponding calculation result of the method in this paper is not given.

It can be seen from Table 3 that the frequency of the structural system solved by the characteristic frequency equation proposed in this paper is in good agreement with the frequency results measured by experiment. Under the three cases, except for the frequency errors of the 1st and 6th orders of case 1, which are −3.4% and −3.9%, respectively, the other order errors are less than 2.0%. Compared with the methods in the literature, except for the 5th and 6th order frequencies, the proposed method is superior to the methods in the literature. Especially for the first four modes, the maximum error in the calculation accuracy of the method proposed in this paper is −3.4% for the first order frequency under working condition 1, with the remaining errors all less than 1.7%. In comparison, the method from the literature yields errors greater than 6.6%, with the maximum error reaching −15.0% for the fourth order frequency in case 1. Additionally, the calculated results from the literature show significant variation, with errors for the 6th to 8th order frequencies being smaller than those for the first four orders. In contrast, the method presented in this paper exhibits consistently smaller errors across all frequency orders, demonstrating its superior stability compared to the literature method.

Table 3. Comparison of frequency calculations in this paper and in the literature [21].

Cases	Order	1	2	3	4	5	6	7	8
1	Measured value	2.93	5.67	8.64	11.57	-	14.80	17.32	20.03
	This paper	2.83	5.68	8.53	11.37	-	14.23	17.03	19.91
	Literature	2.55	5.06	7.49	9.84	-	14.48	16.88	19.33
	Error1 (%)	−3.4%	0.2%	−1.3%	−1.7%	-	−3.9%	−1.7%	−0.6%
	Error2 (%)	−13.0%	−10.8%	−13.3%	−15.0%	-	−2.2%	−2.5%	−3.5%
2	Measured value	2.86	5.75	8.64	11.53	-	14.51	17.33	20.21
	This paper	2.83	5.67	8.51	11.36	-	14.21	17.03	19.88
	Literature	2.62	5.21	7.71	10.04	-	14.47	16.86	19.33
	Error1 (%)	−0.9%	−1.3%	−1.5%	−1.5%	-	−2.0%	−1.7%	−1.6%
	Error2 (%)	−8.4%	−9.4%	−10.8%	−12.9%	-	−0.3%	−2.7%	−4.4%
3	Measured value	2.86	5.75	8.64	11.56	-	14.48	17.36	20.25
	This paper	2.84	5.68	8.53	11.37	-	14.22	17.06	19.91
	Literature	2.67	5.31	7.87	10.23	-	14.50	16.88	19.38
	Error1 (%)	−0.7%	−1.2%	−1.3%	−1.6%	-	−1.8%	−1.7%	−1.7%
	Error2 (%)	−6.6%	−7.7%	−8.9%	−11.5%	-	0.1%	−2.8%	−4.3%

Note: Error1 is the relative error between the measured value and this paper. Error2 is the relative error between the measured value and the literature method.

4.2. Real Bridge Data Verification

.The specimen used in 4.1 is a steel strand, not an actual cable. In order to further verify the accuracy of the proposed method in the application of the cable-damper system in real bridges, this paper selected the cable-damper system of Shaozhou Bridge in Guangdong Province, China for analysis.

The main bridge of Shaozhou Bridge adopts steel–concrete composite beams with a single tower and double cable surface. The span arrangement of the bridge is 33 + 102 + 183 = 318 m, and the whole bridge adopts a half-floating system. The main beam adopts a streamlined flat box beam; the cable tower adopts an arched bridge tower, and the tower column is a reinforced concrete component. The whole bridge has 26 pairs of cables, a total of 52 cables. In the concrete beam section, the distance between the cables is 6 m; in the steel box beam section, the distance between the cables is 12 m, and the vertical distance on the tower between the cables is 2.0 m. The cable-stayed cable adopts the type 250 high-strength epoxy-coated steel strand diagonal cable; the standard strength is 1860 MPa, and the cable group anchor system is adopted. To suppress cable vibrations, an external passive damper is installed near the cable beam section. Figure 5 illustrates the structural layout of the bridge, while Figure 6 shows the installation of the cable dampers

Figure 5. Shaozhou Bridge site photos.

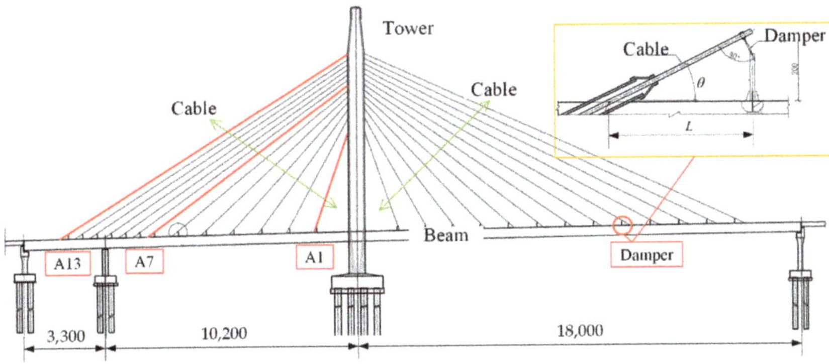

Figure 6. The overall layout of Shaozhou Bridge and external damper installation diagram.

Considering the influence of different cable lengths, the A1, A7, and A13 cables of Shaozhou Bridge are selected for analysis. Table 4 outlines the key parameters of the cable-damper system for the cables analyzed in Shaozhou Bridge. Accelerometers are installed on the above three cables, and the sampling frequency is 50 Hz. In order to reduce the impact of vehicle load, the data at 2:00 a.m. on 1 July 2023 were selected for analysis, and the acceleration data duration was 5 min. The acceleration signal time-history curve and acceleration amplitude spectrogram are illustrated in Figure 7.

Table 4. Parameters of cable-damper system of Shaozhou Bridge.

Cable	l_0(m)	m(kg/m)	D(m)	θ(°)	EI(kN·m^2)	H(kN)	μ_1	c_d(kg/m)
A1	55.2	52.6	0.160	69.746	2509	2348	0.065	200
A7	108.6	67.5	0.160	36.778	2007	3211	0.048	1000
A13	147.4	113.3	0.224	32.375	4627	6694	0.038	2300

Table 5 presents the measured cable frequencies derived from the analysis in Figure 7, along with the cable frequencies calculated using the method proposed in this paper, based on the parameters in Table 4.

Table 5. Cable frequency results of the real bridge.

Cable	Order	1	2	3	4	5	6	7	8
A1	Measured	2.170	4.294	6.348	8.759	11.044	13.471	16.269	18.884
	This paper	2.167	4.256	6.458	8.725	11.090	13.568	16.178	18.937
	Error (%)	−0.1	−0.9	1.7	−0.4	0.4	0.7	−0.6	0.3
A7	Measured	1.085	2.173	3.247	4.312	5.377	6.519	7.562	8.633
	This paper	1.081	2.137	3.211	4.289	5.375	6.470	7.576	8.695
	Error (%)	−0.3	−1.7	−1.1	−0.5	0.0	−0.7	0.2	0.7
A13	Measured	0.879	1.746	2.621	3.503	4.364	5.237	6.125	6.958
	This paper	0.879	1.730	2.598	3.467	4.341	5.219	6.102	6.991
	Error (%)	0.0	−0.9	−0.9	−1.0	−0.5	−0.3	−0.4	0.5

As can be seen from Table 5, the frequency of the cable-damper system calculated by the method in this paper is in good agreement with the measured frequencies, and the calculation errors are less than 1.7%. This verifies that the characteristic frequency equation proposed in this paper has a high precision and can accurately analyze the dynamic characteristics of the cable-damper system.

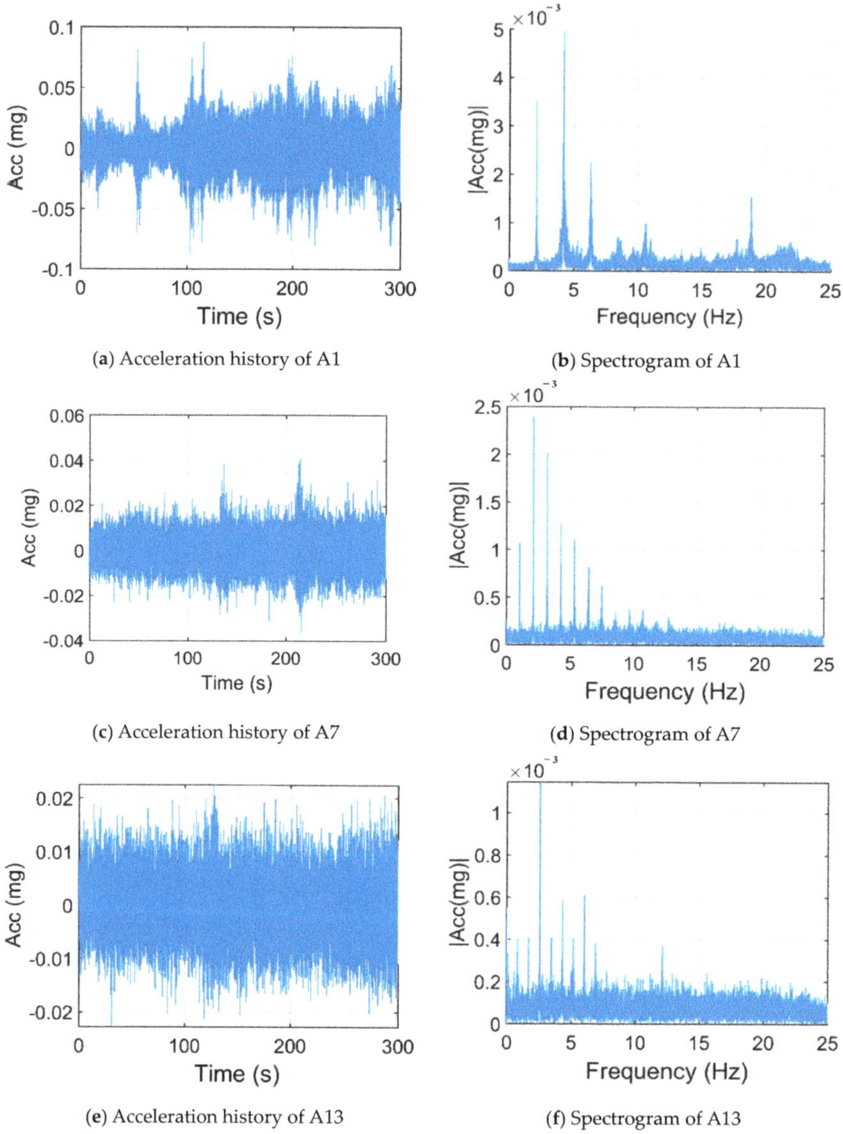

(a) Acceleration history of A1 (b) Spectrogram of A1
(c) Acceleration history of A7 (d) Spectrogram of A7
(e) Acceleration history of A13 (f) Spectrogram of A13

Figure 7. Acceleration history and spectrogram of Shaozhou Bridge cables.

5. Influence of Damper Parameters on Dynamic Characteristics of Cable

It is important to analyze the effect of damper parameters on the frequency of the cable-damper system when designing dampers for vibration modes where controlling the frequency of the cable is the primary concern. This section utilizes the frequency characteristic function of the cable-damper system proposed in this paper to analyze the effects of damper parameters, such as damping coefficient, stiffness, and installation height, on the cable frequencies. The analysis aims to determine how different damper design parameters influence the dynamic characteristics of the system, providing a basis for the design of the cable-damper system.

5.1. Influence of Damper Stiffness and Damping Coefficient on Cable Frequency

In analyzing the effect of the damper stiffness and damping coefficient on cable frequency, the parameters for the two cables are selected from Table 1. The external damper is mainly used to control the low-order vibration of the cable [22]. Therefore, this paper examines how the first four order frequencies of the cable vary with changes in the damper's stiffness and damping coefficient. Among them, the stiffness of the damper varies from 0.5 k_d to 5 k_d, and the change interval is 0.1 k_d. The damping coefficient varies from 0.5 c_d to 5 c_d, and the variation interval is 0.1 c_d. The variation of the first order frequency of the cable-damper system with damper stiffness and damping coefficient is given in Figure 8.

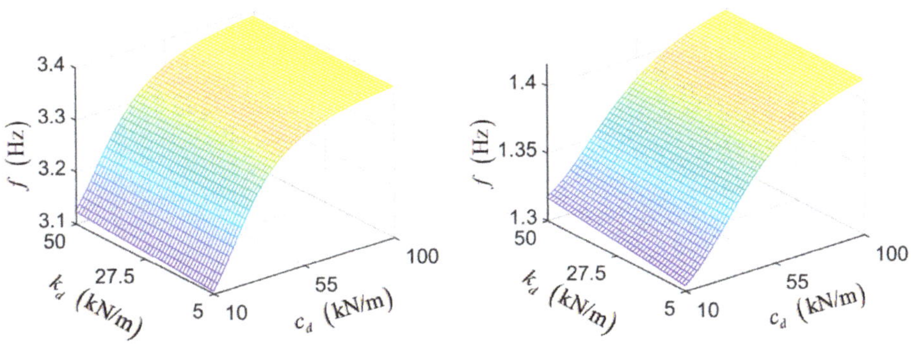

(a) C1 **(b)** C2

Figure 8. Influence of damper stiffness and damping coefficient on cable-damper system frequency.

As can be seen from Figure 8, the influence of the damper damping coefficient on system frequency is much greater than that of damper stiffness. Under different k_d, the damping coefficient has a smaller effect on the cable frequency. As the damping coefficient increases, the system frequency also rises. The influence of damper stiffness on cable frequency is different under different c_d. When c_d is small, k_d has a significant effect on frequency. For the C1 cable, the maximum influence amplitude is 0.6%, and the C2 cable maximum influence amplitude can reach 1%.

To further analyze the effect of the damping coefficient on the first four frequencies of the cable, the relative frequency ratio index (Equation (15)) was defined, and the four frequencies were normalized. The first four frequencies results of the two cables are shown in Figure 9, and the damper stiffness is 10 kN/m. For comparison, this paper examines the first four natural frequencies of the cable when the damping coefficient is infinite, effectively treating the cable as a rigidly supported structure at the damper location. The results are presented in Table 6.

$$Rf = \frac{f_d}{f_c} \tag{15}$$

where f_d is the calculated frequency of the system, and f_c is the calculated frequency of the cable without damper.

Table 6. Cable frequency of cable with stiffness support.

Cable	Order 1	Order 2	Order 3	Order 4
C1	3.396	6.783	10.175	13.568
Rf	1.111	1.111	1.112	1.112
C2	1.402	2.732	4.095	5.461
Rf	1.111	1.111	1.111	1.111

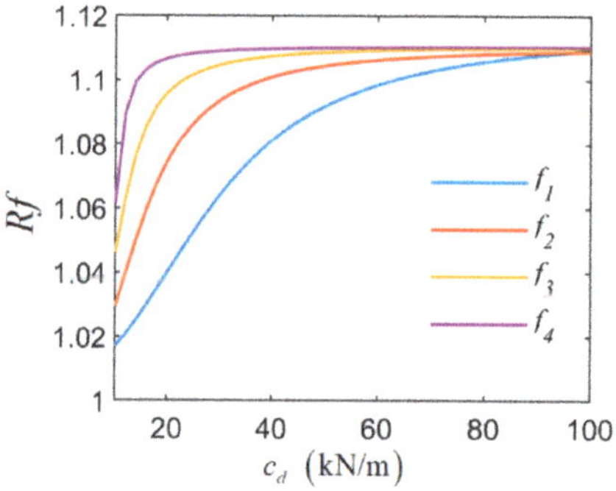

(a) C1

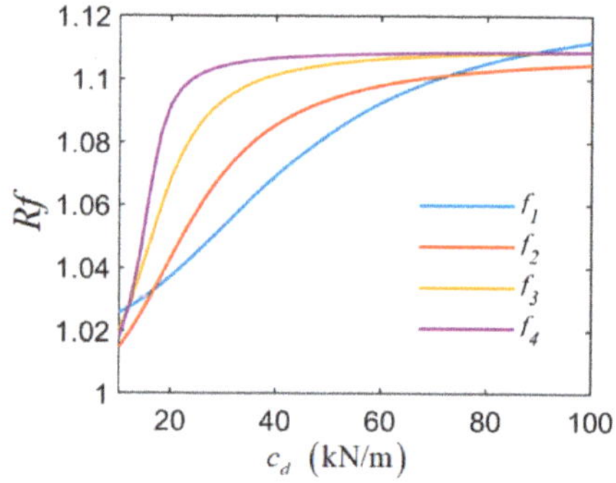

(b) C2

Figure 9. Influence of damping coefficient on cable frequency.

From Figure 9, it can be seen that when the damper stiffness is determined, as the damping coefficient increases, the first four orders of the frequencies of the cable are increased, and the final increases in amplitude are all near 10%. However, when the damping coefficient increases to a certain value, the cable frequency tends to be stable and no longer increases. The value of *Rf* is close to that of the cable frequency when it is rigidly supported. This corresponds to the optimal damping coefficient that enhances the cable frequency. Beyond this point, increasing the damping coefficient further will not significantly increase the stiffness of the cable-damper system. This corresponds to the optimal damping coefficient of the damper at this frequency, representing the minimum damping required to prevent a significant increase in the stiffness of the cable-damper system. With the increase in frequency order, the optimal damping coefficient of cable frequency decreases. When the optimal damping coefficient is reached, increasing the damping coefficient has no significant effect on the adjustment of the cable-damper system stiffness. This shows that the damping coefficient should be selected according to the target order frequency in the actual design of a cable-damper system. The lower the frequency order, the greater the damping coefficient required to achieve the same frequency increase.

As can be seen from Figure 8, the effect of damper stiffness on the cable frequency is influenced by the damping coefficient. Therefore, this paper analyzes the influence of damper stiffness variation on cable frequency when corresponding to different damping coefficients. The frequency difference index Δf (Equation (16)) is defined for analysis. When $\Delta f > 0$, the contribution of damper stiffness to cable stiffness is positive; when $\Delta f < 0$, the contribution of damper stiffness to cable stiffness is negative. Figure 10 shows the variation of frequency difference index with damping coefficient. In Figure 10, Δf reflects the change value of frequency with each change of 0.1 k_d. The farther the frequency deviates from the 0 axis, the greater the influence.

$$\Delta f(n_k, n_c) = f\left(k_{d,n_k+1}, c_{d,n_c}\right) - f\left(k_{d,n_k}, c_{d,n_c}\right) \tag{16}$$

where n_k = 1–45 is the order of values of k_d; n_c = 1–46 is the order of values of c_d; $f\left(k_{d,n_k}, c_{d,n_c}\right)$ is the frequency of the cable-damper system calculated using the n_kth k_d and n_cth c_d.

It can be seen from Figure 10 that with the increase in the damping coefficient, the contribution of k_d to the first 4th-order frequencies of the lasso shows a decreasing (positive contribution)-increasing (negative contribution)-decreasing (negative contribution) trend. When the damping coefficient is between 10 and 40 kN/m, the effect of the damper stiffness on the system frequency is most pronounced. In this paper, c_d is defined as the critical damping coefficient of damper stiffness when crossing the 0 axis. When the damping coefficient is less than c_d, the stiffness of the system increases with the increase in k_d, and the increase in k_d makes a positive contribution to the stiffness of the system. When the damping coefficient is greater than c_d, the stiffness of the system decreases with the increase in k_d, and the increase in dk makes a negative contribution to the stiffness of the system. With the increase in frequency order, the c_d decreases. When c_d exceeds 40 kN/m, the influence of the change in k_d on the first four frequencies of the system tends to be stable and small. This shows that when the damping coefficient is large enough, the influence of the damper stiffness on the stiffness of the system can be ignored.

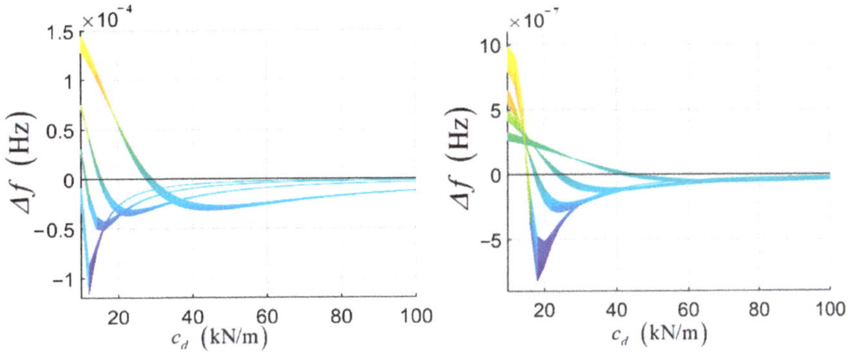

(**a**) C1 (**b**) C2

Figure 10. Influence of damper stiffness on cable frequency.

5.2. Influence of Damper Installation Height on Cable Frequency

Considering that the lower end of the external damper is typically fixed to the main beam, the installation height of the damper is generally kept low when analyzing its impact on the cable frequency. Therefore, this paper analyzes the first four order frequencies of the cable when the installation height of the damper changes from 0.01 l_0 to 0.1 l_0. To assess the influence of the damper's stiffness and damping coefficient on the frequency, three different working conditions are considered: case 1, k_d = 0, c_d = 20 kN/m; case 2, k_d = 10 kN/m, c_d = 0; and case 3, k_d = 10 kN/m, c_d = 20 kN/m. The analysis results of how the first four order frequencies of the two cables change with the installation height of the damper under the three working conditions are shown in Figure 11.

From Figure 11, except for the first order frequency of the C2 cable in Case 2, the first four order frequencies of the cable all increase with the installation height of the damper. This shows that, the higher the damper is installed, the more beneficial it is to increase cable stiffness. As seen in Figure 11a,b, when only the damping coefficient is considered, the impact of installation height on frequency increases exponentially. The effect on high-order frequencies is more significant than that on low-order frequencies. The increase in frequency at 0.1 l_0 installation height is nearly 12% higher than that at 0.01 l_0 installation height. It can be seen from Figure 11c,d that, when only the stiffness of the damper is considered, when the height of the damper is relatively small, k_d will weaken the system stiffness, which is unfavorable to the system. For the short cable C1, when the installation height is around 0.02 l_0, the influence on the first four order frequencies is positive. For the long cable C2, when the installation height is around 0.065 l_0, the influence on the first four order frequencies is positive. It can be observed that, compared to the short cable, the installation height required for the long cable is greater. However, from the perspective of influence amplitude, when k_d is fixed, the influence of μ on the first four order frequencies of the two cables is less than 0.3%. It can be seen from Figure 11a–d that the influence of k_d on system frequency can be ignored compared with the influence of c_d on system frequency. Therefore, in Figure 11e,f, when the damper has both damping coefficient and stiffness, the influence on the cable-damper system is mainly guided by the damping coefficient of the damper, and its variation rule is similar to that of Figure 11a,b.

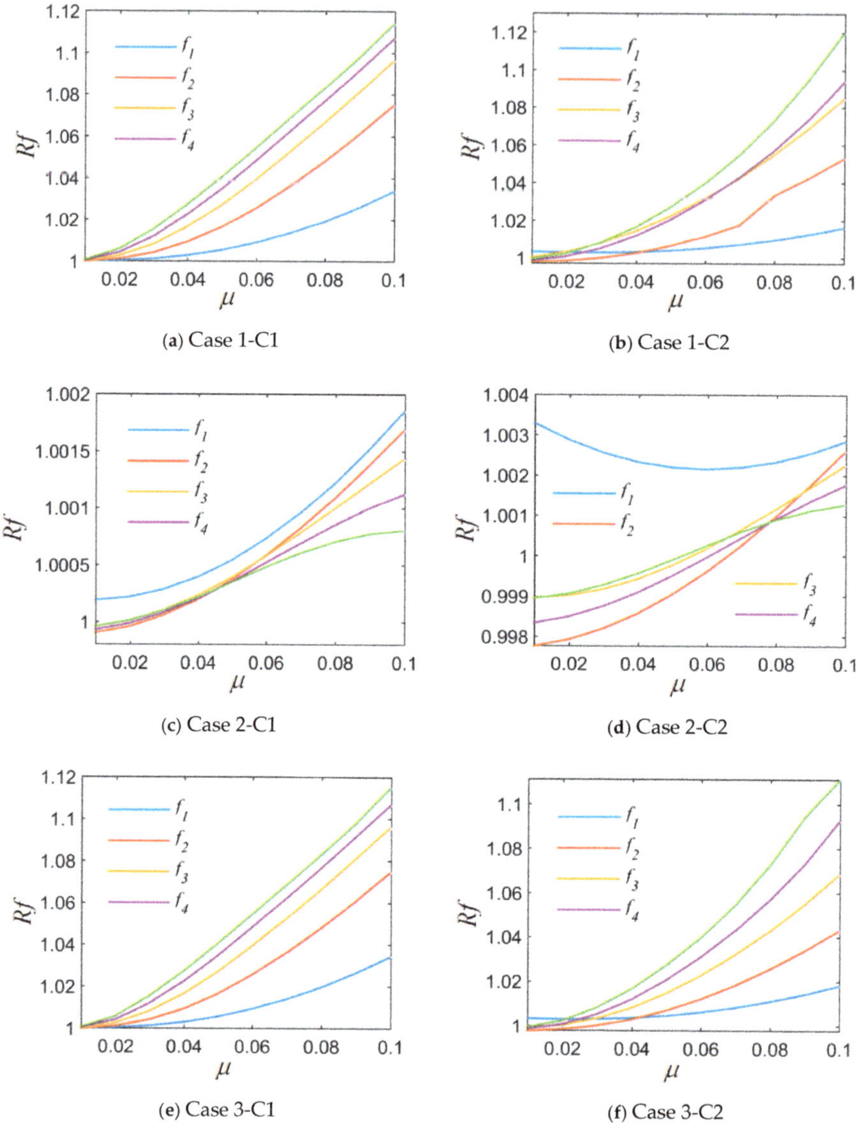

Figure 11. Effect of damper installation height on cable frequency.

In summary, in the system, the damping coefficient of the damper and the installation height of the damper have the most significant influence on the system, while the damper stiffness has relatively little influence on the system. In the design of damper parameters, the damping coefficient and installation height should be taken as the main design parameters, and the damper stiffness should be secondary.

6. Discussion

Based on the dynamic stiffness method, this paper presents an accurate dynamic analysis approach for the cable-damper system. By considering factors such as cable bending stiffness, sag, inclination angle, cable force, and mass per unit length, along with damper stiffness, damping coefficient, and installation height, the proposed method provides a more realistic representation of the cable-damper system as found in actual bridges. The proposed method is more accurate for the frequency analysis of the cable-damper system in comparison to the experimental results of the literature and the results from the real bridge. By analyzing the influence of damper parameters on the dynamic characteristics of the system, the following conclusions are obtained:

(1) The damping coefficient and installation height of the damper significantly influence the dynamic behavior of the cable, whereas the damper stiffness has a relatively minor effect. The analysis indicates that variations in the damping coefficient and installation height can lead to frequency changes exceeding 10%, with the impact increasing at higher vibration modes. The impact of damper stiffness on frequency is less than 1%; the impact on long cable is greater than that of short cable, and the impact on the first order frequency is higher than the impact on other order frequencies.

(2) As the damping coefficient and installation height of the damper increase, the frequency of each mode of the cable also rises. For different order frequencies, there is an optimal damping coefficient. When the optimal damping coefficient is reached, increasing the damping coefficient has little effect on increasing the cable stiffness. As the order increases, the optimal damping coefficient decreases. When the damper is installed within 0.1 L, the damper's contribution to the stiffness of the system increases exponentially.

(3) The influence of the damper stiffness on the dynamic characteristics of the cable is affected by the damping coefficient of the damper, and there is a critical damping coefficient. As the cable order increases, the critical damping coefficient decreases. When the damping coefficient is less than the critical coefficient, increasing the damper stiffness will increase the stiffness of the system. When the damping coefficient exceeds the critical damping value, increasing the damper stiffness reduces the overall stiffness of the cable system. However, once the damping coefficient is sufficiently large, the effect of damper stiffness on the dynamic characteristics of the cable becomes negligible, and its influence on the system frequency stabilizes.

This paper mainly shows the application of the proposed method in the dynamic characteristic analysis of the cable-damper system. This method enhances the design, safety, and longevity of cable-stayed bridges by providing more accurate predictions of dynamic behavior, optimizing damper placement, and supporting better maintenance strategies. In addition, from the proposed cable frequency characteristic equation, it includes cable frequency, cable force, cable parameters, and damper parameters. Therefore, it has great potential to be applied to cable parameter identification and damper effect evaluation. From the analysis results of this paper, it can be seen that the damper has a relatively large impact on the frequency of the cable. If the cable parameter identification method without a damper is used, the identification results will inevitably have large errors. The method proposed in this paper can provide a better solution to this kind of problem, which is also the direction of the authors' next work.

To further strengthen the discussion, future research directions could focus on expanding the dynamic analysis method to account for additional parameters that influence the behavior of cable-damper systems. For example, investigating the effects of varying environmental conditions, such as temperature fluctuations and wind speed changes, on the dynamic response of cables could provide deeper insights into the long-term performance and reliability of the system. Additionally, refining the modeling approach to incorporate more complex nonlinearities or to analyze the interaction between multiple cables and dampers in large-scale bridge networks could enhance the applicability of the method to a wider range of real-world scenarios.

Author Contributions: Conceptualization, H.J. and H.Y.; Methodology, H.J., B.X., Z.J., C.C. and H.Y.; Software, H.J. and Z.J.; Validation, H.J., Z.J. and H.Y.; Formal analysis, C.C. and H.Y.; Investigation, H.Y.; Resources, Z.J.; Data curation, H.J., B.X., Z.J. and H.Y.; Writing—original draft, H.J., B.X., Z.J.

and H.Y.; Writing—review & editing, H.J., B.X., C.C. and H.Y.; Visualization, H.Y.; Supervision, Z.J.; Project administration, Z.J., C.C. and H.Y.; Funding acquisition, C.C. All authors have read and agreed to the published version of the manuscript.

Funding: This research was funded by the National Natural Science Foundation of China (Grant No. 52108290); the China Postdoctoral Science Foundation (Grant No. 2022 M712906); the Foundation of Henan Educational Committee (Grant No. 242102240131).

Data Availability Statement: The data presented in this study are available on request from the corresponding author.

Conflicts of Interest: Authors Hui Jiao and Zhengkai Jiang were employed by the company Investment Company of China Railway Seventh Group Co., Ltd. The remaining authors declare that the research was conducted in the absence of any commercial or financial relationships that could be construed as a potential conflict of interest.

References

1. Larsen, A.; Larose, G.L. Dynamic Wind Effects on Suspension and Cable-Stayed Bridges. *J. Sound Vib.* **2015**, *334*, 2–28. [CrossRef]
2. Fujino, Y.; Siringoringo, D. Vibration Mechanisms and Controls of Long-Span Bridges: A Review. *Struct. Eng. Int.* **2013**, *23*, 248–268. [CrossRef]
3. Sun, P.; Wang, S.; Wang, D. The Numerical Calculation Method Based on Equivalent Frequency-Independent Time-Domain Damping Model. *J. Low Freq. Noise Vib. Act. Control* **2022**, *41*, 1339–1353. [CrossRef]
4. Wang, S.; He, J.; Fan, J.; Sun, P.; Wang, D. A Time-Domain Method for Free Vibration Responses of an Equivalent Viscous Damped System Based on a Complex Damping Model. *Vib. Act. Control.* **2023**, *42*, 1531–1540. [CrossRef]
5. Fournier, J.A.; Cheng, S. Impact of Damper Stiffness and Damper Support Stiffness on the Efficiency of a Linear Viscous Damper in Controlling Stay Cable Vibrations. *J. Bridge Eng.* **2014**, *19*, 04013022. [CrossRef]
6. Pacheco, B.M.; Fujino, Y.; Sulekh, A. Estimation Curve for Modal Damping in Stay Cables with Viscous Damper. *J. Struct. Eng.* **1993**, *119*, 1961–1979. [CrossRef]
7. Krenk, S. Vibrations of a Taut Cable with an External Damper. *J. Appl. Mech.* **2000**, *67*, 772–776. [CrossRef]
8. Lu, Q.; Sun, Z.; Zhang, W. Nonlinear Parametric Vibration with Different Orders of Small Parameters for Stayed Cables. *Eng. Struct.* **2020**, *224*, 111198. [CrossRef]
9. Krenk, S.; Nielsen, S.R.K. Vibrations of a Shallow Cable with a Viscous Damper. *Proc. R. Soc. London. Ser. A Math. Phys. Eng. Sci.* **2001**, *458*, 339–357. [CrossRef]
10. Main, J.A.; Jones, N.P. Vibration of Tensioned Beams with Intermediate Damper. II: Damper near a Support. *J. Eng. Mech.* **2007**, *133*, 379–388. [CrossRef]
11. Sun, L.; Xu, Y.; Chen, L. Damping Effects of Nonlinear Dampers on a Shallow Cable. *Eng. Struct.* **2019**, *196*, 109305. [CrossRef]
12. Mehrabi, A.B.; Tabatabai, H. Unified Finite Difference Formulation for Free Vibration of Cables. *J. Struct. Eng.* **1998**, *124*, 1313–1322. [CrossRef]
13. Fujino, Y.; Hoang, N. Design Formulas for Damping of a Stay Cable with a Damper. *J. Struct. Eng.* **2008**, *134*, 269–278. [CrossRef]
14. Javanbakht, M.; Cheng, S.; Ghrib, F. Control-Oriented Model for the Dynamic Response of a Damped Cable. *J. Sound Vib.* **2019**, *442*, 249–267. [CrossRef]
15. Zhou, Q.; Nielsen, S.R.K.; Qu, W. Stochastic Response of an Inclined Shallow Cable with Linear Viscous Dampers under Stochastic Excitation. *J. Eng. Mech.* **2010**, *136*, 1411–1421. [CrossRef]
16. Sun, Q.; Zhang, N.; Liu, X. A Dynamic Stiffness Matrix Method for Free Vibrations of Partial-Interaction Composite Beams Based on the Timoshenko Beam Theory. *J. Sound Vib.* **2022**, *520*, 116579. [CrossRef]
17. Banerjee, J.R.; Ananthapuvirajah, A. Free Vibration of Functionally Graded Beams and Frameworks Using the Dynamic Stiffness Method. *J. Sound Vib.* **2018**, *422*, 34–47. [CrossRef]
18. Li, X.; Cheng, Z.; Wang, Z.; Fan, B. Free Vibration of a Taut Cable with a Parallel-Connected Viscous Mass Damper and a Grounded Cross-Tie. *J. Low Freq. Noise Vib. Act. Control* **2022**, *41*, 982–995. [CrossRef]
19. Xu, B.; Dan, D.; Zhao, Y. Frequency-Domain Estimation Method for Vibration-Induced Additional Cable Tension Based on Acceleration Monitoring. *J. Vib. Acoust. Trans. ASME* **2019**, *141*, 064502. [CrossRef]
20. Dan, D.-H.; Xu, B.; Chen, Z.-H. Universal Characteristic Frequency Equation for Cable Transverse Component System and Its Universal Numerical Solution. *J. Eng. Mech.* **2016**, *142*, 04015105. [CrossRef]
21. Furukawa, A.; Suzuki, S.; Kobayashi, R. Tension Estimation Method for Cable with Damper Using Natural Frequencies with Uncertain Modal Order. *Front. Built Environ.* **2022**, *8*, 812999. [CrossRef]
22. Tabatabai, H.; Mehrabi, A.B. Design of Mechanical Viscous Dampers for Stay Cables. *J. Bridge Eng.* **2000**, *5*, 114–123. [CrossRef]

buildings

MDPI

Article

Overall Lifting Construction Control Method for Large-Segment Steel Arch Bridges Based on Unstressed State Control Theory

Zhongpei Li [1], Xuetao Dong [2], Hairong Chen [3], Liangjun Chi [3] and Zhicheng Zhang [2,*]

[1] College of Urban Construction, Hangzhou Polytechnic, Hangzhou 311402, China; lzp@mail.hzpt.edu.cn
[2] Department of Civil Engineering, Zhejiang University, Hangzhou 310058, China; dongxuetao@zju.edu.cn
[3] Shaoxing Shangyu Urban Construction Group Co., Ltd., Shaoxing 312399, China; chenhairong2024@126.com (H.C.); chiliangjun2024@126.com (L.C.)
* Correspondence: jszzc@zju.edu.cn

Abstract: The construction method of first splicing the low brackets and then lifting steel arch bridges has become increasingly popular, and its construction control has become a key issue. According to the unstressed state control theory, both the horizontal displacement and rotation angle at the lifted arch segment ends should be approximately 0 during the lifting process. An analytical solution for the position of the lifting points and the tension force in temporary horizontal rods is proposed. The analytical method is verified through a detailed comparison of the analytical results with the finite element model results and the on-site monitoring results. The results show that, for steel arch bridges whose main arch axis is the n-order parabola, the reasonable range of the parameter k is 0.8~0.9 and the specific value is related to the parameter n. The tension force in temporary horizontal rods can be quickly determined. The results meet the requirements of the unstressed state control theory, and the internal forces at the vault and lifting points can be maintained within a reasonable range. This research could be used to effectively control the deformation and stress distributions in the main arch of Shunjiang Bridge during the overall lifting process. The methods proposed provide effective guidance for using the overall lifting method for steel arch bridges.

Keywords: overall lifting construction method; construction control; unstressed state control theory; reasonable lifting point; temporary horizontal tension

1. Introduction

Steel arch bridges have remarkable performance advantages, such as a light weight, high strength, variable shape, and convenient construction [1]. Thus, this kind of arch bridge has been widely applied in bridge engineering, especially in large-scale municipal bridge projects. In the construction of large-scale steel arch bridges, the main arch is mostly prefabricated in segments and assembled on site. The stress distribution characteristics of the segmental prefabricated main arch are significantly different from those of the integrated main arch before closure. Also, temporary facilities or structures, such as framing and cables, are necessary to improve the safety of the main arch's construction. The construction control of the main arch is a key factor during construction [2,3]. Instead of the full-framing method, the swivel method and the cable-hoisting method are commonly used for concrete arch bridges and concrete-filled steel tubular arch bridges [4–6], while the cable-stayed cantilever method and free cantilever method are traditionally used for steel arch bridges. In recent years, the overall lifting method, in which a large, prefabricated steel arch segment is lifted into the designated position using large-scale equipment, has been

 https://doi.org/10.3390/buildings15040523

commonly used [7]. This method was initially applied in the construction of large special steel structures [8] and was further adopted in bridge construction after its successful application in many airports and large ports [9].

In the overall lifting method used in the construction of steel arch bridges, a low support system is first used to assemble each large segment, and then the large segment that will be positioned in the middle is lifted. This method not only has the advantages of framing construction methods, such as a simple operation and high safety level, but also has a short construction period, reduces the required aerial work, and has high stability [10,11]. For example, the Huaihe Bridge of the Shanghang high-speed railway employed the overall lifting construction method for large-section steel pipe arch ribs, which shortened the construction period by nearly one month and reduced the material usage significantly [12].

A schematic diagram of the overall lifting construction method is shown in Figure 1. Firstly, an arc-shaped low support system is erected on a trestle under the longitudinal projection of the designed arch axis, and then the arch elements are vertically assembled to form three segments, namely the two side arch segments and the middle arch segment. Next, two temporary towers are erected to lift the middle arch segment into the designated position. Finally, the three arch segments are drawn together either via welding with a closure section or by directly and accurately trimming a small number of arch plates without a closure section. To control the stability and deformation of the large-segment steel arch during the overall lifting process, especially when using the method without a closure section, temporary auxiliary measures must be implemented. Tensioning temporary horizontal tie rods at the lifting points is a common method, as shown in Figure 1.

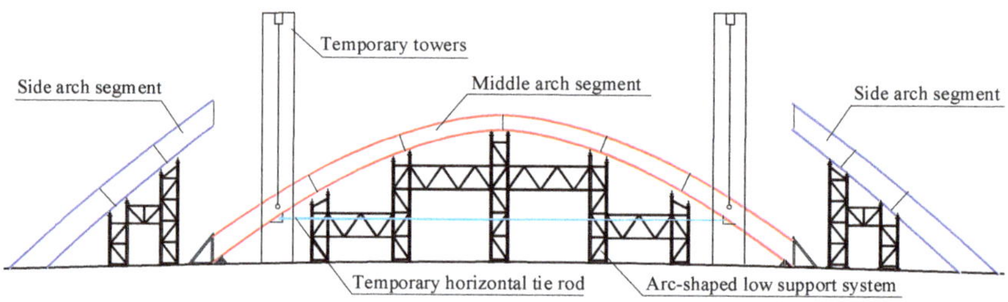

Figure 1. A schematic diagram of the overall lifting construction method for large-segment steel arch ribs.

Due to the strong reliance on lifting facilities [13,14], the theoretical research on and engineering applications of the overall lifting construction method for steel arch bridges are less developed than those of other types of construction methods. Regarding the insufficient theoretical research, numerical analysis is an effective method for studying large steel structures [15]. Most existing studies on the overall lifting method have used practical engineering projects as the research background and used the finite element method for qualitative analysis. For example, Yu Wang et al. [16] studied the overall lifting construction method used for a large cross-sea bridge and developed a monitoring scheme focusing on the control of the arch–rib alignment, the stress, and the temperature, which are key aspects of overall lifting construction control. Focusing on the determination of the lifting point position through a floating crane, Xinge Miao et al. [17] proposed a method that combines mechanical calculation formulas with finite element modeling. The objective of finding a reasonable lifting point position was put forward to ensure a balanced bending moment distribution along the arch rib. Importantly, the bending moments at the vault and lifting point should be equal. Hui Xiang et al. [18] conducted a detailed construction-phase analysis using finite element software, focusing on the overall lifting construction method used for the Xinjiang Jiangxi Branch Bridge of the Changjing–Huang Railway. The specific construction control measures were provided to ensure strict control of the stress and deformation of arch ribs and brackets during construction.

With the development and application of the overall lifting construction method for steel arch bridges, construction control has become an urgent issue. However, the above-mentioned studies based on the finite element method only obtained qualitative conclusions. Studies focused on key technical parameters, such as the optimal lifting point position and the temporary horizontal tension force, are insufficient and cannot provide immediate and accurate references for bridge construction. This study focuses on the key technical parameters for the overall lifting construction method when applied to large-segment steel arch ribs. Based on the unstressed state control theory, a method to determine the optimal lifting point position was proposed and an analytical formula to determine the tension force of the temporary horizontal tie rod was derived. The research findings were applied to the construction of a steel box tied-arch bridge, guiding the overall lifting construction process of large-segment steel arch ribs. This paper will provide a convenient reference for steel arch bridge projects that adopt the overall lifting construction method.

2. Review of Unstressed State Control Theory

The unstressed state control method [19,20] was first proposed by Qin Shunquan, the chief engineer of the China Railway Bridge Bureau, and is a theoretical control method proposed for the phased construction of bridges. This method has been widely used in the construction control of large-scale bridge projects, making significant contributions to high-quality bridge construction [21–23]. This method points out that the following mechanical equilibrium formula should be satisfied during the phased construction of bridges:

$$[K]\{\delta\} = \{P\} + \{L_0\} \tag{1}$$

where $[K]$ is the stiffness matrix composed of the boundary conditions of the structural system; $\{\delta\}$ is the node displacement vector of the structure; $\{P\}$ is the external load vector; and $\{L_0\}$ is the additional generalized load vector, caused by variations in the stress-free states of the elements in the finished bridge state influenced by different construction methods and processes.

Formula (1) indicates that the final internal force and displacement of the bridge using the phased construction method are uniquely solved by four conditions: (1) the position and magnitude of the external loads; (2) the geometric dimensions, positions, and stiffnesses of the structural elements; (3) the boundary conditions of the structure; and (4) the unstressed length and unstressed curvature of each element. According to the unstressed state control theory, once the four specific conditions are met, the internal force and displacement in the finished bridge state are uniquely determined for any bridge structure constructed in stages and are independent of the construction process. That is, the final bridge state achieved through the phrased construction can be theoretically identical to the state achieved through the single-stage construction. Among these conditions, the first three are relatively easy to control, whereas the unstressed length and curvature of the elements change during the phased construction. This is the key and most challenging point in construction control. Linking the intermediate and finished bridge states through the unstressed length and curvature of the elements to control the construction processes is the core concept of the stress-free state control theory.

According to the concept of the unstressed state control theory, to ensure that the main arch of the steel arch bridge can achieve the design target state upon the completion of the phased construction process, the unstressed length and curvature should closely match the design targets. After the middle section of the steel arch is lifted, it should accurately connect with the two side arch segments. To achieve this, the horizontal displacement and rotation angle at both ends of the middle arch segment need to be controlled so that they remain around zero before closure, and the middle arch segment can be seamlessly welded with the side arch segments on the support brackets. However, owing to the fact that the lifted segment is large and the loading conditions are complex, significant deformations may occur during lifting. Furthermore, the structural system is unstable before closure, making it prone to being in an unfavorable stress state. Therefore, it is essential to adopt specific construction measures to control horizontal displacement and rotation angle at the ends of the lifted segment and ensure that the arch is in a favorable internal force state. Here, selecting appropriate lifting points and tensioning temporary horizontal tie rods between them is a low-cost and effective method.

On the basis of the principles of the unstressed state control method, the following sections provide an innovative analytical control method for the determinations of the appropriate positions of the lifting points and the optimal temporary horizontal tension force.

3. Analytical Control Method for the Overall Lifting Method

3.1. Analytical Formula of Temporary Horizontal Tension

In the analytical method, only the self-weight is considered. Other load types, such as the wind load and temperature action, are not considered. The deformation diagram of the lifted steel arch segment during the lifting process is shown in Figure 2. The basic parameters of the lifted steel arch are defined as follows: the longitudinal length of the whole lifted arch segment is l; the sagittal height of the lifted arch segment is f; the axis shape of the arch is assumed to be the n-order parabola; to simplify the tension force calculation of the temporary horizontal tie rod, the section of the steel arch is assumed to be uniform and the cross-sectional area and the moment of inertia are A and I, respectively; the elastic modulus of the arch rib material is E; the ratio of the longitudinal distance between the two lifting points to longitudinal length of the whole lifted arch segment l is k ($0 < k \leq 1$); and the temporary horizontal tension force after the elastic deformation of the horizontal tie rod is P.

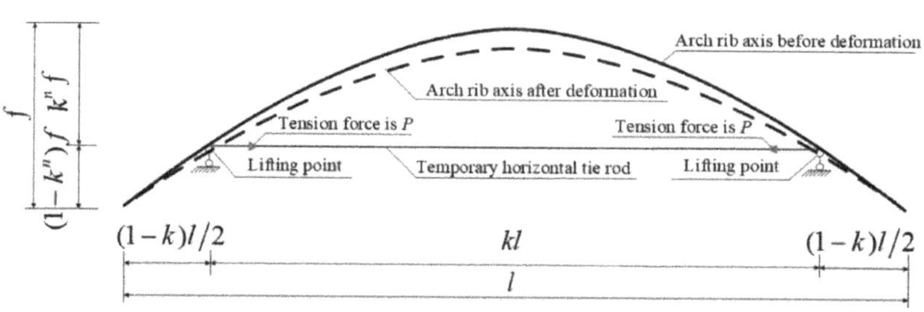

Figure 2. Deformation diagram of the lifted steel arch segment during the lifting process.

Here, since the external forces and bounding conditions acting on the arch rib are symmetrical, only half of the structure was analyzed. The force analysis diagram of the half structure for the overall lifting method of the lifted steel arch segment is shown in Figure 3. In the follow-up analysis, the longitudinal bridge direction is defined as the x-axis direction and the vertical bridge direction is defined as the y-axis direction. The vault is defined as the origin of the coordinates. The function of the arch axis curve is defined as $y(x)$. The weight of the uniform-section steel arch per unit length along the x-axis is marked as q. For the steel arch bridge, the structural self-weight is low and the rise–span ratio of the lifted arch segment f/l is usually small, while the weight of the steel arch per unit length in the longitudinal direction $q_1(x)$ is approximately equal to q, as shown below.

$$q_1(x) = q\sqrt{1 + (y'(x))^2} \approx q \tag{2}$$

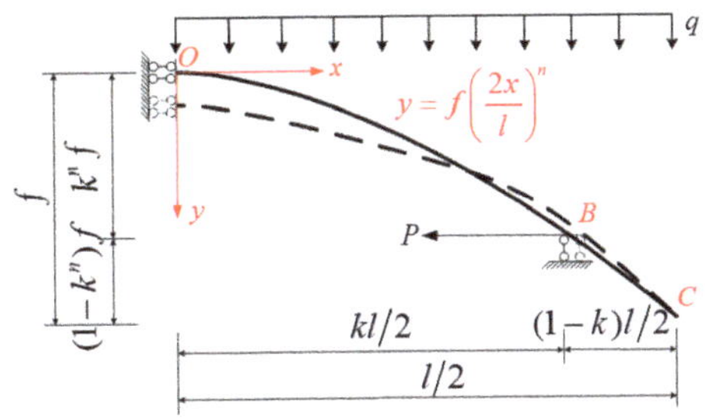

Figure 3. Force analysis diagram of the half structure for the overall lifting method.

Because the lifted steel arch segment is subject to small deformations and remains in the linear and elastic state during construction, according to the linear elastic theory [24], the deformation occurring during the overall lifting of the steel arch is primarily due to the internal bending moments and axial forces, whereas the deformation from shear forces can be ignored. The following conventions for internal forces are established: the bending moment is defined as positive when the lower edge of the steel arch is in tension and negative when the upper edge is in tension. Similarly, the axial force is considered positive when the section is under tension and negative when the section is under compression. By using the unit load method derived from the principle of virtual work [25], the displacement at point C at the end of the lifted steel arch segment can be calculated as follows:

$$\Delta C = \frac{1}{EI}\int_0^{s/2} M(x)\overline{M}(x)ds + \frac{1}{EA}\int_0^{s/2} N(x)\overline{N}(x)ds \tag{3}$$

In Formula (3), since the rise–span ratio of the lifted arch segment f/l is usually small, ds can be approximated as dx. $M(x)$ and $N(x)$ are the bending moment and axial force of the arch rib along the longitudinal bridge direction, respectively, which can be calculated as follows:

$$\begin{gathered} M(x) = \begin{cases} Pf\left(\frac{2x}{l}\right)^n - Pk^n f + \frac{kql^2}{4} - \frac{ql^2}{8} - \frac{qx^2}{2}, 0 \le x \le \frac{kl}{2} \\ \frac{-q\left(\frac{l}{2}-x\right)^2}{2}, \frac{kl}{2} < x \le \frac{l}{2} \end{cases} \\ N(x) = \begin{cases} -P\cdot\sqrt{1+(y')^2} = -P\sqrt{1+\left(\frac{2^n f x^{n-1}}{nl^n}\right)^2}, 0 \le x \le \frac{kl}{2} \\ 0, \frac{kl}{2} < x \le \frac{l}{2} \end{cases} \end{gathered} \tag{4}$$

In Formula (3), $\overline{M}(x)$ and $\overline{N}(x)$ are the bending moment and axial force under the virtual unit load at point C, respectively. The horizontal unit force to the left is assumed to act at point C, and the bending moment and axial force of the arch rib are obtained as follows:

$$\begin{gathered} \overline{M}(x)_x = f\left(\frac{2x}{l}\right)^n - f, 0 \le x \le \frac{l}{2} \\ \overline{N}(x)_x = -\sqrt{1+\left(\frac{2^n f}{nl^n}x^{n-1}\right)^2}, 0 \le x \le \frac{l}{2} \end{gathered} \tag{5}$$

A clockwise unit moment is assumed to act at point C, and the bending moment and axial force of the arch rib are obtained as follows:

$$\begin{aligned}\overline{M}(x)_{\theta} &= -1, 0 \leq x \leq \tfrac{l}{2} \\ \overline{N}(x)_{\theta} &= 0, 0 \leq x \leq \tfrac{l}{2}\end{aligned} \tag{6}$$

By substituting Formulas (4) and (5) into Formula (3) and letting $\Delta Cx = 0$, the following equation can be obtained:

$$P_{\Delta Cx=0} = \frac{ql^2}{8f} \cdot \frac{X(n,k)}{1 + \varphi(n,k,A,I,f,l)} \tag{7}$$

In Formula (6), $X(n,k)$ and $\varphi(n,k,A,I,f,l)$ are the dimensionless coefficients of the arch rib parameters, and the analytical formula is derived as follows:

$$X(n,k) = \frac{\frac{2}{n+3}k^{n+3} - \frac{4n+6}{(n+1)(n+2)}k^{n+2} + \frac{2}{n+1}k^{n+1} - \frac{2}{3}k^3 + 3k^2 - 2k - \frac{2}{(n+1)(n+2)(n+3)} + \frac{1}{3}}{\frac{n(2n+1-k^n)k^{n+1}}{(n+1)(2n+1)}} \tag{8}$$

$$\varphi(n,k,A,I,f,l) = \frac{I}{Af^2} \cdot \frac{(n+1)(2n+1)}{n(2n+1-k^n)k^n} \cdot \left[1 + \frac{k^{2n-2}}{(2n-1)n^2 2^{n-3}} \cdot \frac{f^2}{l^2}\right] \tag{9}$$

In the above formula, $X(n,k)$ is related to the bending moment and $\varphi(n,k,A,I,f,l)$ is related to the axial force. The axial force of the steel arch in the overall lifting process mainly comes from the force of temporary horizontal tension rods and is small compared with the stress state of the completed bridge; thus, the horizontal displacement of the arch rib is generally very small. Thus, the order of magnitude of $\varphi(n,k,A,I,f,l)$ is generally less than 10^{-2}. Therefore, Formula (6) can be further simplified by ignoring the influence of the axial force of the steel arch in the lifting process:

$$P_{\Delta Cx=0} = \frac{ql^2}{8f} \cdot X(n,k) \tag{10}$$

By substituting Formulas (4) and (6) into Formula (3) and letting $\Delta C_{\theta} = 0$, the following equation can be obtained:

$$P_{\Delta C\theta=0} = \frac{ql^2}{8f} \cdot \theta(n,k) \tag{11}$$

In Formula (11), $\theta(n,k)$ is the dimensionless coefficient only about n and k, and the analytical formula is as follows:

$$\theta(n,k) = \frac{(n+1)(3k^2 - 1)}{3nk^{n+1}} \tag{12}$$

3.2. Coefficient Analysis

Equations (10) and (11) indicate that if the axial force in the overall lifting process of the steel arch is ignored, the reasonable tension force of the temporary horizontal tie rods required to ensure that the horizontal displacement and rotation angle at the end of the lifted arch segment are approximately equal to 0 depends only on the values of the coefficients $X(n,k)$ and $\theta(n,k)$. As shown in Figure 4, the variation trends of $X(n,k)$ and $\theta(n,k)$ with the parameter k are the same. Both coefficients are positively correlated with parameter k, and their growth rates decrease as k increases. It can be discovered from Figure 4 that the $X(n,k)$–k curves with different parameter n values converge together within a certain range of k, as well as the $\theta(n,k)$–k curves.

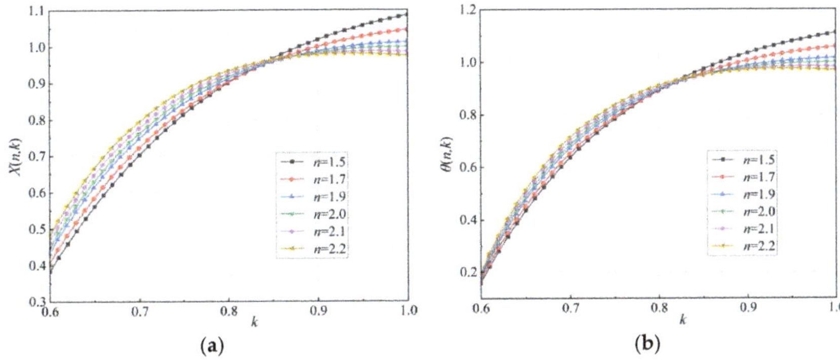

Figure 4. Relationships between the coefficients and the parameter k. (**a**) $X(n,k)$; (**b**) $\theta(n,k)$.

According to the unstressed state control method, the horizontal displacement and the rotation angle at the end of the lifted arch segment should be controlled to be close to 0 at the same time during actual construction, while the force of the temporary horizontal tie rod is unique. Thus, a reasonable lifting point position, which corresponds to a reasonable parameter k, should be determined to solve this contradiction. According to Equations (10) and (11), the ratio of the tension forces when the horizontal displacement and the rotation angle are 0 is equal to the ratio of $X(n,k)$ to $\theta(n,k)$. Figure 5 shows the relationship between the ratio of $X(n,k)$ to $\theta(n,k)$ and the parameter k. It can be found from the figure that when the value of k is in the range of 0.8~0.9, the value of $X(n,k)/\theta(n,k)$ can be controlled within the range 0.99~1.01. This means that the deviation of the tension forces when the horizontal displacement and the rotation angle are 0 is within 1%, which fully meets the requirements in engineering use. Furthermore, the specific value of k gradually increases as the parameter n increases.

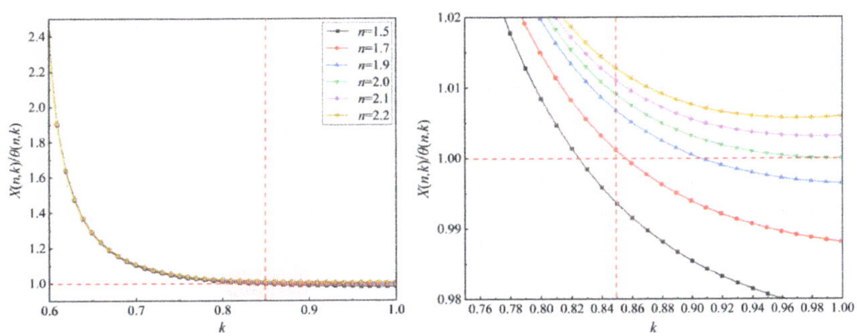

Figure 5. Relationships between the ratio of $X(n,k)$ to $\theta(n,k)$ and the parameter k.

3.3. Internal Force Analysis of Key Sections of Steel Arch Segment

During the lifting process, the vault section and the sections at lifting points are the key sections [18]. According to Formulas (3) and (6), the internal forces at the vault section can be obtained as follows:

$$\begin{cases} M(0) = \frac{ql^2}{8}\cdot\left[2k - 1 - k^n\cdot\frac{X(n,k)}{1+\varphi(n,k,A,I,f,l)}\right] \approx \frac{ql^2}{8}\cdot[2k - 1 - k^n\cdot X(n,k)] \\ N(0) = -P \approx -\frac{ql^2}{8f}\cdot X(n,k) \end{cases} \tag{13}$$

The internal force of the section at the lifting point is as follows:

$$\begin{cases} M\left(\frac{kl}{2}\right) = \frac{-ql^2}{8}\cdot(1-k)^2 \\ N\left(\frac{kl}{2}\right) = -P\sqrt{1+\frac{4f^2}{l^2}k^{2n-2}} \approx -\frac{ql^2}{8f}\cdot X(n,k)\cdot\sqrt{1+\frac{4f^2}{l^2}k^{2n-2}} \end{cases} \quad (14)$$

According to Formulas (13) and (14), the bending moments and axial forces of the key sections of the steel arch are plotted in Figures 6 and 7, respectively. In Figure 7, for the convenience of comparison, the rise–span ratio of the lifted arch segment is taken as 1:4. Figure 6a shows that the absolute value of the bending moment of the vault section decreases at first and then increases as the parameter k increases for different n parameter values. It reaches the minimum value when k is approximately 0.85. Figure 6b shows that the absolute value of the bending moment at the lifting point decreases as the parameter k increases and is independent of the parameter n. Furthermore, the rate of decrease begins to reduce when k is approximately 0.85. Relevant studies [17,18] have pointed out that, when the peak values of positive and negative bending moments of arch ribs are closed, it is easier to control the internal force level of arch ribs within a reasonable range. Figure 6c indicates that when k is in the range of 0.8~0.9, the bending moments of the vault and the lifting point are approximately equal and the internal force of the bending moment of the arch rib during lifting is in a reasonable state.

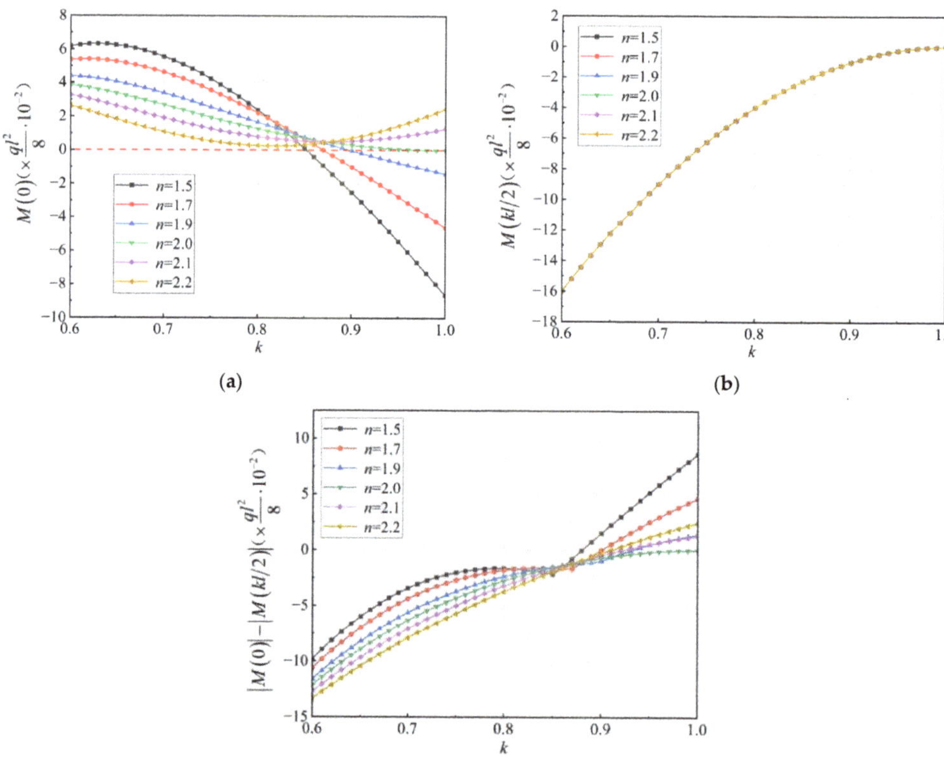

Figure 6. Relationships between bending moments of key sections and the parameter k. (**a**) Vault section; (**b**) lifting point; (**c**) difference between the absolute values.

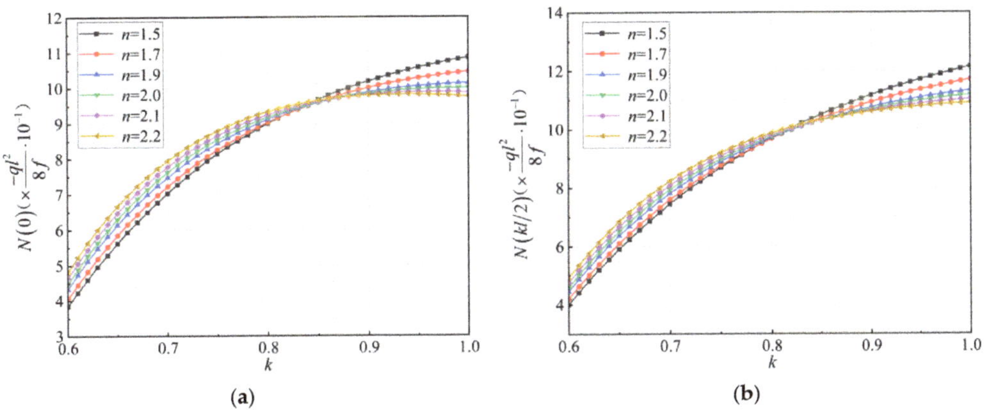

Figure 7. Relationships between axial forces of key sections and the parameter k. (**a**) Vault section; (**b**) lifting point (take f/l = 1:4).

It can be seen from Figure 7 that the cross-sectional axial forces at the vault and the lifting point gradually increase as the parameter k increases, and the rate of increase rate begins to decrease when k is approximately 0.85. For different parameter n values, the cross-sectional axial forces at the vault and the lifting point are similar when k is approximately 0.85.

In summary, when the value of k is in the range of 0.8~0.9, the bending moment and axial force of the key section of the steel arch are within a reasonable range, and the peak value of the bending moment of the key section of the arch rib is relatively low. This means that the recommended range of k obtained by controlling the deviation of $X(n,k)$ and $\theta(n,k)$ at the end of the lifted arch segment based on the unstressed state control method meets the internal force requirements in key sections of the steel arch.

3.4. Selection of Lifting Point and Calculation of Temporary Horizontal Tension

It can be seen from Sections 3.2 and 3.3 that, when $X(n,k)/\theta(n,k) = 1$ and the internal forces in the key sections are taken as the optimization objectives, the recommended range of k is 0.8~0.9. According to Figure 5, the specific value of k is related to the arch rib parameter n. Assuming that the value of k is determined as k_0 through Figure 5, the temporary tension force in the horizontal tie rod P_0 can be calculated as follows:

$$P_0 = \frac{ql^2}{8f} \cdot X(n, k_0) \tag{15}$$

Notably, if the influence of the axial force generated by the horizontal tension cables during the lifting process is considered, the value of P_0 needs to be reduced. According to Equations (7) and (9), the temporary tension force in the horizontal tie rod considering the influence of the axial force P_0' is as follows:

$$P_0' = \frac{ql^2}{8f} \cdot \frac{X(n, k_0)}{1 + \varphi(n, k_0, A, I, f, l)} \tag{16}$$

If the actual influence of the axial force on the arch rib is considered, the vertical coordinate in Figure 5 should actually be $X/[(1 + \varphi)\cdot\theta]$, and the reasonable value of k should be smaller. Therefore, it is recommended that the value of k_0 should be determined through Figure 5 by taking the upper limit value of X/θ, which is 1.01 in this paper. If so,

k_0 is recommended to be approximately 0.815 when n = 1.5, while it is recommended to be approximately 0.890 when n = 2.0.

To facilitate the use of the proposed analytical construction control method in practical projects, the recommended lifting point position and the tension force of the temporary horizontal tie rod for different values of the parameter n are calculated and summarized in Table 1.

Table 1. Quick calculation table of arch rib lifting point selection and temporary horizontal tension.

n	k_0	$X(n,k_0)$	$\varphi(n,k_0,A,I,f,l)$	P_0	P_0'
1.5	0.815	0.9224	$\frac{2.78I}{Af^2}\cdot\left(1+\frac{0.51f^2}{l^2}\right)$	$\frac{ql^2}{8f}\cdot X$	$\frac{ql^2}{8f}\cdot\frac{X}{1+\varphi}$
1.6	0.830	0.9401	$\frac{2.66I}{Af^2}\cdot\left(1+\frac{0.37f^2}{l^2}\right)$		
1.7	0.845	0.9548	$\frac{2.55I}{Af^2}\cdot\left(1+\frac{0.28f^2}{l^2}\right)$		
1.8	0.860	0.9664	$\frac{2.45I}{Af^2}\cdot\left(1+\frac{0.21f^2}{l^2}\right)$		
1.9	0.875	0.9749	$\frac{2.35I}{Af^2}\cdot\left(1+\frac{0.17f^2}{l^2}\right)$		
2.0	0.890	0.9801	$\frac{2.25I}{Af^2}\cdot\left(1+\frac{0.13f^2}{l^2}\right)$		
2.1	0.895	0.9796	$\frac{2.20I}{Af^2}\cdot\left(1+\frac{0.10f^2}{l^2}\right)$		
2.2	0.900	0.9788	$\frac{2.15I}{Af^2}\cdot\left(1+\frac{0.08f^2}{l^2}\right)$		

3.5. Example Verification

To verify the accuracy of the proposed construction control method, a certain steel arch segment to be lifted was considered as an example and finite element analysis was conducted. The structural and material parameters were as follows: n = 2, l = 103.2 m, f = 22.6 m, E = 2.06 × 10^5 MPa, I = 0.8545 m^4, A = 0.4331 m^2, and q = 34.0 kN/m.

From Table 1, the suggested value of k_0 is 0.890. The calculated result of P_0 is 1963.0 kN, while the calculated value of P_0' is 1946.0 kN. The relative deviation of P_0 from P_0' is 0.9%. It is verified that the actual axial force has little influence on the reasonable value of the temporary horizontal tension force.

Based on the recommended values of k_0 and P_0', the beam element model of the lifted steel arch segment established through the finite element software Midas Civil 2021 is shown in Figure 8. The deformation of the steel arch before and after the tie rod is tightened is shown in Figure 9 and Figure 10, respectively. The results show that the horizontal displacement and rotation angle at the end of the lifted arch segment are close to 0 when k_0 is 0.890 and the temporary horizontal tension is 1946.0 kN. This indicates that the derived formula can provide a reasonable and accurate solution to meet the requirements of the unstressed state control method under the given conditions.

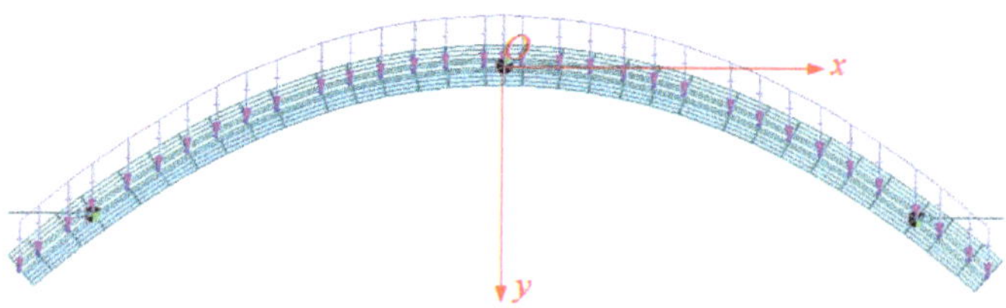

Figure 8. Finite element model of the steel arch example.

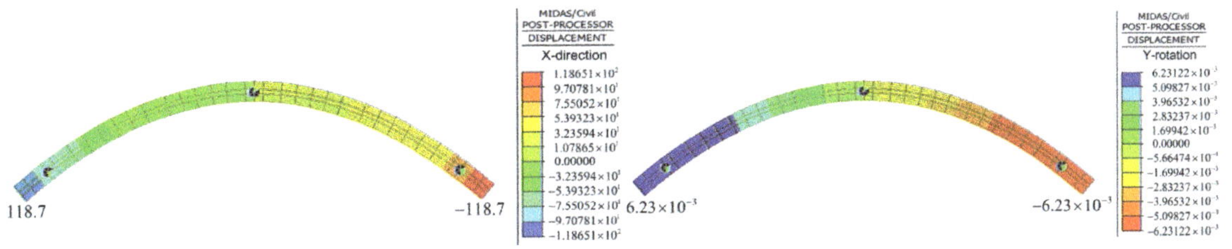

(a) (b)

Figure 9. Deformation of the arch rib before the tie rod is tightened. (**a**) Horizontal displacement (unit: mm); (**b**) rotation angle (unit: rad).

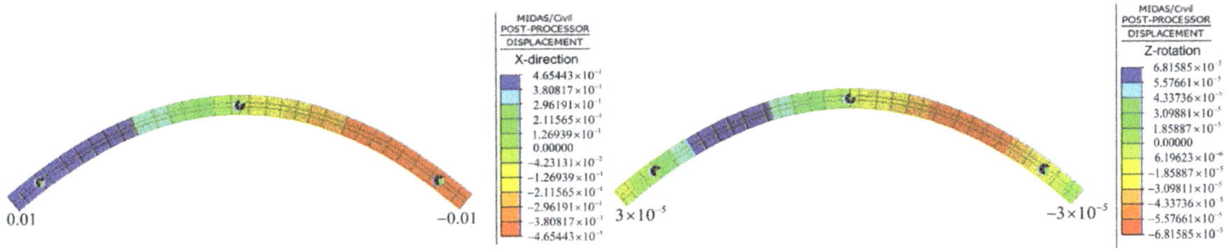

(a) (b)

Figure 10. Deformation of the arch rib after the tie rod is tightened. (**a**) Horizontal displacement (unit: mm); (**b**) rotation angle (unit: rad).

4. Application of Proposed Construction Control Method

The proposed construction control method was applied in the engineering project of Shunjiang Bridge. This application case is introduced as follows.

4.1. Design Overview

Shunjiang Bridge is located in Shangyu District of Shaoxing City, which is in Zhejiang Province of China. It is a bottom-supported box-section steel arch bridge with a standard span of 176 m. The completed state of the bridge is shown in Figure 11a. The calculated span of Shunjiang Bridge is 170 m, while the arch height is 50 m. The theoretical rise–span ratio of the main arch is 0.294. The longitudinal projection lines of the main arches are a 1.5-order parabola followed by a 1.7-order parabola, as shown in Figure 11b. The transverse projection lines of the main arches are the oblique straight lines, as shown in Figure 11c. The cross-section of the main arch is octagonal. The cross-sectional sizes of the arch springing and the vault are 4.5 m × 2.2 and 2.5 m × 2.2 m with linear transition. The arch vault section and the arch springing section are shown in Figure 11d,e. The standard cross-sectional width in the middle span of the bridge is 22.5 m. There are 26 pairs of suspenders along the whole bridge, which employ parallel steel wire stay cables with a standard strength of 1670 MPa. The maximum length of these suspenders is 47 m. The spacing between the upper lifting points of the suspenders is 4 m, and the spacing between the lower lifting points of the suspender is 6 m. The suspenders are attached to the beam via an anchor pipe and are connected to the arch through lifting lugs. A cantilever sidewalk is set on one side of the main girder to help pedestrians on the east bank levee and the west bank beach to cross the bridge. The cantilever and the main arch form a loop on the facade to enhance the overall look of the main bridge.

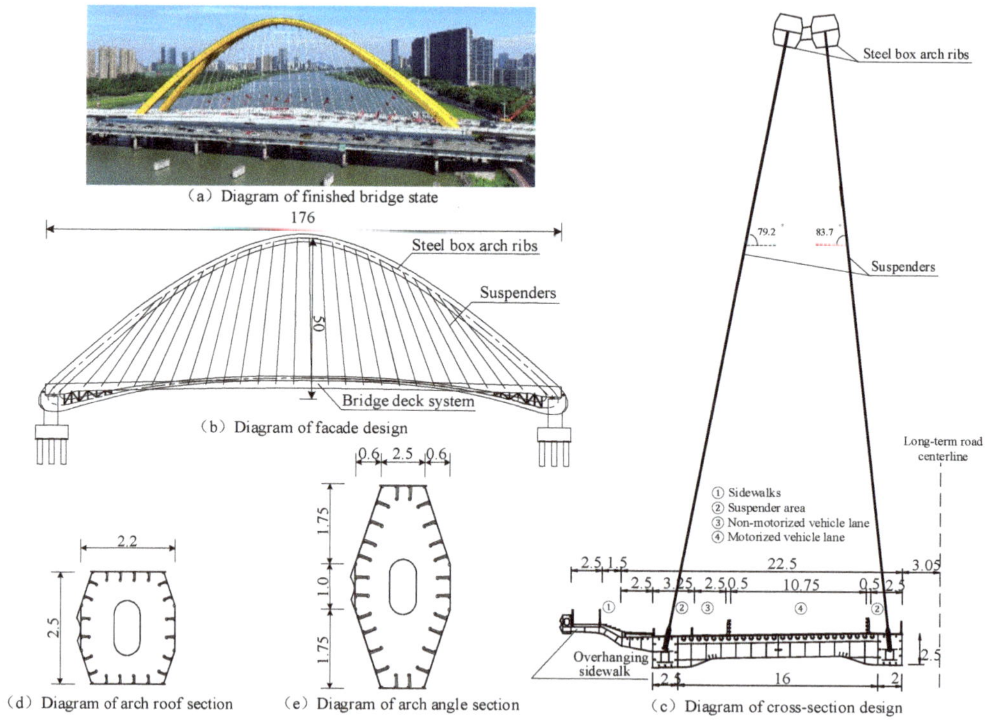

Figure 11. Design drawing of Shunjiang Bridge (unit: m). (**a**) Completed bridge state; (**b**) facade design; (**c**) cross-sectional design; (**d**) arch vault section; (**e**) arch springing section.

4.2. Construction Method Overview

Shunjiang Bridge was constructed with the sequence of "beam first and arch second". All the segmented steel components were fabricated in a centralized manner in a factory and transported to the site for lifting and assembly. Once the main girder installation was completed, arch rib welding and overall lifting construction were carried out, which was the critical construction phase. The arch ribs were divided into three longitudinal sections. The main arch segments at both ends were hoisted and welded into place by gantry cranes, which were temporarily supported and fixed through steel pipe tire frames. All of the segments were assembled vertically on the tire frame through a gantry crane. The middle main arch segment, which is made of Q345 steel, has a longitudinal length of 103.7 m and a dead weight of 858.3 t. Two portal lifting frames were employed to elevate the assembled middle main arch segment to the designated installation position. When they were welded to the adjacent arch segments, the structural conversion of the steel arch was completed. The construction site for the overall lifting construction of Shunjiang Bridge is illustrated in Figure 12, and the relevant design parameters are detailed in Table 2.

The overall lifting arch segment of Shunjiang Bridge is a 1.5-order parabola arch rib. According to Table 1, the recommended value of k_0 is 0.815. Then, the reasonable lifting point spacing $k_0 l$ is calculated as 84.5 m, and the equivalent uniform load is determined as q = 31.7 kN/m. According to the data in Table 1, the recommended temporary horizontal tension P_0' is 1734.2 kN, and each main arch is tensioned with one horizontal tie rod. Since the arch rib of Shunjiang Bridge is a type of oblique arch rib with non-uniform cross-sections, and the two steel arch ribs are slightly asymmetrical in the transverse direction, the reasonable tension force of the temporary horizontal tie rod calculated above needs to be further iteratively optimized through finite element analysis.

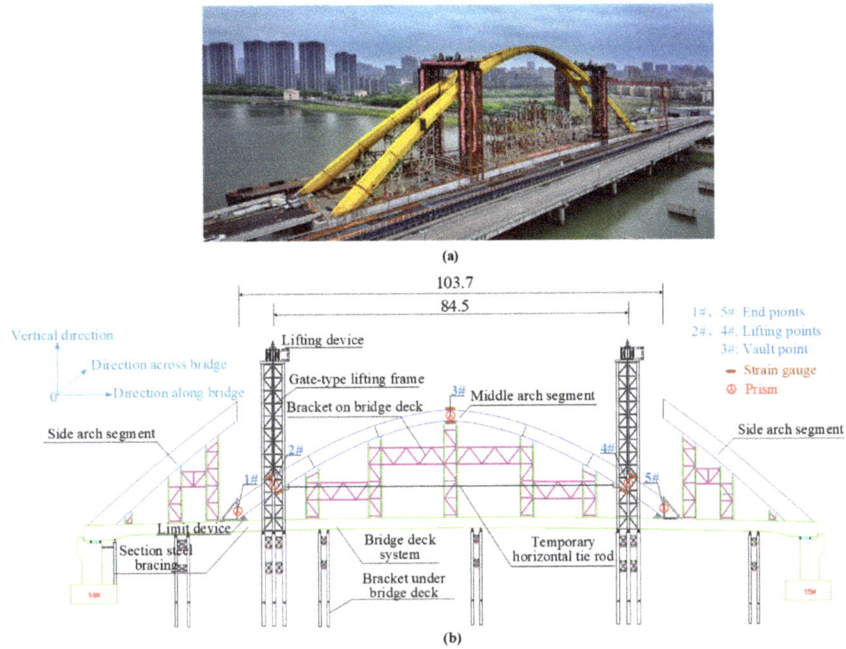

Figure 12. Drawing of the overall lifting construction of Shunjiang Bridge. (**a**) Construction state; (**b**) overall lifting construction (unit: m).

Table 2. Design parameters of the lifted main arch segment.

Parabola Shape Parameter n	Arch Longitudinal Length l (m)	Arch Rise f (m)	Elastic Modulus E (kN/m^2)	Area of Arch Vault Section A (m^2)	Inertia Moment of Arch Vault Section I (m^4)	Volumetric Weight γ (kN/m^3)
1.5	103.8	22.6	2.06×10^8	0.2621	0.2257	78.50

4.3. Finite Element Analysis Model

MIDAS Civil 2021 software was used to establish the beam element analysis model of the lifted middle main arch segment by setting a reasonable lifting point position and inputting the initial force of the temporary horizontal tie rod, as shown in Figure 13a. The whole lifting section model was constructed by the beam element B31 and the truss element T3D2. The model contained 214 beam elements, 2 truss elements, and 224 nodes. Here, the value of k_0 was set as 0.815 through the proposed analytical method. Based on the recommended temporary horizontal tension P_0', the iterative optimization with the goal of controlling the horizontal deformation and rotation angle of the arch end so that they remained around 0 at the same time was performed. The optimized inner arch cable force was 1575.0 kN, and the optimized outer arch cable force was 1915.0 kN, and these values were used in actual construction. The vertical lifting cable force of 2171 kN was one quarter of the weight of the lifted middle main arch segment. In addition, to analyze the local mechanical performance around the arch lifting point, the local shell element model of the segment near the arch lifting point was established in ABAQUS 6.14 software, as shown in Figure 13b. The local model was constructed by the shell element S4R. The mesh size was not unique and was more refined in the lift anchor components. The local model contained 62,464 elements and 61,535 nodes. In this model, the simply supported coupling points were set at the end of the lifted arch segment. The horizontal tension force and vertical lifting force were applied at the lifting points, respectively. The internal forces transmitted on the ends of the middle main arch segment were added into the model.

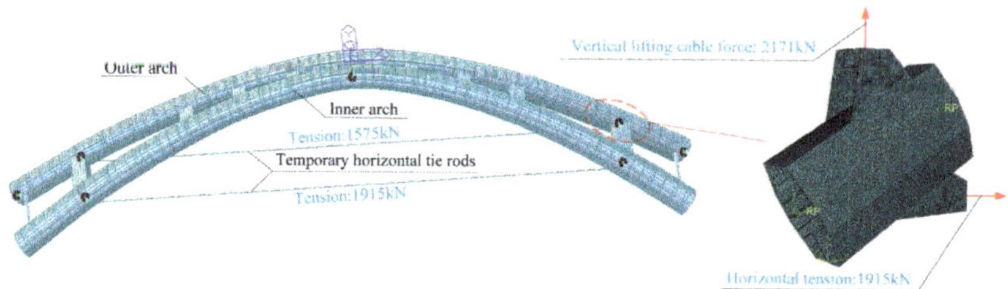

(a) (b)

Figure 13. Calculation model of the lifted middle main arch segment. (**a**) Beam element model of the lifted middle main arch segment; (**b**) local shell element model of the segment near the arch lifting point.

4.4. Finite Element Analysis Results

4.4.1. Calculation Results of Beam Element Model

After setting the reasonable lifting point position through the proposed analytical method and the tension force of the temporary horizontal tie rod through the finite element analysis, the deformation of the middle main arch segment of Shunjiang Bridge in overall lifting construction is shown in Figure 14. It can be seen from Figure 14 that both the horizontal displacement and the rotation angle at the end of the lifted arch segment are close to 0 simultaneously.

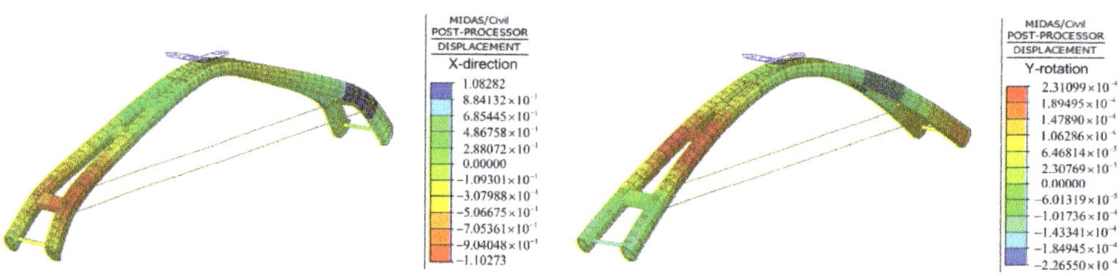

(**a**) (**b**)

Figure 14. Deformation of middle arch segment in overall lifting construction. (**a**) Horizontal displacement (unit: mm); (**b**) rotation angle (unit: rad).

The stress distribution of the middle arch segment during the overall lifting process of Shunjiang Bridge is depicted in Figure 15. The normal stress at the upper edge of the middle main arch segment ranges from −11.8 to 9.4 MPa, whereas the stress at the lower edge ranges from −19.7 to 0 MPa. The maximum absolute values of the stresses at the vault section are −5.3 MPa and −7.9 MPa for the upper and lower edges, respectively. At the lifting point, the maximum stresses for the upper and lower edges of the arch rib are 9.4 MPa and −19.7 MPa, respectively. The stress in the key sections is relatively low, which indicates that the lifted arch segment is in a secure state with a significant safety margin.

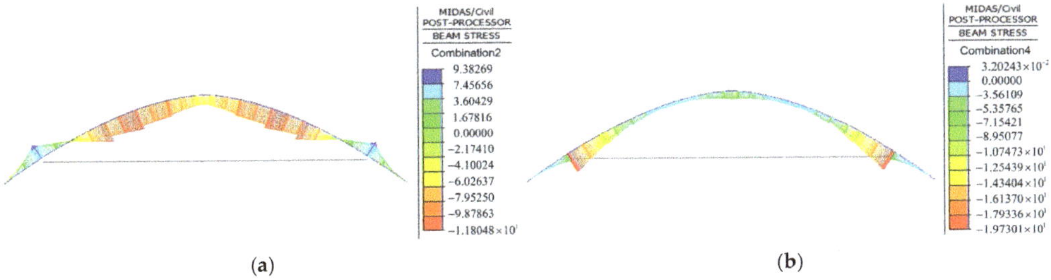

Figure 15. Stress distribution of middle arch segment in overall lifting construction (unit: MPa). (**a**) Upper edge; (**b**) lower edge.

4.4.2. Calculation Results of Local Shell Element Model

Owing to the constructional complexity near the lifting point, the overall calculation results cannot accurately reflect the stress distribution and stability performance of steel plates in local areas. Further analysis based on the local shell element model is necessary.

Figure 16a shows the von Mises stress distribution for the established segment near the lifting point of the steel arch. The maximum von Mises stress in the main arch near the lifting point is 51.9 MPa, as shown in Figure 16b. The maximum stress in the temporary tension and lift anchor components is 138.9 MPa, as shown in Figure 16c. Both values are less than the design strength of Q345 steel, which is 270 MPa, thus meeting the specification requirements. The maximum von Mises stress at the lifting point occurs on the steel member at the anchor point of the arch rib. The first-order instability mode for the established segment near the lifting point is depicted in Figure 17. It can be seen that the critical load coefficient is 21.5, which exceeds the limiting value of 4.0 in the design code [26], thus meeting the specification requirements. The most instability-susceptible area is located at the diaphragm near the lifting point of the main arch.

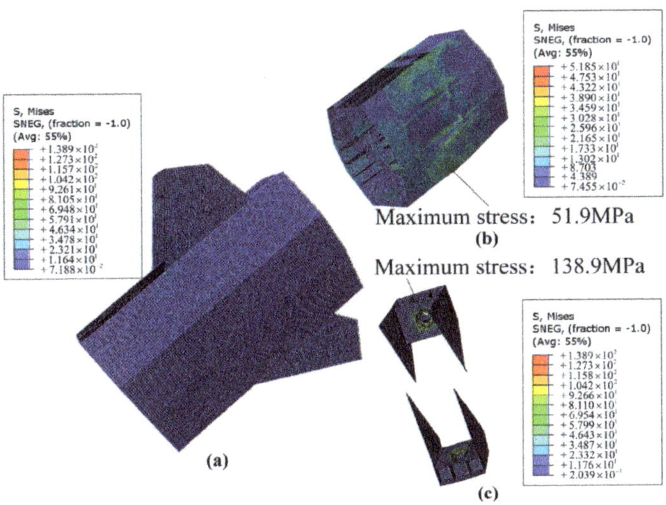

Figure 16. Von Mises stress nephogram of the established arch segment near the lifting point (unit: MPa). (**a**) Whole established arch segment; (**b**) main body of the steel arch segment; (**c**) temporary tension and lift anchor components.

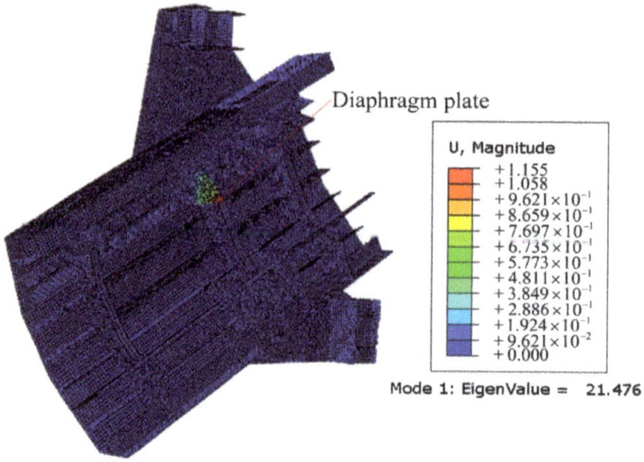

Figure 17. First-order buckling mode of the established segment near the lifting point.

4.5. On-Site Monitoring Results

During the actual lifting process for the middle arch segment, the inclinometers were positioned at the wind support near the lifting point and the inclination angle of the arch segment was monitored throughout the whole lifting process. Figure 18 shows the measured inclination angle of the arch rib lifting section in both the longitudinal and transverse directions during the lifting process. Table 3 presents the changes in the coordinates of the measuring points at the end of the lifted arch segment. The locations of the monitoring points in Table 3 are marked in Figure 12.

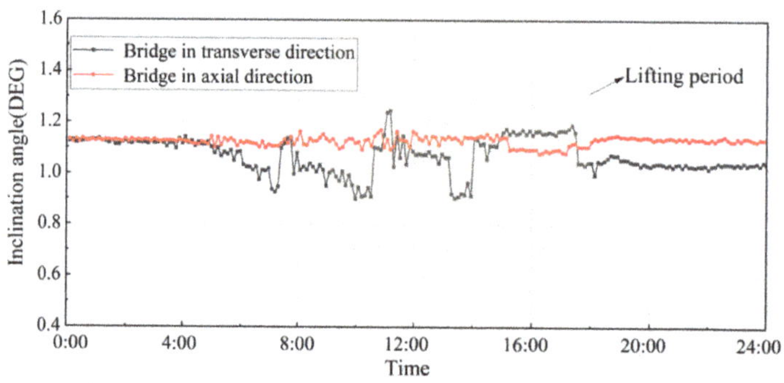

Figure 18. Measurement results of the inclination angle in the lifting section.

Table 3. Coordinate measurement results at the end of the lifted arch segment.

Number	Measuring Result (m)			Coordinates Before Lifting (m)			Deviation Value (mm)		
	Direction Along Bridge	Direction Across Bridge	Vertical Direction	Direction Along Bridge	Direction Across Bridge	Vertical Direction	Direction Along Bridge	Direction Across Bridge	Vertical Direction
1#	36.18377	6.47273	45.36168	36.1736	6.4644	23.05662	10.17	8.33	15.06
2#	51.09771	5.51281	54.87641	51.09866	5.51012	32.57208	−0.95	2.69	14.33
3#	88.13294	4.23741	67.08195	88.13951	4.24834	44.78131	−6.57	−10.93	10.64
4#	123.31742	5.36685	55.75667	123.31981	5.36909	33.45448	−2.39	−2.24	12.19
5#	139.54647	6.47551	45.56474	139.55481	6.45929	23.26053	−8.34	16.22	14.21

The measurements indicated that the inclination angle of the lifted arch segment remained stable throughout the lifting process, the posture of the arch rib in the overall lifting section was in a controllable state. The maximum offset between mileage and each measuring point was 10.17 mm and reached 15.06 mm after accounting for the lifting height, indicating that the lifted middle arch segment was essentially in place.

Based on the finite element analysis results, the key stress measuring points, which included the vault section and the lifting points of the lifted arch segment, were selected for real-time monitoring. A wireless automatic comprehensive testing system was employed for stress monitoring. The sampling frequency was every 10 min.

Figure 19 displays the stress changes of the measuring points on the day when the middle arch segment was lifted. The lifting period of the middle arch segment was from 12:00 to 16:00. The real-time stress monitoring data from key sections of the arch demonstrated that the stress variation in the lifting sections was small throughout the lifting process, and the variation was similar to the theoretical calculation results of the beam element model. Numerical analysis and on-site monitoring results show that the stress of key sections of the steel arch rib was in the range of −25 MPa to 10 MPa, which also verifies the accuracy of the finite element model. This confirmed that the arch lifting section remained essentially stable during the overall lifting process.

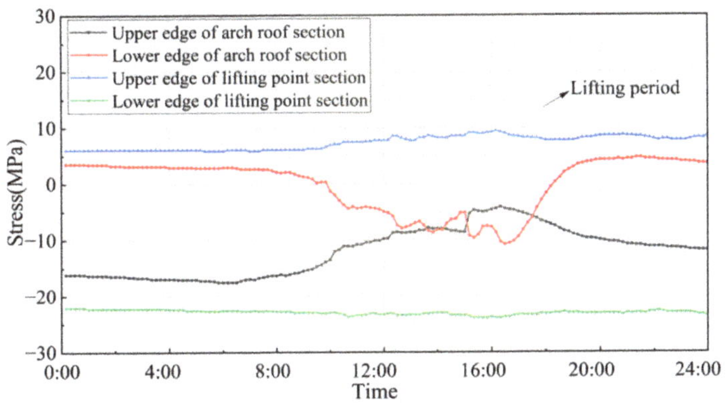

Figure 19. Stress monitoring results in the arch lifting process.

Overall, the on-site monitoring results indicate that the deformation and stress distributions of the main arch during the overall lifting process can be effectively controlled when employing the proposed construction control method. As a result, the middle arch segment of Shunjiang Bridge was successfully lifted, and joint formation was completed in one day.

5. Conclusions and Discussion

5.1. Conclusions

The overall lifting construction method for large-segment steel arch bridges is commonly used. The lifting point positions and the temporary horizontal tension force are key technical parameters for this construction method. On the basis of the unstressed state control theory, this study proposed the analytical control method to ensure that both the horizontal displacement and rotation angle at the lifted arch segment ends are approximately 0 during the lifting process. This method provides an analytical solution for the positioning of lifting points and determination of tension force in temporary horizontal rods. Furthermore, the application case of the proposed construction control method in

Shunjiang Bridge was deeply analyzed. The following conclusions have been drawn in this study:

(1) In the proposed overall lifting construction control method, the position of the lifting point significantly impacts the deformation and internal forces of the steel arch. According to the proposed analytical control method, for steel arch bridges whose main arch axis is an n-order parabola, the reasonable range of the parameter k is 0.8~0.9 under the action of uniformly distributed dead loads. The specific value is related to the parameter n.

(2) Once the reasonable lifting points are determined, the tension force in the temporary horizontal rods can be quickly determined by using the theoretical formula provided in this study. By using the recommended lifting point positions and temporary horizontal tension force, the horizontal displacement and rotation angle at the end of the lifted arch segment can be controlled at around 0 simultaneously, and the bending moments at the vault and lifting points can be maintained within relatively low ranges and at similar values. The axial force of the arch rib is also within a reasonable range.

(3) In the whole lifting process of the middle arch segment of Shunjiang Bridge, the lifting point parameter k was set as 0.815 based on the proposed analytical control method, and the tension force of the temporary horizontal tie rod was further determined through the finite element analysis. On-site monitoring indicates that the deformation and stress distributions of the main arch during the overall lifting process are effectively controlled. The methods proposed in this paper provide direct and effective guidance for steel arch bridge construction using the overall lifting construction method.

5.2. Discussion

In this paper, to simplify the tension force calculation of the temporary horizontal tie rod, only the uniformly distributed dead load is considered in the analytical process. However, arch ribs often have non-uniform cross sections in practical engineering, and they also bear temperature and wind loads, causing the real-world situation to be much more complicated. Therefore, there is some deviation between the simplified conditions presented in this paper and the practical construction site. It is necessary to use the finite element method for some theoretical corrections and dynamic adjustments in the application of this technique. In the future, more complex load conditions will be considered, such as the influence of non-uniform loads and temperature fields.

Author Contributions: Conceptualization, methodology, and writing—original draft, Z.L.; software and formal analysis, X.D.; visualization, data curation, and supervision, H.C.; resources and investigation, L.C.; writing—review and editing and project administration, Z.Z. All authors have read and agreed to the published version of the manuscript.

Funding: This research was funded by the research project of Hangzhou Urban and Rural Construction Committee (No. 2023071).

Data Availability Statement: The original contributions presented in the study are included in the article; further inquiries can be directed to the corresponding author.

Conflicts of Interest: Authors Hairong Chen and Liangjun Chi were employed by the company Shaoxing Shangyu Urban Construction Group Co., Ltd. The remaining authors declare that the research was conducted in the absence of any commercial or financial relationships that could be construed as a potential conflict of interest.

Abbreviations

l	Horizontal projection length of the axis of the whole lifting arch rib segment
f	Sagittal height of the axis of the whole lifting arch rib segment
n	Sub-parabolic axis shape parameter of the arch rib
A	Cross-sectional area of the whole lifting arch rib segment
E	Elastic modulus of the arch rib material
I	Moment of inertia of the arch rib section around the transverse bridge direction
k	Ratio of the horizontal distance between the two lifting points to the horizontal projected length of the arch rib
q	Arch rib bears a uniformly distributed vertical load
P	Temporary horizontal tension
M	Bending moment
N	Axial force

References

1. Xie, K.; Wang, H.; Guo, X.; Zhou, J. Study on the safety of the concrete pouring process for the main truss arch structure in a long-span concrete-filled steel tube arch bridge. *Mech. Adv. Mater. Struct.* **2019**, *6*, 731–740. [CrossRef]
2. Chen, B.C.; Liu, J.P. Review of construction and technology development of arch bridges in the world. *J. Traffic Transp. Eng.* **2020**, *20*, 27–41. (In Chinese)
3. Qi, L.; Zhao, Y.Y. Stress optimization analysis of long-span concrete-flled steel tube arch bridge based on midas full-bridge fnite element analysis model. *Proc. Indian Natl. Sci. Acad.* **2023**, *89*, 267–278. [CrossRef]
4. Zheng, J.L.; Wang, J.J. Concrete-flled steel tube arch bridges in China. *Engineering* **2018**, *4*, 143–155. [CrossRef]
5. Sun, J.; Li, J.; Jiang, Y.; Ma, X.; Tan, Z.; Zhufu, G. Key construction technology and monitoring of long-span steel box tied arch bridge. *Int. J. Steel Struct.* **2023**, *23*, 191–207. [CrossRef]
6. Feng, Y.; Qi, J.; Wang, J.; Zhang, W.; Zhang, Q. Rotation construction of heavy swivel arch bridge for high-speed railway. *Structures* **2020**, *26*, 755–764. [CrossRef]
7. Shi, Y.C. Theoretical Analysis and Model Test Study on Integral Lifting of Six Thousands Tons' Steel Box Arch Bridge. Master's Dissertation, Southwest Jiaotong University, Chengdu, China, 2018.
8. Ruan, R.; Lai, M.; Jiang, C.; Wang, J.; Lin, Y. Integral Lifting of Steel Structure Corridor between Two Super High-Rise Buildings under Wind Load. *Buildings* **2023**, *13*, 2441. [CrossRef]
9. Li, G.; Mo, T.; Xu, N.; Zhang, W.; Lu, H.; Deng, D. Improvement of overall lifting construction technology of large span steel structure hangar. In Proceedings of the 2nd International Conference on Internet of Things and Smart City, Xiamen, China, 18–20 February 2022; SPIE: Bellingham, WA, USA, 2022; Volume 12249, pp. 923–928. [CrossRef]
10. Zhou, G.J. Construction technology of main bridge superstructure of Cuijiaying Hanjiang river bridge on Han-Shi High-speed railway. *World Bridges* **2021**, *49*, 10314. (In Chinese)
11. Wang, J.F.; Xiang, H.W.; Zhang, J.T.; Wu, T.M.; Xu, R.Q. Geometric state transfer method for construction control of a large-segment steel box girder with hoisting installation. *J. Zhejiang Univ.-Sci. A* **2020**, *21*, 382–391. [CrossRef]
12. Zhang, Z.; Pan, W. Lift planning and optimization in construction: A thirty-year review. *Autom. Constr.* **2020**, *118*, 103271. [CrossRef]
13. Song, W.H. Study on Mechanical Performance and Key Issues of Long-span Steel Box Arch Bridge During Integral Lifting and Incremental Launching Construction. Master's Dissertation, Southeast University, Nanjing, China, 2021.
14. Zheng, J. Recent Construction Technology Innovations and Practices for Large-Span Arch Bridges in China. *Engineering* **2024**, *41*, 110–129. [CrossRef]
15. Nassiraei, H. Probabilistic Analysis of Strength in Retrofitted X-Joints under Tensile Loading and Fire Conditions. *Buildings* **2024**, *14*, 2105. [CrossRef]
16. Wen, F.; Liang, X.; Chen, C.; Xu, L.; Feng, Q. Construction Control and Monitoring Platform of a Large-Segment Steel Box Girder with Hoisting Installation. *Appl. Sci.* **2023**, *13*, 9573. [CrossRef]
17. Miao, X.; Yuan, A.; Wang, X.; Chen, Q. Arch First and Beam Later Technology: An Accelerated and Safe Construction for Tied Arch Bridges. *Struct. Eng. Int.* **2024**, 1–9. [CrossRef]
18. Xiang, H.; Xia, D.; Wang, S.; Zhang, Z.; Luo, S.; Chen, W. Stress and deformation analysis of the steel pipe arch integral lifting construction process of a continuous beam arch bridge. *Sci. Rep.* **2024**, *14*, 10314. [CrossRef] [PubMed]
19. Qin, S.Q. *Theory and Practice of Stress-Free-State Control Method*; People's Communications Press: Beijing, China, 2006. (In Chinese)

20. Qin, S.Q. Stress-free-state control method for bridges constructed in stages bridge construction. *Bridge Constr.* **2008**, *1*, 8–14. (In Chinese)
21. Qin, S.; Wei, K.; Qin, J.; Yuan, R.; Xu, L.; Dan, Q. Stress-free-state based structural analysis and construction control theory for staged construction bridges. *Adv. Bridge Eng.* **2020**, *1*, 1. [CrossRef]
22. Du, M.; Tang, H.; Xue, X.; Wang, H.; Wu, G.; Gao, Q. Intelligent control method of large-span arch bridge based on stress-free state. In Proceedings of the International Conference on Mathematics and Machine Learning, Nanjing, China, 24–26 November 2023; pp. 287–291. [CrossRef]
23. Wei, S.Y.; Gong, W.F.; Wu, X.G.; Zhang, Z.H. Nonlinear Stress-Free-State Forward Analysis Method of Long-Span Cable-Stayed Bridges Constructed in Stages. *Buildings* **2023**, *13*, 1735. [CrossRef]
24. Xanthakos, P.P. *Theory and Design of Bridges*; John Wiley & Sons: Hoboken, NJ, USA, 1994. [CrossRef]
25. Strydom, J.H.; Elvin, A. Optimizing the design of steel bridges using the principle of virtual work. In *Insights and Innovations in Structural Engineering, Mechanics and Computation*; CRC Press: Boca Raton, FL, USA, 2016. [CrossRef]
26. *JTG D64-2015*; Industry Standards of the People's Republic of China. Code for Design of Highway steel Bridges. Ministry of Transport of the People's Republic of China: Beijing, China, 2015.

buildings

MDPI

Article

Multi-Hazard Vibration Control of Transmission Infrastructure: A Pounding Tuned Mass Damper Approach with Lifelong Reliability Analysis

Zhuoqun Zhang [1], Lizhong Qi [1], Jingguo Rong [1], Yaping Zhang [1], Peijie Li [1] and Ziguang Jia [2,*]

[1] State Grid Economic and Technological Research Institute Co., Ltd., Beijing 102209, China; zhangzhuoqun@chinasperi.sgcc.com.cn (Z.Z.); qilizhong@chinasperi.sgcc.com.cn (L.Q.); rongjingguo@chinasperi.sgcc.com.cn (J.R.); zhangyaping@chinasperi.sgcc.com.cn (Y.Z.); lipeijie@chinasperi.sgcc.com.cn (P.L.)

[2] School of Infrastructure Engineering, Dalian University of Technology, Dalian 116026, China

* Correspondence: jiaziguang@dlut.edu.cn

Abstract: Power transmission tower-line systems are exposed to various dynamic hazards, including wind and earthquakes, among others. Despite the multitude of dampers proposed to mitigate vibrations, the dual control effect on both seismic and wind-induced vibrations has rarely been addressed. This paper introduces a comprehensive methodology for evaluating the reliability of power transmission towers under a range of dynamic disasters, encompassing both earthquakes and wind loads. Subsequently, a lifelong reliability approach was employed to assess the efficacy of a pounding tuned mass damper (PTMD). The proposed algorithm leverages the incremental dynamic analysis (IDA) method to compute structural fragility with regard to each type of disaster and integrates these findings with hazard functions to determine the probability of overall failure. The results conclusively demonstrate that the PTMD substantially diminished the towers' dynamic response to both earthquakes and wind loads, thereby enhancing their overall reliability. Specifically, the PTMD reduced the vibration reduction ratio by 10% to 30% under wind loads and by 20% to 80% under seismic actions, with more pronounced effects at higher wind speeds and peak ground accelerations (PGAs). Furthermore, the reliability index (β) of the transmission tower increased from 2.1849 to 2.4295 when the PTMD was implemented, highlighting its effectiveness in dual-hazard scenarios. This study underscores the potential for reliability to be considered as a key metric for optimizing damping devices in power transmission structures, particularly in the context of multi-hazard scenarios.

Keywords: dual control; reliability analysis; fragility analysis; pounding tuned mass damper; vibration control; damper

1. Introduction

As the foundational support structures of electrical power systems, transmission towers are indispensable in guaranteeing stable power transmission, enhancing system safety and reliability, and integrating renewable energy sources. They not only facilitate the spanning of power lines across extensive distances but also constitute crucial infrastructure for the efficient distribution and utilization of energy.

Due to their inherent flexibility, transmission towers are prone to damage from a variety of dynamic hazards [1,2], which can result in significant consequences. Over the past few decades, numerous reports have chronicled instances ranging from minor damage to catastrophic failures. Two important types of hazard are as follows:

- **Wind-induced hazards:** Wind loads represent one of the most prevalent types of disaster leading to the collapse of power transmission towers [3,4]. When exposed to extreme wind loads, such as those from hurricanes, typhoons [5], galloping [6,7] or downbursts, the primary structural members of the transmission tower experience a substantial surge in stress levels, which may result in yielding, buckling, and ultimately, complete structural collapse. Furthermore, wind-driven particles, such as wind-blown raindrops [8] or sand [9], can exacerbate the dynamic responses of power transmission towers, and prolonged exposure to wind-induced excitations can induce fatigue damage in operational transmission towers, progressively diminishing the structural capacity of their components. In severe cases, this cumulative damage can escalate to collapse and result in catastrophic failure;
- **Earthquake activity:** As potent natural disasters, earthquakes generate intense ground motions that can cause immediate physical damage, such as the collapse of transmission towers and the fracture of conductors [10,11]. Additionally, these seismic events have the potential to trigger secondary geological hazards, such as landslides or structural failures, which can further compound the damage to transmission lines. Moreover, the electrical equipment situated along these lines can be adversely affected by earthquakes, thereby jeopardizing the stability of the entire power system. The ensuing disruptions not only lead to widespread power outages that inflict substantial economic losses on society but also have the potential to initiate a chain of consequences that disrupts the normal functioning of other vital infrastructure.

In order to reduce the losses caused by these dynamic hazards, a wide array of dampers have been developed and tested both in laboratory settings and in real-world applications.

Wind load is typically considered the dominant lateral load in the design of transmission towers, leading to the proposal of various dampers to mitigate wind-induced vibration and reduce structural response under such conditions. In 2021, Roy and Kundu [12] summarized state-of-the-art vibration control techniques with regard to the response of transmission towers under wind load. In early efforts, tuned mass dampers (TMDs) were initially utilized because of their simplicity and effectiveness. Stockbridge dampers, which employ a damping mechanism similar to that of a TMD, are widely used to suppress the vibration of long-span conductors [13]. Recently, in order to enhance damping performance or minimize space requirements, numerous supplementary devices have been integrated into the traditional TMD. Upgraded TMD-type dampers include tuned mass damper inerters (TMDIs) [14], shape memory alloy-spring pendulums (SMA-SPs) [15], etc. In addition, Song et al. [16] explored the use of electromagnetic inertial mass dampers (EIMDs) for control of vibration in wind-disturbed transmission towers. They presented an analytical model, control approach, and parametric studies that demonstrated the effectiveness and robustness of EIMDs in reducing structural dynamic responses under varying intensities of wind load. Lei et al. [17] developed a new device named ECD-TMD, incorporating eddy current damping (ECD) and characterized by its non-contact ECD and structural simplicity. Field tests verified the significant vibration suppression achieved with this damper. Lou et al. [18] conducted a wind-tunnel test to examine the damping effect of an omnidirectional cantilever-type eddy current tuned mass damper (ECTMD). Furthermore, Grotto et al. [19] proposed a methodology to optimize the stiffness and damping coefficients of connection dampers in lattice steel towers subjected to wind action, aiming to reduce the dynamic response and horizontal displacements of the structure. Numerical results demonstrated that through incorporating semi-rigid connections with optimized connection dampers, a significant reduction in horizontal displacements can be achieved without significantly altering the fundamental frequency of the vibration, thereby enhancing the structural performance against wind-induced vibrations.

Regarding mitigation of earthquake-induced vibration, while wind load is commonly considered the primary lateral load in the design of transmission towers, numerous tall transmission towers in regions with high seismic intensity are susceptible to damage or even collapse. To mitigate earthquake-induced vibration, Wu et al. [20] utilized the energy dissipation property of SMA wires, due to their superelasticity, and developed a tube-type SMA damper. Zhang et al. [21,22] introduced impact into the traditional TMD and proposed a pounding tuned mass damper (PTMD). Lin and Liu [23] conducted an exhaustive parametric study of a TMD to meticulously examine its damping effect against near-fault earthquakes. Their paper further introduced an innovative multi-objective optimization framework, leveraging the NSGA-III algorithm to concurrently optimize the vibration reduction ratios and the cost of the TMD. Additionally, an artificial neural network (ANN)-based predictive framework was established to efficiently forecast the intricate relationship between TMD parameters and vibration reduction ratios. Miguel et al. [24] introduced a method for optimizing the placement and friction forces of dampers in structures under seismic loading, employing a recently developed optimization algorithm. Their findings revealed substantial decreases in structural response, showcasing the potential of the proposed methodology as an efficient tool for the design of friction dampers.

Despite the considerable advancements in research on vibration control for power transmission towers, a significant limitation is evident. The majority of existing dampers have been tailored to address only a single type of hazard, such as earthquakes or wind load, but not both. This narrow focus stems from the absence of evaluation criteria that specifically address multi-hazard scenarios. As a result, the application of these dampers in mitigating the combined effects of multiple dynamic hazards remains limited. To bridge this gap, this paper introduces a comprehensive methodology for assessing the reliability of power transmission towers under a wide range of dynamic disasters (shown in Figure 1), including both earthquakes and wind load. We employ a lifelong reliability approach to rigorously evaluate the effectiveness of a PTMD. Our proposed algorithm utilizes incremental dynamic analysis (IDA) to calculate structural fragility with regard to each type of disaster. These fragility estimates are then integrated with hazard functions to determine the overall failure probabilities of the towers.

The results of this study demonstrate that the PTMD can significantly reduce the dynamic response of towers to both earthquakes and wind loads, thereby enhancing the towers' overall reliability. This research highlights the potential of reliability as a crucial metric for optimizing damping devices in power transmission structures, particularly in the context of multi-hazard scenarios. By providing a comprehensive framework for evaluating and optimizing damping solutions, this study contributes to the advancement of vibration control technologies for power transmission towers.

The remainder of this paper is structured as follows: Sections 2–6 elucidate the detailed procedure of the proposed methodology. Specifically, Section 2 provides an explanation of the simplified model of the power transmission tower-line system. Section 3 subsequently introduces the damping mechanisms facilitated by the PTMD and derives the corresponding motion equation for the tower–PTMD system. In Section 4, we address considerations of structural uncertainties and uncertainties that may affect the PTMD. Section 5 details the method for generating seismic and wind excitations, and Section 6 outlines the calculation of the reliability of the tower–PTMD system through the integration of fragility and hazard functions. This is followed by Section 7, which presents numerical results and discussions. Finally, Section 8 concludes this paper with remarks on the current findings and their implications.

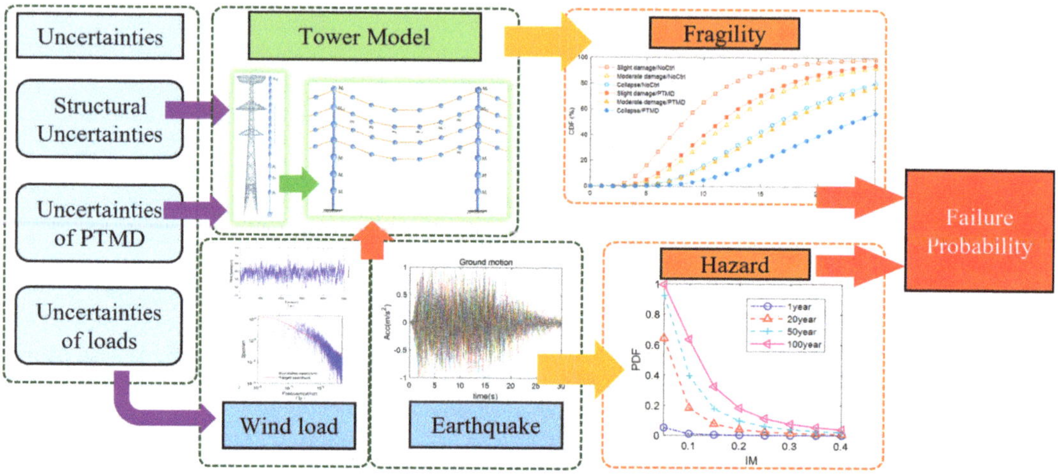

Figure 1. Flow chart of the proposed algorithm.

2. Numerical Model of Power Transmission Tower

2.1. FE Model

This study simulated a 35.71 km long transmission line as the engineering background. The voltage level of this project was 500 kV. The transmission line utilized a double-circuit, four-bundle conductor configuration, with a horizontal span of 400 m. The conductors were composed of 4 $\times$ LGJ-400/35 aluminum conductor steel reinforced (ACSR) cables, while the ground wires were LGJ-95/55 ACSR.

The selected power transmission tower was a tangent-type transmission tower standing at a total height of 53.9 m, with a nominal (structural) height of 30 m. It featured a base width of 8.36 m and a top width of 2.0 m, incorporating three strategically positioned crossarms at the top, middle, and bottom. These crossarms extended outwards by 8.7 m, 10.45 m, and 10 m, respectively. The primary structural components, including the main members, diagonal members, and auxiliary materials, were constructed of Q345 and Q235 angle steel. For visual reference, a photograph of this type of tower is provided in Figure 2a. Figure 2b–e illustrates the finite element (FE) model of the transmission tower from different angles. The whole FE model consisted of 618 nodes and 1588 elements.

2.2. Lumped Mass Model

Although a finite element (FE) model can facilitate time-domain analysis of a power transmission tower-line system, considering the complexity of multi-hazards, a simplified numerical model remains necessary for the following reasons. The proposed PTMD is intricate and challenging to simulate using FE software. Additionally, conducting time-domain analysis with an FE model is computationally intensive and time-consuming. Therefore, this study established a lumped mass model to simulate the behavior of the transmission tower under wind load and seismic action.

The lumped mass model represents a simplification of the detailed FE model. It conceptualizes the power transmission tower as a series of concentrated masses interconnected by rods. Figure 3 provides a visualization of the lumped mass model applied to the tangent tower. To obtain the mass, stiffness, and damping matrices for this model, the following steps were followed:

Step 1 The power transmission tower was divided into n nodes based on its structural characteristics. The positions of the crossarms and crossbraces were considered as a single node layer. In Figure 3, the power transmission tower considered in this study is represented by a lumped mass model of 15 mass components.

Step 2 The mass of the steel members between two adjacent nodes n_i and n_{i+1} was calculated. Here, n_i represents the *i*th node, and n_{i+1} represents the *i* + 1th node. Then, half of the calculated mass was assigned to node n_i and the other half to node n_{i+1}. Thus, the mass matrix $\boldsymbol{M}_\mathrm{t}$ was obtained.

Step 3 A unit force was aaplied at the *j*th node. The deformed shape under this load case was considered the *j*th flexibility vector, δ_j. The total flexibility matrix $\boldsymbol{\Delta}$ was then obtained as followed:

$$\boldsymbol{\Delta} = [\delta_1, \delta_2, \ldots, \delta_n] \quad (1)$$

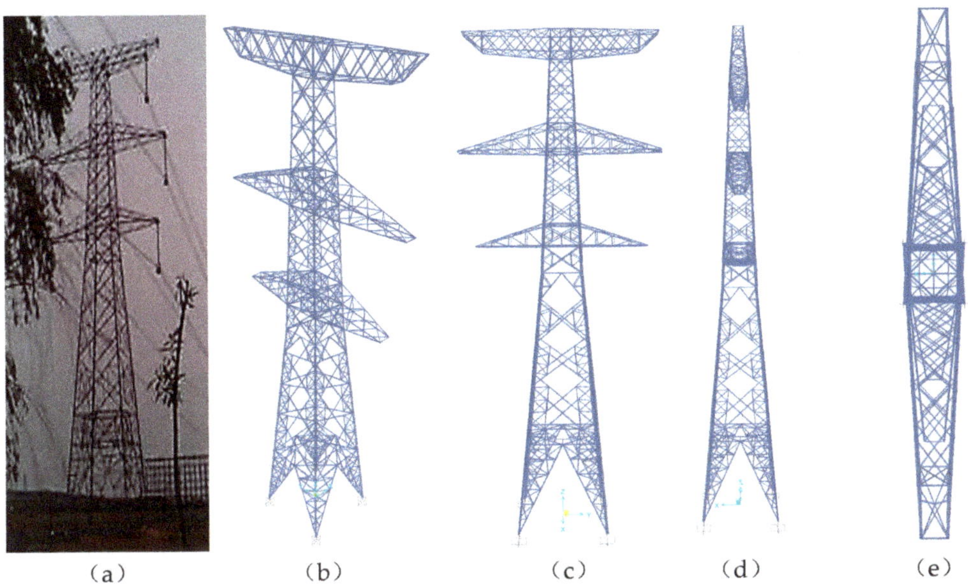

Figure 2. Established FE model of the power transmission tower: (**a**) onsite photo; (**b**) 3D view; (**c**) front view; (**d**) side view; (**e**) top view.

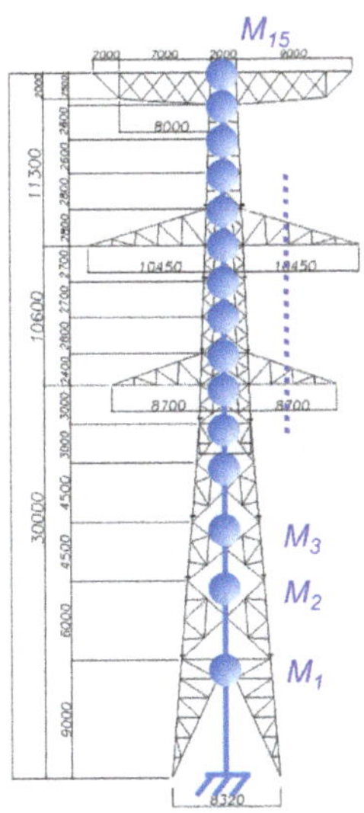

Figure 3. Lumped mass model of a power transmission tower.

Then, the stiffness matrix $\boldsymbol{K}_t$ was calculated as the inverse of the flexibility matrix (as shown in Figure 4):

$$\boldsymbol{K}_t = \boldsymbol{\Delta}^{-1} \tag{2}$$

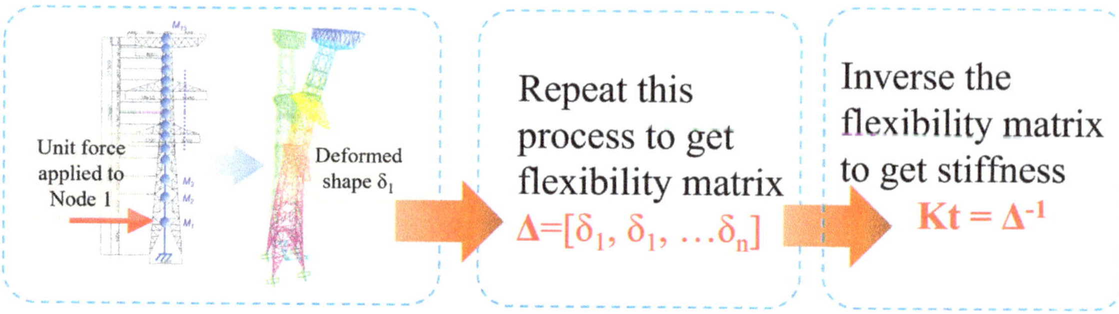

Figure 4. Process for calculating the stiffness matrix.

Step 4 Assuming Rayleigh damping, the damping matrix was then calculated as a superposition of the mass matrix and stiffness matrix, as follows:

$$C_t = a_1 + \boldsymbol{K}_t \tag{3}$$

where

$$\begin{aligned} a_1 &= \frac{2(\xi_j\omega_i - \xi_i\omega_j)}{\omega_i{}^2 - \omega_j{}^2}\omega_i\omega_j \\ a_2 &= \frac{2(\xi_i\omega_i - \xi_j\omega_j)}{\omega_i{}^2 - \omega_j{}^2} \end{aligned} \tag{4}$$

where ξ_i is the damping ratio of the ith order, and ω_i is the circular frequency of the i-th mode of vibration.

As shown in Figure 3, the FE model was simplified to a lumped mass model of 15 degrees of freedom (d.o.f.). The lumped mass and the height of each node are presented in Table 1.

Table 1. Mass and height of each node of the simplified model.

Node No.	Height (m)	Mass (kg)
1	9.00	3163.85
2	15.0	1033.85
3	19.5	1708.71
4	24.0	1123.44
5	26.8	767.98
6	30.0	1393.82
7	32.4	1413.27
8	35.2	478.67
9	37.9	448.92
10	40.6	1441.11
11	43.4	1202.66
12	46.2	289.97
13	48.8	203.59
14	51.4	1506.49
15	53.9	891.70

In order to validate the effectiveness of the lumped mass model in predicting the dynamic response of the power transmission tower, the first-order modal shapes computed using the finite element (FE) model and the lumped mass model were compared, as shown in Figure 5. The results revealed a striking similarity between the two modal shapes. Furthermore, the first-order frequencies obtained from the FE model and the lumped mass model were 1.877 Hz and 1.872 Hz, respectively, with a minimal relative error of only 0.26%, demonstrating the suitability of the lumped mass model for numerical analysis purposes.

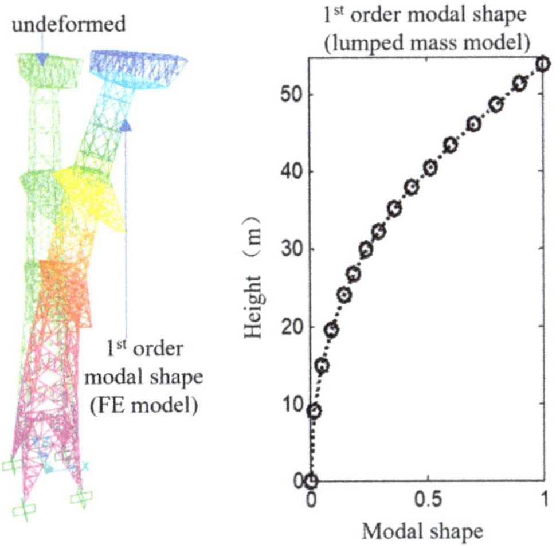

Figure 5. Comparison of first-order modal shape.

To further validate the accuracy of the lumped mass model for predicting the dynamic response of the transmission tower, the time-history responses of the tower under dynamic loading were computed and the results obtained from the lumped mass model and the finite element model were compared. As shown in Figure 6, the displacement responses from the lumped mass model closely matched those from the finite element model, with minimal deviations observed. The maximum displacement calculated with the lumped mass model and the FE model was 0.1622 m and 0.1595 m, respectively, with a relative difference of only 1.66%. This demonstrates that the lumped mass model provided a high level of computational accuracy, making it a reliable and efficient alternative for dynamic analysis of transmission towers, especially when computational efficiency is a priority.

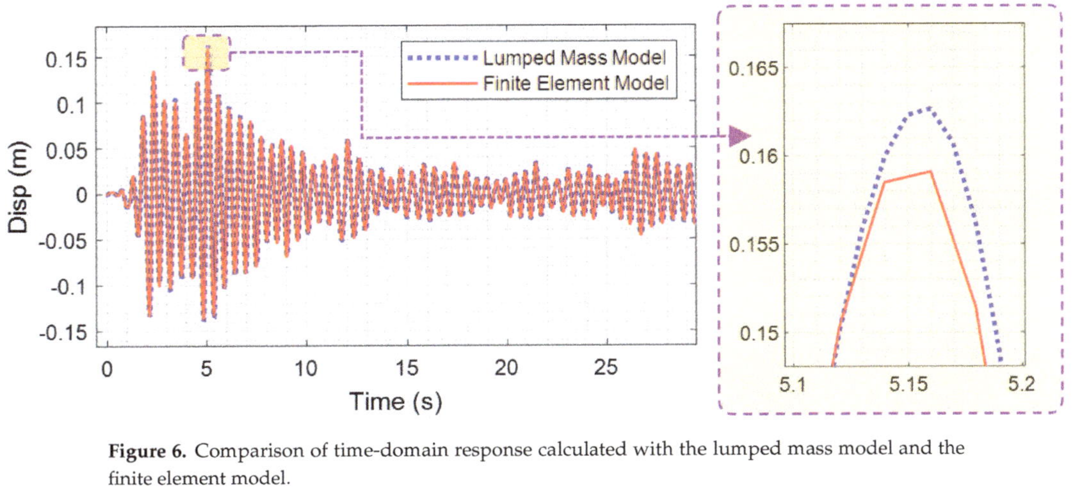

Figure 6. Comparison of time-domain response calculated with the lumped mass model and the finite element model.

2.3. Lumped Mass Model Considering Tower-Line Coupling Effect

In actual high-voltage power transmission projects, multi-split conductors are often used for transmission lines, with the mass per unit length reaching several kilograms. Considering that the span of the conductors often extends to hundreds of meters, the total mass of the conductors borne by the transmission tower is substantial, sometimes exceeding half of the tower's own mass. Therefore, the dynamic coupling effect of the conductors on the transmission tower's structure cannot be ignored.

This study employed the classical coupling model, as delineated in the literature [25], to compute the dynamic response of the power transmission tower-line system. A schematic of this model is presented in Figure 7. Within this framework, the conductor is approximated as a series of lumped masses interconnected with rigid bars. Previous experimental studies have validated both the accuracy and efficiency of this model.

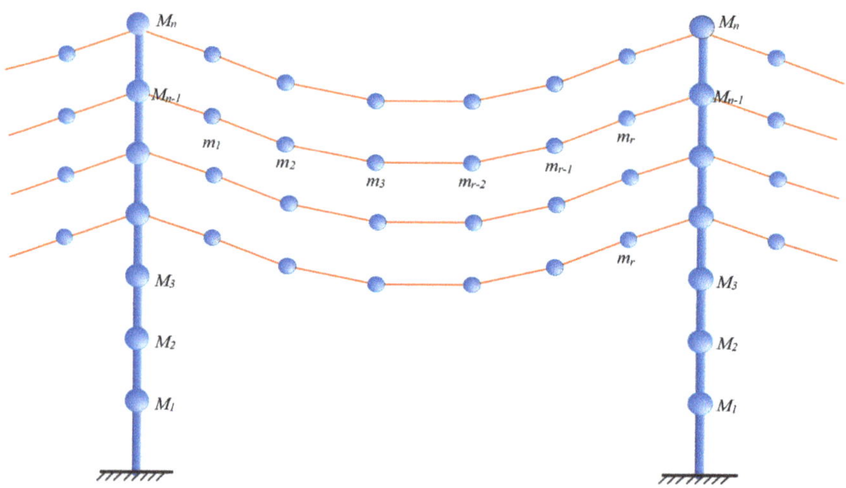

Figure 7. Power transmission tower-line system.

3. Power Transmission Tower Controlled with Pounding Tuned Mass Damper

3.1. Dampin Mechanism of Pounding Tuned Mass Damper

The pounding tuned mass damper (PTMD) is rooted in the principle of pounding damping. Extensive research has shown that PTMDs surpass traditional tuned mass dampers (TMDs) in terms of efficient vibration reduction, providing a wider damping bandwidth and exceptionally resilient performance. Furthermore, the model is distinguished by its straightforward design and minimal maintenance needs.

Figure 8 illustrates a schematic of the PTMD. Its damping mechanism operates in two distinct phases. In instances where the structural amplitude is minor, the amplitude of m_2 remains similarly low, causing the PTMD to function like a conventional TMD, mitigating vibration by absorbing energy. Under these conditions, m_2 generates an inertial force that aligns with the primary structure m_1, consistently directed towards the equilibrium position and thereby effectively diminishing its dynamic responses, including displacement and acceleration. Conversely, when the primary structure experiences substantial vibrations, m_2 within the PTMD collides with the delimiter on either side, further dissipating the seismic kinetic energy through impact. In contrast to a traditional TMD, the innovative PTMD offers multiple avenues for energy dissipation, leading to superior damping capabilities.

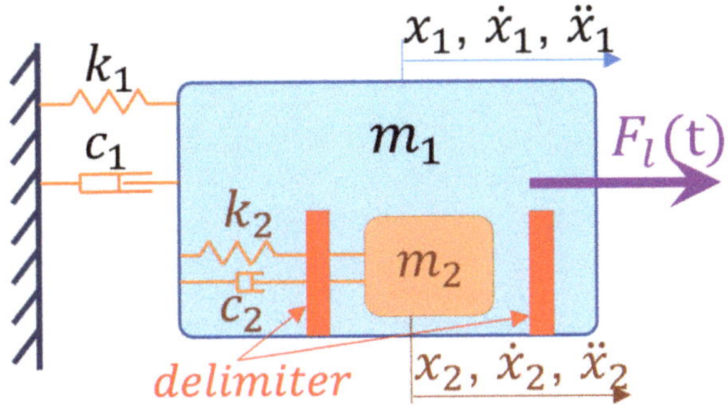

Figure 8. Schematic of the PTMD.

3.2. Motion Equation of Structure–PTMD System

The motion equations of the structure–PTMD damping system illustrated in Figure 8 are as follows:

$$\begin{cases} m_1\ddot{x}_1(t) + c_1\dot{x}_1(t) + k_1x_1(t) = F_{l1} + k_2(x_2 - x_1) + c_2(\dot{x}_2 - \dot{x}_1) + F_{impact} \\ m_2\ddot{x}_2(t) + c_2\dot{x}_2(t) + k_2x_2(t) = F_{l2} - k_2(x_2 - x_1) - c_2(\dot{x}_2 - \dot{x}_1) - F_{impact} \end{cases} \tag{5}$$

where m_1, c_1, and k_1 are the mass, damping, and stiffness of the primary structure, respectively. $\ddot{x}_1(t)$, $\dot{x}_1(t)$, and $x_1(t)$ are its acceleration, velocity, and displacement response. Similarly, m_2, c_2, and k_2 denote the mass, damping, and stiffness of the PTMD, and $\ddot{x}_2(t)$, $\dot{x}_2(t)$, and $x_2(t)$ are the corresponding acceleration, velocity, and displacement. F_{impact} denotes the nonlinear impact force when auxiliary mass m_2 impacts the delimiter. F_{l1} and F_{l2} are the loading forces added onto m_1 and m_2.

Suppose the distance between m_2 and the delimiter is d_{gap}. Then, the impact deformation can be calculated as follows:

$$\delta = x_2 - x_1 - d_{gap} \tag{6}$$

Additionally, the velocity of the impact deformation can be calculated as follows:

$$\dot{\delta} = \dot{x}_2 - \dot{x}_1 \tag{7}$$

Then, the impact force F_{impact} can be calculated as follows:

$$\begin{cases} F_{impact} = \beta_{impact}\delta + c_{impact}\dot{\delta} & \text{(Phase I)} \\ F_{impact} = \beta_{impact}\delta & \text{(Phase II)} \end{cases} \tag{8}$$

where β_{impact} and c_{impact} denote the impact stiffness and impact damping. Their values can be obtained from experimental tests. As shown in Equation (8), the impact force is calculated in two phases. Phase I refers to the colliding phase where m_2 penetrates into the delimiter, whereas Phase II is the restitution phase during which m_2 leaves the delimiter.

4. Consideration of Uncertainties

This study considers two types of uncertainties, namely, uncertain structural parameters and uncertain damper parameters. Since the tower was made mainly of Q235 and Q345 angle steel, structural uncertainties were considered regarding the material properties and geometry of the members, including the elastic modulus, Poisson's ratio, mass density, yield strength, web thickness, and the width of the angle steel. Table 2 lists the stochastic parameters of these factors.

Table 2. Probability distribution of random variables.

Random Variable	Average Value	Standard Deviation	Distribution Type
Elastic modulus	206 GPa	6.18 GPa	Lognormal
Poisson's ratio	0.3	0.009	Lognormal
Mass density	7800 kg/mm^3		Deterministic
Yield strength for Q235	263.7 MPa	18.46 MPa	Lognormal
Yield strength for Q345	387.1 MPa	27.10 MPa	Lognormal
Web thickness	0.985	0.031	Normal
Width	1.001	0.008	Normal

The parameters of the PTMD constituted another source of uncertainty. Specifically, impact stiffness, denoted as β_{impact}, fluctuates with changes in environmental temperature. Additionally, repeated impacts also modify this value. Nevertheless, this parameter was excluded from the present study because prior parametric analysis indicated that vibration control performance is relatively insensitive to β_{impact}. Conversely, the gap between the auxiliary mass m_2 and the delimiter significantly impacts vibration control performance. As such, it was recognized as the key uncertain parameter in this study. However, due to the limited application of PTMDs in practical engineering, stochastic data on this parameter (d_{gap}) are currently unavailable. Essentially, d_{gap} is a geometric parameter; thus, this paper assumes that it follows the same probability distribution and coefficient as the web thickness.

5. Simulation of Dynamic Hazards

5.1. Simulation of Wind Loads

The time-varying wind speed is composed of a mean speed and a fluctuating speed, expressed as follows:

$$v(z,t) = \overline{v}(z) + v_f(z,t) \tag{9}$$

where $\overline{v}(z)$ is the mean wind speed, $v_f(z,t)$ is the fluctuating speed, and z represents the height. The mean wind speed is considered a function of height. It can be calculated with the following formula:

$$\overline{v}(z) = v_{10}\left(\frac{z}{10}\right)^{\alpha} \tag{10}$$

where v_{10} is the mean wind speed at 10 m above the ground, and α is the power-law exponent, which is normally set to 0.15.

The fluctuating speed $v_f(z,t)$ can be simulated using the Davenport spectrum and the method proposed by Shinozuka et al. [26]. More details of this method can be found in [27]. Figure 9 shows the simulated wind speed at the top of the tower, when the basic wind speed $v_{10} = 25$ m/s. Comparison of the simulated spectrum and the target spectrum validated the accuracy of this method.

Given the total wind speed at a simulation point of height z, the wind load at this point can be calculated using the following equation:

$$F_w = \frac{\mu_s A v(z,t)^2}{1.6} \tag{11}$$

where F_w represents the time-varying wind load, μ_s is the shape coefficient, which is taken to be 1.3, and A is the projected area in the windward direction.

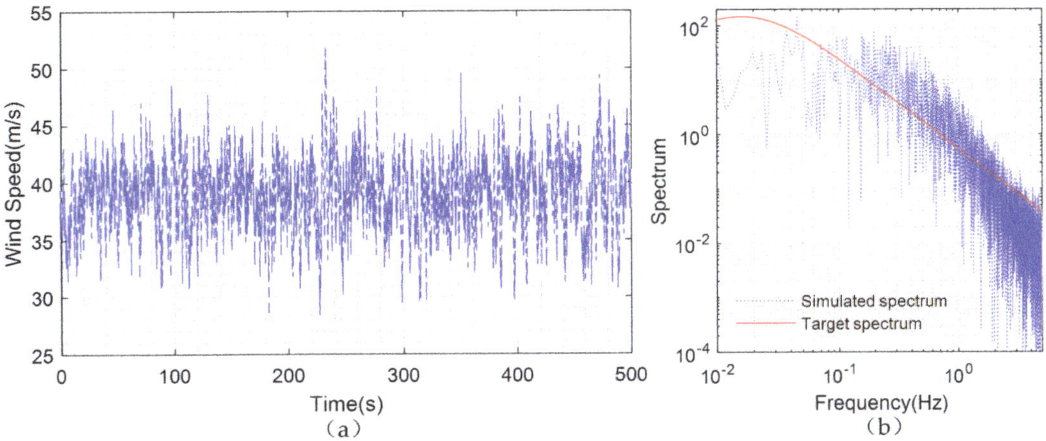

Figure 9. Simulated wind speed: (**a**) time history of wind speed at z = 53.9 m with basic wind speed $v_{10} = 25\frac{\mathrm{m}}{\mathrm{s}}$; (**b**) comparison between simulated spectrum and target spectrum.

5.2. Simulation of Earthquake Actions

Earthquakes are highly random processes. To consider uncertainty in seismic load, this study simulated 20 ground motions using the following parameters: (1) characteristic period of the site = 0.4 s; (2) structural damping ratio = 5%; (3) duration of the earthquake = 30 s. Figure 10 shows the simulated ground motion record and the response spectrum.

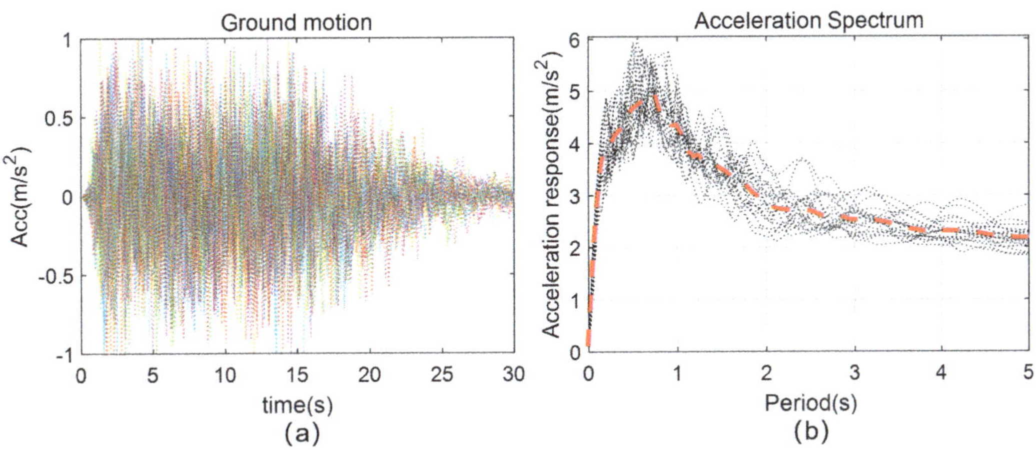

Figure 10. Simulated seismic action: (**a**) time history of ground motion; (**b**) response spectra.

6. Reliability Analysis

In this study, the probability of failure of a power transmission structure incorporating a PTMD was assessed by integrating its fragility curve with the corresponding seismic and wind hazard functions, thus enabling evaluation of its structural reliability. The fragility analysis was conducted using the incremental dynamic analysis (IDA) method. The hazard function comprehensively considered both seismic activity and wind load. The detailed procedure is outlined below.

6.1. Fragility Analysis

The fragility of the power transmission tower with and without a PTMD was calculated using incremental dynamic analysis (IDA) [28]. The core principle of IDA involves modulating the amplitude of excitations, such as earthquakes or wind load, and subjecting the structure to nonlinear time-history analysis. By progressively increasing the amplitude, this method allows comprehensive observation of the structure's response, ranging from elastic deformation to potential collapse. The culmination of the IDA process is the generation of IDA curves, wherein each data point represents the outcome of an individual nonlinear time-history analysis. A solitary IDA curve indicates correlation between the excitation intensity measure (IM) and the structural damage measure (DM). Conversely, a collection of multiple IDA curves serves to illustrate variability in structural responses under diverse earthquake excitation or wind load.

The IDA method assumes that the engineering demand parameter (*EDP*) and the *IM* of the excitation can be defined by the following a log-linear relationship:

$$\ln(EDP) = \ln(b) + a\ln(IM) \tag{12}$$

where a and b are the regression coefficients obtained from linear regression. For a power transmission tower structure, the *EDP* normally refers to the maximum displacement at the top of the tower. The conditional probability for a transmission tower reaching a specific damage state is as follows:

$$P[EDP \geq LS|IM] = \Phi\left(\frac{\ln(EDP) - \ln(LS)}{\beta_{(EDP|IM)}}\right) \tag{13}$$

$$\beta_{EDP|IM} = \sqrt{\frac{\sum[lnEDP_i - (a(ln(IM_i))^2 + bln(IM_i)]^2}{n-2}} \tag{14}$$

where *LS* represents engineering demand in a given damage state. For slight damage, it is set to 50% of the buckling displacement; for moderate damage, it is set to 75% of the buckling point; and collapse occurs at 100% of the buckling point. The determination of the buckling point involves conducting static nonlinear pushover analysis on the tower model. During the pushover analysis, the displacement at the top of the tower with the increasing load is recorded. By examining the pushover curve, a distinct buckling point can be identified, marking the transition from elastic deformation to plastic buckling of the tower structure (Figure 11).

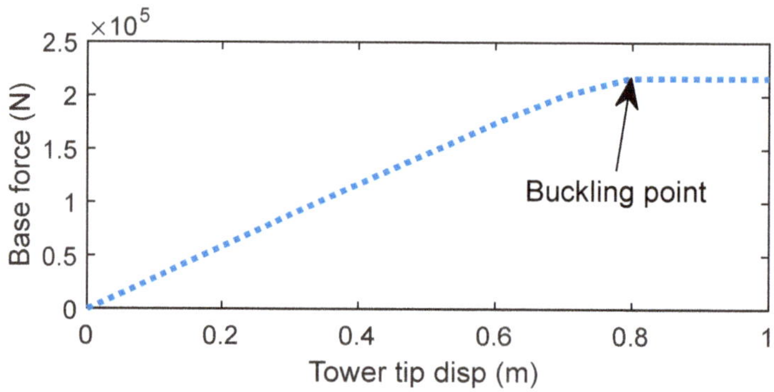

Figure 11. The pushover curve and the buckling point.

6.2. Consideration of Hazards

Quantification of the hazard function of earthquakes and wind load is necessary to enable the reliability analysis of the power transmission tower-PTMD system.

Based on [29], the hazard function for seismic action can be expressed as follows:

$$H_q(IM) = k_0 IM^{-k} \tag{15}$$

where IM denotes the intensity of the ground motion, and k_0 and k are shape factors that can be obtained using the following equations:

$$k = \frac{\ln(p_1/p_2)}{\ln(IM_2/IM_1)} \tag{16}$$

$$\ln(k_0) = \frac{\ln(IM_1)\cdot\ln(p_2) - \ln(IM_2)\cdot\ln(p_1)}{\ln(IM_1/IM_2)} \tag{17}$$

In Equation (15), the annual mean occurrence probabilities of ground motion under IM_1 and IM_2 are denoted as p_1 and p_2, respectively. In this study, p_1 and p_2 were determined in accordance with the current seismic design code for steel structures.

The hazard function of the wind load can also be determined following a similar procedure. According to the current design code, the basic wind speed is defined as the wind speed determined using probability statistics based on the average wind speed observed at a height of 10 m on open and flat ground for 10 min (i.e., v_{10}), with a maximum value determined by a 50-year return period. v_{10} follows a Gumbel distribution, and its cumulative probability distribution function is expressed as follows:

$$H_w(v_{10}) = \exp\{\exp[-\alpha(x-u)]\} \tag{18}$$

$$\alpha = \frac{1.28255}{\sigma} \tag{19}$$

$$u = \mu - \frac{0.57722}{\alpha} \tag{20}$$

where u represents the location parameter of the distribution, which is the mode of the distribution, α is the scale parameter of the distribution, and μ and σ are the mean value and standard deviation of the sample, respectively.

The probability density function in relation to Equation (18) is as follows:

$$d[H_w(v_{10})] = (1-\alpha)\cdot\exp\left[-\frac{x-u}{\alpha} - \exp\left(-\frac{x-u}{\alpha}\right)\right] \tag{21}$$

where have the same meanings as those employed in Equation (18).

6.3. Reliability Analysis Based on Fragility and Harzard Analysis

The probability of a structure's failure can be calculated by integrating the hazard function and cumulative density function of the fragility curve. Specifically, for seismic induced failure, the following applies:

$$P_{f,q} = \int F_q d\left[H_q(x)\right] \tag{22}$$

and for wind induced failure:

$$P_{f,w} = \int F_w d[H_w(x)] \tag{23}$$

where $P_{f,q}$ and $P_{f,w}$ denote the failure probability regarding earthquake activity and wind load, respectively. F_q and F_w are the fragility functions, determined using Equation (13), and $H_q(x)$ and $H_w(x)$ are the hazard functions of earthquake and wind load.

Based on Equations (22) and (23), the overall failure probability of the power transmission tower during its whole service life can be determined as follows:

$$P_f = 1 - \left[\left(1 - P_{f,q}\right) \cdot \left(P_{f,w}\right)\right] \tag{24}$$

where P_f denotes the probability of failure, which can be further applied to express reliability as follows:

$$\beta = 1 - \Phi^{-1}\left[1 - P_f\right] \tag{25}$$

7. Results and Discussion

The primary findings of this study indicate that installation of PTMDs can significantly reduce the dynamic response of transmission towers subjected to both seismic and wind load. This reduction in dynamic response is reflected in the numerical results presented in this paper, demonstrating a substantial decrease in the displacement and acceleration of the tower under various loading conditions.

The significance of these findings lies in their implications for the design and maintenance of transmission towers. By effectively mitigating vibrations induced by earthquakes and wind, PTMDs have the potential to enhance the structural safety and durability of transmission towers, thereby reducing the risk of damage and failure. This is particularly important given the critical role transmission towers play in ensuring the reliable delivery of electrical power.

In comparison to existing research, the present study introduces a novel approach for evaluating the effectiveness of PTMDs in multi-hazard contexts. By considering both seismic activity and wind load, this study provides a more comprehensive assessment of PTMD performance than previous studies, which have typically focused on a single type of hazard.

It should be noted that the proposed reliability-driven design framework currently relies on numerical simulations without systematic experimental validation. This limitation stems primarily from the challenges in replicating the strongly nonlinear structural behaviors under extreme loads (e.g., seismic and wind-induced dynamic effects) and the stochastic nature of real-world environmental conditions in controlled laboratory settings. Future efforts will prioritize three experimental validation pathways: (1) collaboration with power grid operators to conduct long-term monitoring of in-service towers to capture real-world responses under extreme loads; (2) destructive testing of critical components (e.g., buckling-prone members, high-stress connections) at laboratory scale; and (3) probabilistic modeling of manufacturing tolerances and material variability to enhance the comparability of numerical and experimental data. This phased hybrid validation strategy aims to systematically improve the method's engineering applicability while maintaining cost-effectiveness.

7.1. Performace Evaluation Based on Time-Domain Response

The displacement at the apex of the power transmission tower is frequently chosen as the key parameter for assessing structural safety. Although acceleration is occasionally taken into account to evaluate occupants' comfort during moderate earthquakes or wind events, this was not a consideration in the current study because power transmission towers are not designed for human habitation.

Figure 12 presents an overview of the displacement responses of all structure samples subjected to varying levels of wind. In Figure 12a, a box plot compares the displacements with and without control, showing that the maximum displacement increases as the basic wind speed v_{10} rises from 25 m/s to 75 m/s. Across all windspeed conditions, the responses observed with the implementation of a PTMD were consistently smaller than those in the non-control case. Figure 12b describes the vibration reduction ratio. The data reveal that the PTMD effectively reduced wind-induced vibrations. Specifically, the reduction in vibration fluctuated between 10% and 30%, with a more pronounced reduction effect at higher wind speeds. This further underscores the significant efficacy of PTMDs in controlling structural vibrations, particularly under high-windspeed conditions.

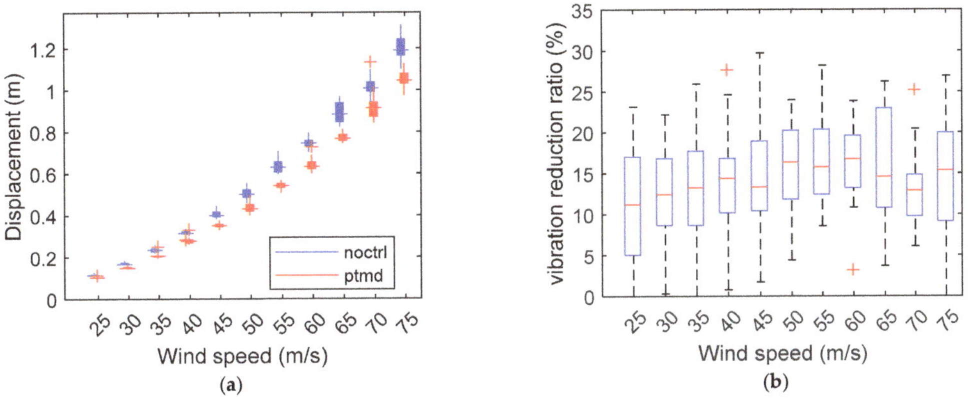

Figure 12. Wind-induced response with and without control: (**a**) displacement at the top of the power transmission tower; (**b**) vibration reduction ratio (+ denotes outliers in the boxplot).

Similarly, Figure 13 showcases the results pertaining to seismic-induced vibrations. Panel (a) features a box plot that compares the displacement of structures with and without control measures. As the peak ground acceleration (PGA) increases from 0.1 m/s^2 to 25 m/s^2, the maximum displacement also escalates. However, across all PGA conditions, the responses observed with the implementation of the PTMD were consistently lower than those without control. Panel (b) of Figure 13 presents the vibration reduction ratio, demonstrating that the PTMD effectively dampened earthquake-induced vibrations. The reduction in vibration ranged from 20% to 80%, with a more pronounced reduction effect at higher PGAs. Comparing Figures 12b and 13b, it is evident that the reduction ratio relating to seismic activity is greater than for wind load, suggesting that the PTMD is more effective in mitigating the effects of earthquakes.

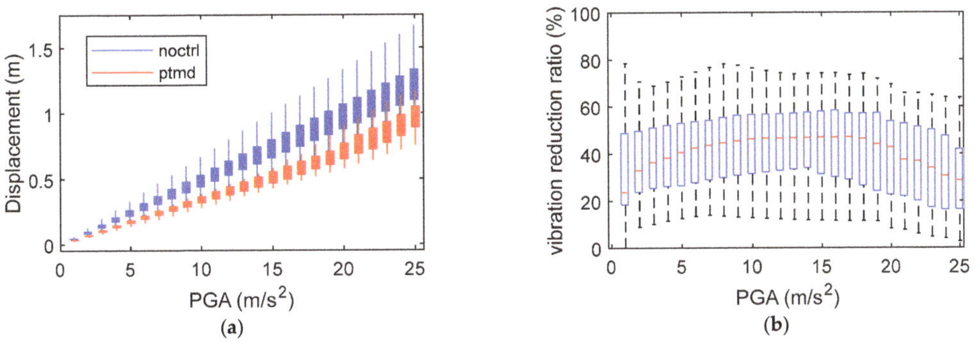

Figure 13. Earthquake-induced response with and without control: (**a**) displacement at the top of the power transmission tower; (**b**) vibration reduction ratio.

7.2. Fragility Curves with and Without Control

Figure 14 shows the IDA curves for the power transmission tower with and without the PTMD, under wind load. The blue symbols and lines denote the no-control conditions, revealing a positive correlation between increasing wind speed and structural displacement. In contrast, the introduction of PTMD control, represented by red symbols and lines, significantly mitigated this displacement. Figure 15 presents the fragility curves computed using Equation (11), quantifying the probability of different damage states as a function of wind speed. Both figures collectively demonstrate the effectiveness of the PTMD in mitigating wind-induced failure. For instance, at a wind speed of 75 m/s, the power transmission tower without the PTMD had a high 75.0% probability of experiencing collapse (Collapse/NoCtrl curve). However, when the PTMD was installed, the failure probability dropped significantly to 70.1% (Collapse/PTMD curve), indicating a notable reduction in the likelihood of structural failure.

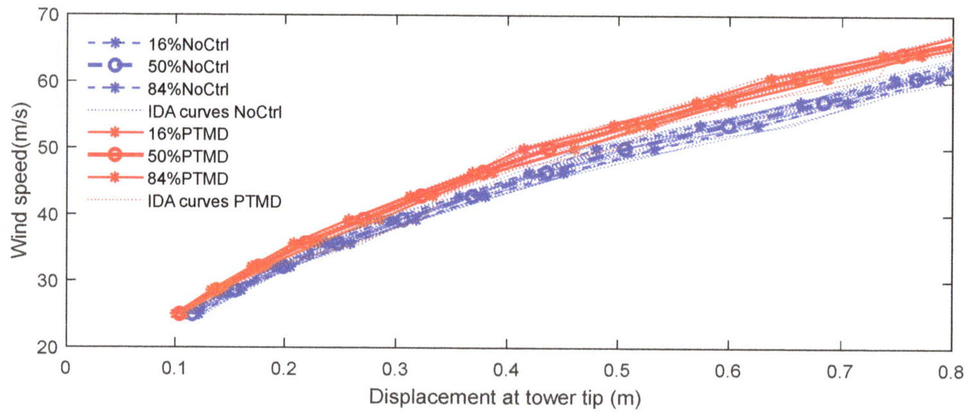

Figure 14. IDA curves of the power transmission tower with and without control under wind load.

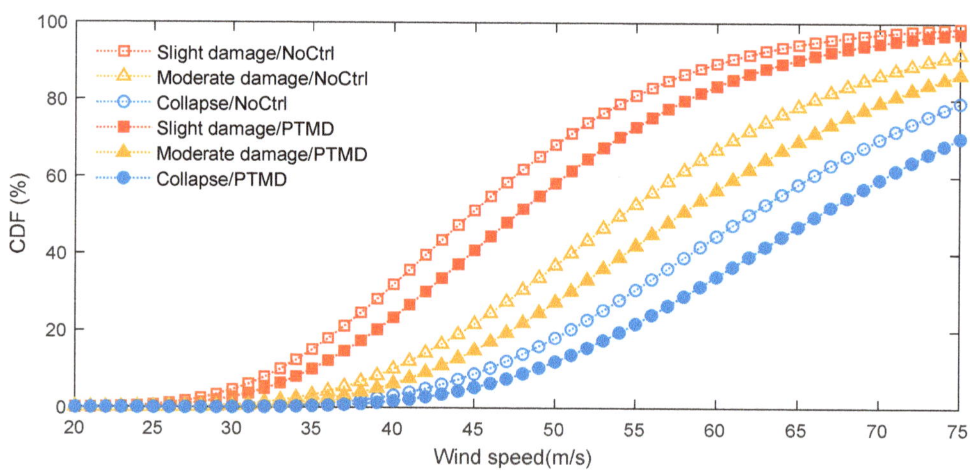

Figure 15. Fragility curves of the power transmission tower with and without control under wind load.

Similar results are evident in both Figures 16 and 17, which present the calculated outcomes of the earthquake-induced vibrations. The dashed red lines represent the IDA curves for the tower with the PTMD, all of which are elevated compared with the blue dashed lines, thereby highlighting the damping capability of the PTMD during seismic events. Compared with Figure 14 (depicting the wind load scenario), the IDA curves under seismic activity exhibit greater dispersion, suggesting that the uncertainty associated with earthquake activity significantly influences the seismic response. Therefore, the selection of earthquake records should be given careful consideration in structural design.

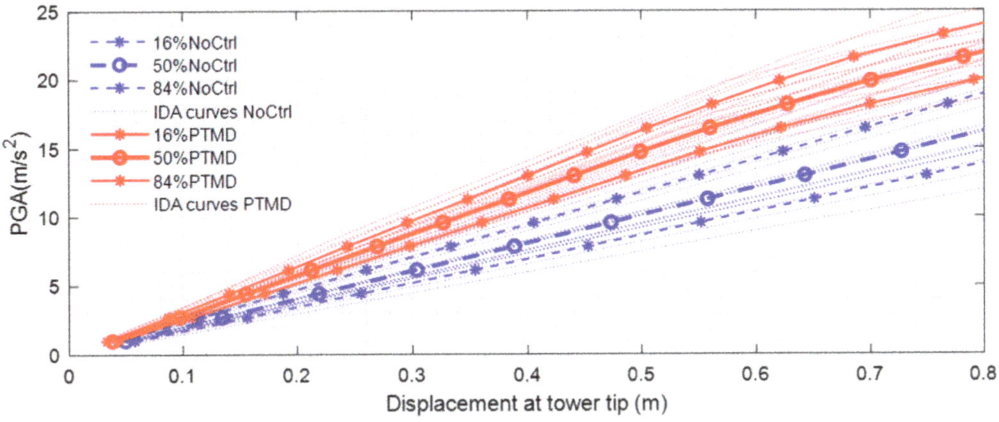

Figure 16. IDA curves of the power transmission tower under seismic actions, with and without PTMD.

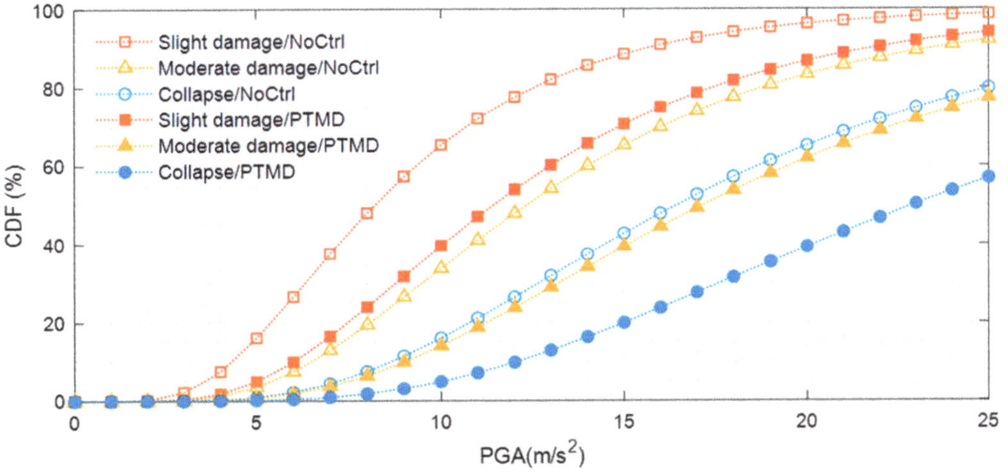

Figure 17. Fragility curves of the power transmission tower under wind load with and without control (under seismic activity).

Figure 17 presents the fragility curves of the power transmission tower subjected to earthquakes, both with and without the incorporation of the PTMD; these results imply that a PTMD can effectively reduce the likelihood of failure during seismic events. For instance, when the tower is subjected to a ground motion of 25 m/s^2, the probability of failure soars to 79.7%; however, with the addition of a PTMD, this probability drops to just 56.8%.

7.3. Structural Reliability with and Without Control

This study considers a power transmission tower-line system with the following wind load and seismic activity parameters: 50-year recurrence interval; basic wind speed $v_{10} = 25.3$ m/s; mean value $\mu = 14.591$ m/s; standard deviation $\sigma = 4.115$ m/s; scale parameter $\alpha = 3.2085$ m/s; location parameter $u = 12.739$ m/s; and shape factors for seismic hazard $k = 2.3219$ and $k_0 = 4.7651 \times 10^{-5}$.

Table 3 presents a summary of the reliability indicator β for the power transmission tower, both with and without a PTMD. The following conclusions can be drawn from the table:

- As the service life extends from 30 to 100 years, a decrease in reliability is observed. Specifically, in the absence of control measures, β is 3.4399 for a 1-year service life. However, when the tower is expected to operate for 100 years, β drops significantly to 1.9005;
- Comparing wind load to earthquakes, wind load emerged as the dominant factor. For instance, with a 50-year service life, the reliability β is 2.1851 when only wind load is considered. Conversely, the tower's reliability under earthquake conditions reaches 4.3830, indicating that the tower is more susceptible to damage from wind load than from earthquakes. Considering both wind and earthquake load, the reliability factor β is slightly lower at 2.1849 compared with the wind-only case (β = 2.1851);
- In all scenarios, the PTMD exhibited a satisfactory damping effect. For example, it enhanced the reliability from 2.1849 to 2.4295 under combined wind and earthquake load, demonstrating its effectiveness in providing dual control against vibrations and seismic activity.

Table 3. Reliablility of the power transmission tower with and without PTMD, considering both wind load and seismic activity.

Service Life (Year)	Wind		Seismic		Wind + Seismic	
	No Ctrl	PTMD	No Ctrl	PTMD	No Ctrl	PTMD
1	3.4400	3.6124	5.1700	5.3467	3.4399	3.6123
30	2.3786	2.6089	4.4930	4.6934	2.3784	2.6088
50	2.1851	2.4296	4.3830	4.5879	2.1849	2.4295
100	1.9006	2.1686	4.2296	4.4410	1.9005	2.1685

8. Conclusions

Power transmission tower-line systems face significant challenges from various dynamic hazards, notably, wind and earthquakes. Despite the proposal of numerous damping solutions, effective dual control systems for both seismic and wind-induced vibrations have been largely unexplored. This study introduces a comprehensive framework for assessing the reliability of power transmission towers under a spectrum of dynamic disasters, including earthquakes and wind load. Through employing a lifelong reliability approach, the efficacy of a PTMD was rigorously evaluated. The following conclusions can be drawn from the results of this study:

(1) While deterministic structural analysis demonstrated the damping ability of the PTMD using the vibration reduction ratio, this study further verified the effectiveness of the PTMD when considering structural uncertainties and excitation uncertainties;

(2) This paper proposes utilizing reliability as a metric to assess the effectiveness of dampers in enhancing structural safety. With the implementation of the PTMD, the lifelong reliability of the power transmission tower increased from 2.1849 to 2.4295;

(3) The proposed method takes into account the influence of multiple dynamic disasters. The results indicate that the PTMD is effective for dual control of seismic and wind-induced vibrations;
(4) The PTMD utilized in this study was not optimized. However, reliability can serve as a valuable target for optimizing PTMDs and other damping devices, with the potential to further improve performance in dual seismic and vibration control.

The present study focused on wind- and earthquake-induced vibrations. Meanwhile, the proposed methodology shows potential for assessing hazards from other excitation sources such as traffic, machinery, or human activities. Additionally, the reliability-based optimization framework is not restricted to the investigated transmission tower configuration; it can be extended to other tower types and structural systems by adjusting the design parameters and performance criteria.

The current PTMD design parameters represent an initial stage, requiring further optimization through advanced simulations and experiments to maximize the dual-control system's effectiveness. Future work will refine these parameters to enhance resilience against earthquakes and wind-induced vibrations, with practical implementations. Subsequent studies will integrate real-time structural health monitoring (SHM) capabilities into PTMD control frameworks, enabling adaptive damping strategies under operational uncertainties. This integration will embed sensor-driven feedback into adaptive control algorithms, advancing responsive vibration regulation in dynamic environments.

Author Contributions: Conceptualization, Z.J. and Z.Z.; methodology, Z.J. and P.L.; software, P.L.; validation, Z.Z., L.Q. and Z.J.; formal analysis, Z.Z., J.R. and Y.Z.; investigation, P.L. and Z.J.; resources, Z.Z., J.R. and Y.Z.; data curation, Z.Z. and L.Q.; writing—original draft preparation, Z.Z., L.Q. and Z.J.; writing—review and editing, Z.Z., L.Q. and Z.J.; visualization, P.L. and Z.J.; supervision, Z.Z., L.Q. and Z.J.; project administration, Z.Z. and L.Q.; funding acquisition, Z.Z. and L.Q. All authors have read and agreed to the published version of the manuscript.

Funding: This work is supported by the State Grid science and technology project under Grant No. 5200–202356685A-3–3-JC. This support is greatly appreciated.

Data Availability Statement: Data are contained within the article.

Conflicts of Interest: Z.Z., L.Q., J.R., Y.Z. and P.L. are employed the by State Grid Economic and Technological Research Institute Co., Ltd. The remaining authors declare that the research was conducted in the absence of any commercial or financial relationships that could be construed as a potential conflict of interest. The authors declare that this study received funding from State Grid. The funder was not involved in the study design, collection, analysis, interpretation of data, the writing of this article, or the decision to submit it for publication.

References

1. He, X.; Ma, H.; Zhang, S.; Wang, W.; Zhang, L. Study on the Multi-Hazard Responses of Transmission Tower-Line Systems Under Fire and Wind Loads Using ABAQUS. *Appl. Sci.* **2025**, *15*, 255. [CrossRef]
2. Wang, W.; Wang, G.; Sheng, H.; Gao, Y.; Lv, X. Research on multi-hazard response of transmission tower-line systems under rare earthquake-wind combined actions. *World Earthq. Eng.* **2023**, *39*, 165–173.
3. Azzi, Z.; Elawady, A.; Irwin, P.; Chowdhury, A.G.; Abi Shdid, C. Aeroelastic modeling to study the wind-induced response of a self-supported lattice tower. *Eng. Struct.* **2022**, *245*, 112885.
4. Vu, Q.-V.; Lee, H.-J.; Kim, J.-M. Nonlinear Pushover Analysis of Retrofitted Transmission Towers Under Static Wind Loads. *Buildings* **2025**, *15*, 654. [CrossRef]
5. Zhou, X.; Li, J. Risk Assessment of High-Voltage Power Grid Under Typhoon Disaster Based on Model-Driven and Data-Driven Methods. *Energies* **2025**, *18*, 809. [CrossRef]
6. Zhao, M.; Li, M.; Li, S.; Wan, Y.; Hai, Y.; Li, C. Galloping Performance of Transmission Line System Aeroelastic Model with Rime Through Wind-Tunnel Tests. *Energies* **2025**, *18*, 1203. [CrossRef]

7. Wu, Y.; Chen, X. Prediction and characterization of three-dimensional multi-mode coupled galloping of multi-span ice-accreted transmission conductors. *J. Wind Eng. Ind. Aerodyn.* **2023**, *241*, 105516.
8. Soares Fernandes, Y.M.; Rodrigues Machado, M.; Dutkiewicz, M. The Spectral Approach of Love and Mindlin-Herrmann Theory in the Dynamical Simulations of the Tower-Cable Interactions under the Wind and Rain Loads. *Energies* **2022**, *15*, 7725. [CrossRef]
9. Ma, X.; Lu, J.; Li, B.; Tian, W.; Zhang, Y.; Zhang, P. Fragility Analysis of Power Transmission Tower Subjected to Wind–Sand Loads. *Energies* **2024**, *17*, 6339. [CrossRef]
10. Chang, M.; Chen, B.; Xiao, X.; Chen, Y. Nonstationary Stochastic Responses of Transmission Tower-Line System with Viscoelastic Material Dampers Under Seismic Excitations. *Materials* **2025**, *18*, 1138. [CrossRef]
11. Nie, P.; Liu, H.; Wang, Y.; Han, S. Seismic Response and Collapse Analysis of a Transmission Tower Structure: Assessing the Impact of the Damage Accumulation Effect. *Buildings* **2024**, *14*, 2243. [CrossRef]
12. Roy, S.; Kundu, C.K. State of the art review of wind induced vibration and its control on transmission towers. *Structures* **2021**, *29*, 254–264. [CrossRef]
13. Li, Z.; Hu, Y.; Tu, X. Wind-Induced Response and Its Controlling of Long-Span Cross-Rope Suspension Transmission Line. *Appl. Sci.* **2022**, *12*, 1488. [CrossRef]
14. Bian, Y.; Liu, X.; Sun, Y.; Zhong, Y. Optimized Design of a Tuned Mass Damper Inerter (TMDI) Applied to Circular Section Members of Transmission Towers. *Buildings* **2022**, *12*, 1154. [CrossRef]
15. Zhou, M.; Meng, X.; Tian, L.; Jin, Q.; Liu, J. Shape Memory Alloy-Spring Pendulum for Vibration Suppression of Wind-Induced Transmission Tower-Line Systems. *Int. J. Struct. Stab. Dyn.* **2023**, *23*, 2350168. [CrossRef]
16. Song, X.; Chen, Y.; Bo, C.; Wu, J. Response control of a transmission tower-line system under wind excitations by electromagnetic inertial mass dampers. *Adv. Struct. Eng.* **2022**, *25*, 3334–3348. [CrossRef]
17. Lei, X.; Shen, L.; Niu, H.; Chen, J.; Xie, W.; Zhang, X. Development and Application of Eddy Current Damping-Based Tuned Mass Damper for Wind-Induced Vibration Control of Transmission Tower. *Int. J. Struct. Stab. Dyn.* **2024**, *24*, 2450053. [CrossRef]
18. Lou, W.; Wen, Z.; Chen, Y.; Huang, M. Wind Tunnel Study on Bidirectional Vibration Control of Lattice Towers with Omnidirectional Cantilever-Type Eddy Current TMD. *Appl. Sci.* **2019**, *9*, 2978. [CrossRef]
19. Grotto, L.G.; Miguel, L.F.F.; Kaminski Junior, J. Optimization of stiffness and damping coefficients of connection dampers to reduce the dynamic response of transmission line steel towers subjected to wind action. *Adv. Steel Constr. Int. J. Hong Kong* **2023**, *19*, 254–261.
20. Wu, J.; Chen, B.; Zhi, L.; Song, X. Seismic Response Mitigation of a Television Transmission Tower by Shape Memory Alloy Dampers. *Materials* **2021**, *14*, 6987. [CrossRef]
21. Zhang, P.; Song, G.; Li, H.; Lin, Y. Seismic control of power transmission tower using pounding TMD. *J. Eng. Mech.* **2013**, *139*, 1395–1406. [CrossRef]
22. Tian, L.; Rong, K.; Bi, K.; Zhang, P. A Bidirectional Pounding Tuned Mass Damper and Its Application to Transmission Tower-Line Systems under Seismic Excitations. *Int. J. Struct. Stab. Dyn.* **2019**, *19*, 1950056. [CrossRef]
23. Lin, Y.; Liu, T. Seismic Vibration Control and Multi-Objective Optimization of Transmission Tower with Tuned Mass Damper Under Near-Fault Pulse-like Ground Motions. *Buildings* **2024**, *14*, 3572. [CrossRef]
24. Miguel, L.F.F.; Miguel, L.F.F.; Lopez, R.H. Simultaneous optimization of force and placement of friction dampers under seismic loading. *Eng. Optim.* **2016**, *48*, 582–602. [CrossRef]
25. Li, H.; Xu, J. Simplified aseismic calculation model for transmission tower-line system considering pile-soil dynamic interaction. *Chin. J. Geotech. Eng.* **2009**, *31*, 1763–1767.
26. Shinozuka, M.; Deodatis, G. Simulation of Stochastic Processes by Spectral Representation. *Appl. Mech. Rev.* **1991**, *44*, 191.
27. Eisenhut, C.; Krug, F.; Schram, C.; Klockl, B. Wind-Turbine Model for System Simulations Near Cut-In Wind Speed. *IEEE Trans. Energy Convers.* **2007**, *22*, 414–420. [CrossRef]
28. Li, J.-X.; Zhang, C.; Cheng, J.-P. Seismic Fragility Analysis of the Transmission Tower-Line System Considering Bolt Slippage. *Appl. Sci.* **2024**, *14*, 10909. [CrossRef]
29. Cornell, C.A.; Jalayer, F.; Hamburger, R.O.; Foutch, D.A. Probabilistic Basis for 2000 SAC Federal Emergency Management Agency Steel Moment Frame Guidelines. *J. Struct. Eng.* **2002**, *128*, 526–533. [CrossRef]

buildings

MDPI

Article

A Comparative Study of Inertial Mass Dampers and Negative Stiffness Dampers for the Multi-Mode Vibration Control of Stay Cables

Zhaowei Shen [1], Xiaohong Sun [1], Zhipeng Cheng [2,*] and Ruisheng Ma [2,*]

[1] Shandong Electric Power Engineering Consulting Institute Corp., Ltd., Jinan 250014, China; shenzhaowei@spic.com.cn (Z.S.); sunxiaohong@sdepci.com (X.S.)

[2] State Key Laboratory of Bridge Safety and Resilience, Beijing University of Technology, Beijing 100124, China

* Correspondence: chengzhipeng@emails.bjut.edu.cn (Z.C.); ruishengma@bjut.edu.cn (R.M.)

Abstract: Previous studies have demonstrated that two representative passive control devices, including inertial mass dampers (IMDs) and negative stiffness dampers (NSDs), exhibit superior control performance in single-mode vibration control of stay cables. However, observations in recent years have increasingly reported rain–wind-induced multi-mode vibrations of stay cables on actual bridges. Therefore, it is of considerable significance to investigate the control effectiveness of the two representative passive dampers in mitigating multi-mode cable vibrations. For this reason, this study presents a comparative study of the IMD and NSD for the multi-mode vibration control of stay cables. The mechanical models of typical IMDs and NSDs are first introduced, followed by the numerical modeling of the two cable-damper systems using the finite difference method. Subsequently, the effectiveness of three multi-mode optimization strategies is comprehensively assessed, and the most effective strategy is selected for the optimal design of the IMD and NSD. Finally, the effectiveness of the control of the IMD and NSD in suppressing harmonic, white noise and wind-induced multi-mode vibrations of a 493.72 (m) long ultra-long cable is systematically evaluated. The numerical results indicate that the NSD significantly improves the cable damping ratios for multiple vibration modes as its negative stiffness coefficient increases, while IMD performs well only within a small inertia coefficient. Moreover, the NSD outperforms the IMD in suppressing multi-mode cable vibrations induced by harmonic, white noise and wind excitations.

Keywords: stay cable; inertial mass damper; negative stiffness damper; control effectiveness; damping ratio

1. Introduction

Stay cables, as essential load-carrying components in cable-stayed bridges, are highly susceptible to excessive vibrations caused by environmental factors such as wind or combined wind–rain effects. This vulnerability stems from their inherent characteristics, including low damping and high flexibility [1–3]. Such adverse vibrations not only pose a risk of immediate structural damage but can also lead to fatigue failure at the cable anchorages over time. These issues significantly compromise the durability and lifespan of the cables, while also undermining public confidence in the safety and reliability of the bridge structure [4–6]. In response to these challenges, a wide range of vibration control measures have been developed and implemented. According to the working mechanism, these control measures can be roughly grouped into the following categories: active control systems, which rely on external energy input to counteract vibrations [7,8]; semi-active control systems, which combine the high efficiency of active control with the cost-effectiveness of passive methods [9,10]; and passive control systems, which utilize energy-dissipating devices without external power, such as magnetorheological damper [11], dynamic vibration absorber [12], and tuned mass damper (TMD) [13–15]. Each approach has its unique advantages and limitations, making the selection of an appropriate control strategy critical for ensuring the long-term performance and safety of cable-stayed bridges.

Among these control measures, transversely attached passive viscous dampers (VDs) are one of the most commonly used. In this regard, the effectiveness of the control of VDs in suppressing cable vibrations has been extensively researched. Some previous studies have analyzed the available damping ratio of a cable with length L, provided by a VD at a specific installation position x_d, which is approximately $x_d/2L$ [16–18]. In addition, the effects of various non-ideal conditions on the dynamic characteristics of the cable–VD system have been investigated, such as internal stiffness and support stiffness [19–21], damper nonlinearity [22–24], cable inclination [25,26], and sag and flexural rigidity [27,28]. To design VDs for multi-mode cable vibration mitigation, Wang et al. [29] developed a novel approach for optimizing VD parameters to mitigate these adverse vibrations. Weber et al. [30] proposed a practical multi-mode design procedure for the optimal design of VDs. Yang et al. [31] presented a novel control strategy to improve the multi-mode cable vibration mitigation performance of two optimized VDs installed at the lower cable anchorage. The literature review indicates that adopting appropriate control strategies is conducive to improving the cable vibration control performance of VDs. However, the effectiveness of VDs is constrained by installation height, which results in a limited damping ratio provided to the cable, making it difficult to satisfy the multi-mode vibration mitigation requirements of ultra-long cables with lengths over 450 (m) [32–34].

Previous investigations have indicated that the enhanced control performance of active and semi-active systems is attributed to the negative stiffness behavior of their control dampers, which allows these control devices to undergo larger displacements and dissipate more energy [35–37]. However, the successful on-site implementation of active and semi-active systems relies heavily on a complex control scheme and a dependable power source. Compared to active and semi-active systems, passive dampers are favored in practical applications for their simplicity and reliability [38–40]. In this aspect, lots of efforts have been made to develop practical passive control devices incorporating negative stiffness mechanisms. In particular, two representative passive dampers, namely inertial mass dampers (IMDs) and negative stiffness dampers (NSDs), offer a promising solution for addressing the challenges of vibration control in stay cables subjected to wind or wind–rain excitations. The viscous inertial mass dampers (VIMDs) [41–43] and electromagnetic inertial mass dampers (EIMDs) [44–46] are common types of IMDs, whose hysteresis curves of inertial force and displacement exhibit equivalent negative stiffness characteristics. Theoretical and experimental studies have indicated that they provide superior additional damping to stay cables compared to conventional VDs. The pre-stressed spring NSDs [47,48] and magnetic NSDs [49–51] are typical NSDs capable of generating true negative stiffness, which has been demonstrated theoretically and experimentally to offer larger damping to stay cables than conventional VDs. Although some papers have investigated the cable vibration control performance of IMD and NSD, most studies are limited to single-mode cable vibration mitigation, with limited attention paid to multi-mode cable vibration mitigation. Specifically, Chen et al. [52] conducted a unified analysis of NSD and inerter-based dampers for multi-mode cable vibration control. Nevertheless, the IMD is the most typical inerter-based damper; no comparison is made between the IMD and NSD.

Given the above facts, this study conducts a comprehensive comparative analysis of IMDs and NSDs for multi-mode vibration control of stay cables. The primary objective is to evaluate their control performance in mitigating vibrations across multiple modes, which is critical for ensuring the safety and longevity of cable-stayed bridges. The remaining paper is organized as follows: in Section 2, the mechanical models of typical IMD and NSD are first introduced, and the numerical models of the two cable-damper systems are established utilizing the finite difference method; the effectiveness of three multi-mode optimization strategies is comprehensively assessed in Section 3, and the most effective strategy is selected for the optimal design of the IMD and NSD; Section 4 evaluates the performance of the IMD and NSD in suppressing harmonic, white noise and wind-induced multi-mode vibrations of a 493.72 (m) long ultra-long cable; and finally, key conclusions are drawn in Section 5.

2. Dynamic Formulation of a Stay Cable Equipped with an IMD or a NSD

2.1. Mechanical Models of Typical IMD and NSD

Figure 1 shows the mechanical models of two representative passive dampers, i.e., the IMD and NSD. As shown in Figure 1a, the IMD comprises an inerter element and a damping element in parallel, and its axial force F_{d1} can be expressed as follows:

$$F_{\mathrm{d1}} = b_{\mathrm{d}}\ddot{u} + c_{\mathrm{d}}\dot{u} \tag{1}$$

where b_{d} and c_{d} denote the inertia coefficient and damping coefficient of the IMD, respectively.

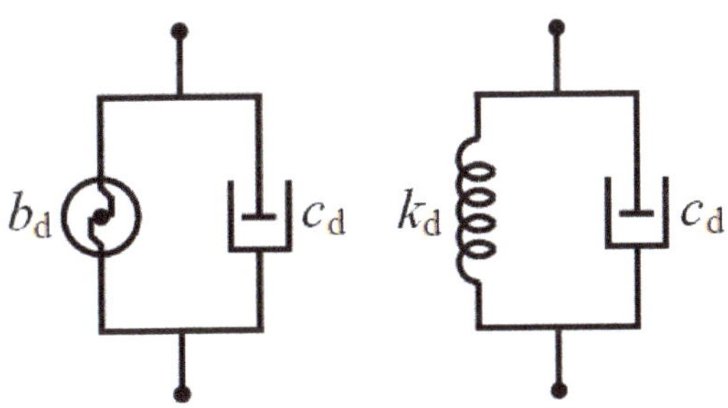

Figure 1. Mechanical models of two representative passive dampers. (**a**) IMD. (**b**) NSD.

It can be found from Figure 1b that the NSD consists of a negative stiffness element and a damping element in parallel, with its axial force F_{d2} mathematically expressed as follows:

$$F_{\mathrm{d2}} = k_{\mathrm{d}}u + c_{\mathrm{d}}\dot{u} \tag{2}$$

in which k_{d} represents the negative stiffness coefficient of the NSD.

Notably, the IMD contains an inerter element that generates an acceleration-dependent force, while the NSD incorporates a negative stiffness element capable of generating a displacement-dependent force. These differences result in distinct mechanical behaviors of the IMD and NSD, namely equivalent negative stiffness for the IMD with a frequency of f (i.e., $k_{\mathrm{eq}} = -b_{\mathrm{d}}(2\pi f)^2$) and real negative stiffness for the NSD.

In recent years, various practical prototypes of typical IMDs [7,42,53] and NSDs [47,50,51] have been developed for cable vibration control. Given the differences in mechanical performance between the IMD and NSD and their potential applications in multi-mode cable vibration mitigation, there is an urgent need for a comparative study on the control effectiveness of these two representative passive dampers in suppressing multi-mode cable vibrations.

2.2. Numerical Modeling of Two Cable-Damper Systems

As illustrated in Figure 2, an external damper is transversely equipped to an inclined cable at a distance x_d from the lower cable anchorage. The assumptions are as follows [28]: (1) the cable static profile follows a second-order parabola; (2) the cable vibrations are confined to the x-u plane, with negligible motion along the x-direction; (3) the sag-to-span ratio is small enough; (4) the cable maintains a consistent cross-sectional area throughout its length; and (5) the nonlinear characteristics of the cable, including large deformations, material nonlinearities, and temperature-induced nonlinearity, are neglected. On this basis, the equation governing the motion of the cable-damper system can be formulated as [42,50]

$$EI\frac{\partial^4 u(x,t)}{\partial x^4} - T\frac{\partial^2 u(x,t)}{\partial x^2} + \frac{\lambda^2 T}{l^3}\int_0^l u(x,t)dx + c\frac{\partial u(x,t)}{\partial t} + m\frac{\partial^2 u(x,t)}{\partial t^2} + F_\mathrm{d}\delta(x-x_d) = f_\mathrm{ext}(x,t) \quad (3)$$

where EI, T, and l denote the flexural rigidity, the chord tension force, and the length of the cable, respectively; $u(x,t)$ and $f_\mathrm{ext}(x,t)$ represent the transverse cable displacement and external load applied on the cable, respectively; c is the inherent damping of the cable; m represents the mass per unit length of the cable; F_d denotes the axial force of the external damper, which can be expressed as Equation (1) for the IMD and Equation (2) for the NSD, respectively; $\delta(\cdot)$ is the Dirac delta function.

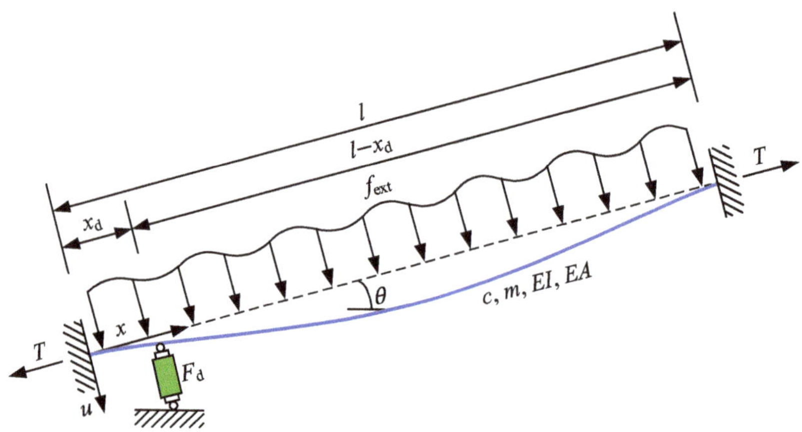

Figure 2. An inclined cable equipped with an external damper.

In Equation (3), the sag parameter λ^2 is represented as [50]:

$$\lambda^2 = \left(\frac{mgl\cos\theta}{T}\right)^2\frac{EAl}{TL_\mathrm{e}} \quad (4)$$

where g denotes the gravitational acceleration, which is 9.8 $\mathrm{m/s^2}$; θ is the cable inclination angle; and $L_\mathrm{e} = l\left[1 + (mgl\cos\theta)^2/(8T^2)\right]$ represents the static length of the cable under gravity.

Equation (3) can be discretized utilizing the finite difference method provided by Mehrabi and Tabatabai [54], whose effectiveness has been validated in Refs. [11,41,42,50]. Notably, the boundary condition for fixed ends is taken into account, which can be mathematically expressed as

$$\begin{cases} u(0,t) = u(l,t) = 0 \\ \frac{\partial u(0,t)}{\partial x} = \frac{\partial u(l,t)}{\partial x} = 0 \end{cases} \tag{5}$$

It is assumed that the cable is discretized into $N + 1$ equal segments with N internal nodes ($N = 199$ in the present study). Accordingly, Equation (3) can be formulated in matrix form:

$$\mathbf{M}\ddot{\mathbf{u}} + \mathbf{C}\dot{\mathbf{u}} + \mathbf{K}\mathbf{u} = \mathbf{f} - \boldsymbol{\gamma} F_{\mathrm{d}} \tag{6}$$

in which **M**, **C**, and **K** represent the mass matrix, damping matrix, and stiffness matrix of the cable, respectively, with the corresponding mathematical expressions given as follows:

$$\mathbf{M} = ma\mathbf{I}_N \tag{7}$$

$$\mathbf{C} = c\mathbf{I}_N \tag{8}$$

$$\mathbf{K} = \frac{\lambda^2 Ta}{l^3}\mathbf{B}_{N\times N} + \frac{T}{a}\varepsilon\chi_{N\times N} \tag{9}$$

where $\mathbf{I}_N$ is a N-dimensional identity matrix; $\boldsymbol{B}_{N\times N}$ represents a full unit matrix; $\varepsilon = EIT^{-1}l^{-2}$ denotes the dimensional flexural rigidity; $a = l/N$ is the segment length; and

$$\chi_{N\times N} = \begin{bmatrix} Q & U & W & & & 0 \\ & S & U & W & & \\ & & S & U & \ddots & \\ & & & \ddots & \ddots & W \\ & \text{sym} & & & S & U \\ & & & & & Q \end{bmatrix} \tag{10}$$

where

$$Q = 7 + \frac{2}{\varepsilon} \tag{11}$$

$$S = 6 + \frac{2}{\varepsilon} \tag{12}$$

$$U = -4 - \frac{1}{\varepsilon} \tag{13}$$

$$W = 1 \tag{14}$$

in which **u**, **f**, and $\boldsymbol{\gamma}$ represent the displacement, external load, and damper location vector, respectively, with the corresponding mathematical expressions given as follows:

$$\mathbf{u} = \begin{bmatrix} u_1 & u_2 & u_3 & \cdots & u_N \end{bmatrix} \tag{15}$$

$$\mathbf{f} = \begin{bmatrix} f_1 & f_2 & f_3 & \cdots & f_N \end{bmatrix} \tag{16}$$

$$\boldsymbol{\gamma} = \begin{bmatrix} \gamma_1 & \gamma_2 & \gamma_3 & \cdots & \gamma_N \end{bmatrix} \tag{17}$$

where γ_i is determined by the damper location, with its installation at the jth node given by

$$\gamma_i = \begin{cases} 0 & i \neq j \\ 1 & i = j \end{cases} \tag{18}$$

Considering the damper as an intrinsic element of the cable-damper system, Equation (6) can be rewritten as follows:

For a cable–IMD system,

$$\mathbf{M}_\mathrm{d}\ddot{\mathbf{u}} + \mathbf{C}_\mathrm{d}\dot{\mathbf{u}} + \mathbf{K}u = \mathbf{f} \tag{19}$$

in which $\mathbf{M}_\mathrm{d} = \mathbf{M} + \gamma b_\mathrm{d}\gamma^\mathrm{T}$ and $\mathbf{C}_\mathrm{d} = \mathbf{C} + \gamma c_\mathrm{d}\gamma^\mathrm{T}$, respectively.

For a cable–NSD system,

$$\mathbf{M}\ddot{\mathbf{u}} + \mathbf{C}_\mathrm{d}\dot{\mathbf{u}} + \mathbf{K}_\mathrm{d}u = \mathbf{f} \tag{20}$$

where $\mathbf{C}_\mathrm{d} = \mathbf{C} + \gamma c_\mathrm{d}\gamma^\mathrm{T}$ and $\mathbf{K}_\mathrm{d} = \mathbf{K} + \gamma k_\mathrm{d}\gamma^\mathrm{T}$.

An equivalent state-space form of Equations (19) and (20) can be represented as

$$\dot{\mathbf{Z}} = \mathbf{AZ} + \mathbf{Bf} \tag{21}$$

in which **Z**, **A**, and **B** denote state vector, system matrix, and input matrix, respectively, with their mathematical expressions provided as follows

$$\mathbf{Z} = \begin{bmatrix} \mathbf{u} \\ \dot{\mathbf{u}} \end{bmatrix} \tag{22}$$

For a cable–IMD system,

$$\mathbf{A}_1 = \begin{bmatrix} \mathbf{0}_{N\times N} & \mathbf{I}_N \\ -\mathbf{M}_\mathrm{d}^{-1}\mathbf{K} & -\mathbf{M}_\mathrm{d}^{-1}\mathbf{C}_\mathrm{d} \end{bmatrix} \tag{23}$$

$$\mathbf{B}_1 = \begin{bmatrix} \mathbf{0}_{N\times 1} \\ \mathbf{M}_\mathrm{d}^{-1} \end{bmatrix} \tag{24}$$

For a cable–NSD system,

$$\mathbf{A}_2 = \begin{bmatrix} \mathbf{0}_{N\times N} & \mathbf{I}_N \\ -\mathbf{M}^{-1}\mathbf{K}_\mathrm{d} & -\mathbf{M}^{-1}\mathbf{C}_\mathrm{d} \end{bmatrix} \tag{25}$$

$$\mathbf{B}_2 = \begin{bmatrix} \mathbf{0}_{N\times 1} \\ \mathbf{M}^{-1} \end{bmatrix} \tag{26}$$

Based on Equation (21), the dynamic characteristics of the cable-damper system, including frequencies, damping ratios, and mode shapes, can be obtained using complex eigenvalue analysis [42]. Notably, the numerical model employed in the present study was developed using the finite difference method. However, the finite difference method is not the only viable approach, and alternative methods such as the Galerkin method [55,56] and the finite element method [57] may also be applicable.

3. Design IMD and NSD for Controlling Multi-Mode Cable Vibrations

In Section 2, the mechanical models of typical IMD and NSD are introduced in detail, and the numerical models of the two cable-dampers systems are established to obtain their dynamic characteristics. In this section, the effectiveness of three multi-mode optimization strategies is comprehensively assessed, and the most effective strategy is selected for the optimal design of the NSD and IMD. On this basis, the design parameters of IMD and NSD are then determined to suppress the multi-mode vibrations of a 493.72 (m) long ultra-long cable.

3.1. Effectiveness of Multi-Mode Optimization Strategies

In this subsection, the effectiveness of three multi-mode optimization strategies is first assessed. Prior to this, a detailed description of the optimization process is first presented in Figure 3.

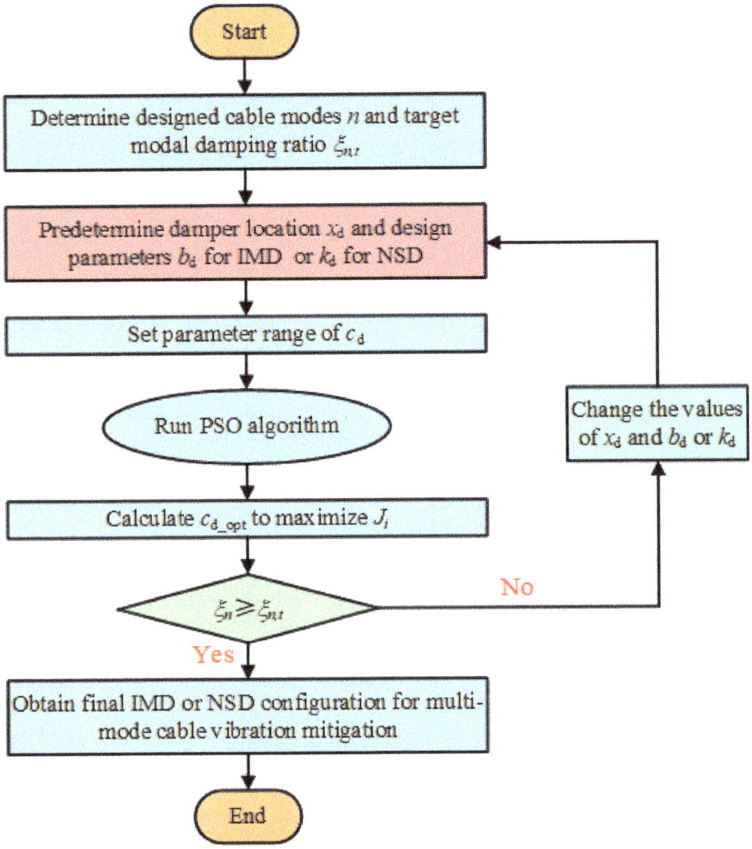

Figure 3. Optimization flowchart.

Notably, the optimization problem can be regarded as determining the optimal value c_{d} (namely $c_{\mathrm{d\_opt}}$) under given constraints of x_{d}, b_{d} for IMD or k_{d} for NSD, and $\xi_n \geq \xi_{n,t}$ to maximize the performance index J_i, with the corresponding mathematical formulation provided as follows:

$$\underset{c_{\mathrm{d}}}{\mathrm{maximize}} J_i \tag{27}$$

subjected to given values of x_{d}, b_{d} for IMD or k_{d} for NSD, and ξ_n

$$0.005l \leq x_{\mathrm{d}} \leq 0.05l, -\frac{T}{x_{\mathrm{d}}} < k_{\mathrm{d}} \leq 0 \text{ or } b_{\mathrm{d}} \geq 0, \xi_n \geq \xi_{n,t} \tag{28}$$

in which J_i is the performance index obtained using the ith optimization strategy; $\xi_{n,t}$ denotes the target modal damping ratio.

To address the aforementioned optimization problem, the particle swarm optimization (PSO) algorithm in MATLAB (version R2022a) is utilized in the present study.

Subsequently, to achieve the optimal design of IMD or NSD for mitigating cable multi-mode vibrations, the effectiveness of three multi-mode optimization strategies is comprehensively assessed. The three optimization strategies are described in detail below.

Strategy 1: The optimization of the minimum cable damping ratio in the first mode, with the constraint that the cable damping ratios are provided by the conventional VD in the designed cable modes, is considered [57]. By employing this strategy, the performance index J_1 in the designed cable modes can be defined as

$$J_1 = \min\xi_1 \tag{29}$$

$$\xi_1 > \min\xi_n^{\mathrm{VD}} \tag{30}$$

in which n is the designed cable modes; ξ_n^{VD} denotes the minimum cable damping ratio provided by VD in the designed cable modes.

Strategy 2: The optimization of the minimum cable damping ratio provided by the equipped damper in the designed cable modes [58] and the mathematical expression of the performance index J_2 is provided as follows:

$$J_2 = \min\begin{pmatrix} \xi_1 & \xi_2 & \xi_3 & \cdots & \xi_n \end{pmatrix} \tag{31}$$

Strategy 3: The average value $\mu(c_{\mathrm{d}})$ of cable damping ratios provided by an equipped damper in the designed cable modes is maximized, and in the meantime, the standard deviation $\sigma(c_{\mathrm{d}})$ is minimized. $\mu(c_{\mathrm{d}}) - \sigma(c_{\mathrm{d}})$ is optimized to assess the control effectiveness of the equipped damper, with the performance index J_3 mathematically expressed as follows [50,59]:

$$J_3 = (\mu(c_{\mathrm{d}}) - \sigma(c_{\mathrm{d}})) \tag{32}$$

where

$$\mu(c_{\mathrm{d}}) = \frac{1}{\kappa}\sum_{n=1}^{\kappa} \xi_n(c_{\mathrm{d}}) \tag{33}$$

$$\sigma(c_{\mathrm{d}}) = \sqrt{\frac{\sum_{n=1}^{\kappa}(\xi_n(c_{\mathrm{d}}) - \mu(c_{\mathrm{d}}))^2}{\kappa}} \tag{34}$$

The frequencies at which stay cables experienced rain-wind-induced vibrations (RWIVs) are generally below 3 Hz [60], covering multiple vibration modes. A case study is performed on the No. A30U stay cable of the Sutong Bridge (see Figure 4), with its main properties shown in Table 1. Numerical simulations of the ultra-long stay cable indicate that the first eleven vibration modes must be considered to effectively suppress the cable RWIVs. In the numerical simulations, it is assumed that the IMD or NSD is installed at the location x_{d} = 2%l from the lower anchorage of the ultra-long cable.

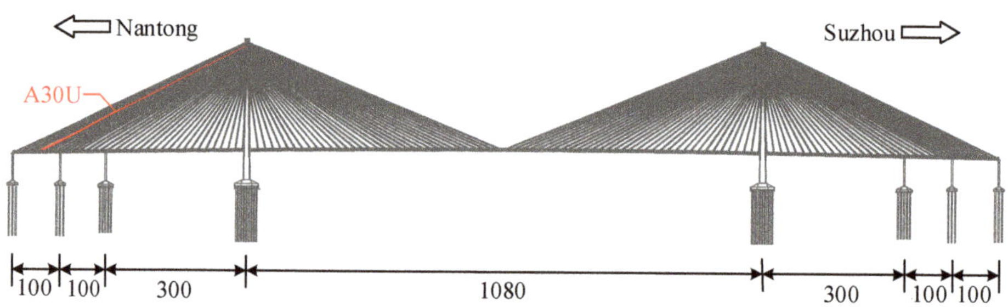

Figure 4. No. A30U stay cable of the Sutong Bridge (unit: m).

Table 1. Primary properties of the No. A30U stay cable of the Sutong Bridge.

Parameter	Value
Cable length (l)	493.72 (m)
Mass per unit length (m)	78.50 (kg/m)
Diameter (D)	0.142 (m)
Tension force (T)	5347 (kN)
Flexural rigidity (EI)	3.460×10^5 (N·m^2)
Inclination angle (θ)	25.84 (°)
Sag parameter (λ^2)	1.30

To facilitate designing the IMD and NSD to mitigate the multi-mode vibrations of the ultra-long cable, defining the following dimensionless parameters:

$$\bar{b}_{\mathrm{d}} = \frac{b_d}{ml} \tag{35}$$

$$\bar{k}_{\mathrm{d}} = \frac{k_{\mathrm{d}} x_{\mathrm{d}}}{T} \tag{36}$$

where $\bar{b}_{\mathrm{d}}$ and $\bar{k}_{\mathrm{d}}$ are the dimensionless inertia and negative stiffness coefficients, respectively.

Notably, the dimensionless inertia coefficient $\bar{b}_{\mathrm{d}}$ of the IMD is assumed to be in the range of 0 to 0.25. This is because the cable possesses a length of 493.72 (m) and a mass per unit length of 78.50 (kg), with the corresponding $\bar{b}_{\mathrm{d}}$ of the IMD ranging from 0 to 9.69 (tons), a value that is relatively easy to achieve in practice. In addition, the dimensionless negative stiffness coefficient $\bar{k}_{\mathrm{d}}$ of the NSD ranges from −0.9 to 0, preventing stability issues in the cable–NSD system as $\bar{k}_{\mathrm{d}}$ approaches −1.

Figure 5 compares the performance index J and optimal damping coefficient $c_{\mathrm{d\_opt}}$ of the IMD obtained using the three optimization strategies with respect to its dimensionless inertia coefficient $\bar{b}_{\mathrm{d}}$. As shown in Figure 5a, J initially increases and then decreases as $\bar{b}_{\mathrm{d}}$ increases for the three optimization strategies. It can be observed from Figure 5b that $c_{\mathrm{d\_opt}}$ initially decreases and then increases as $\bar{b}_{\mathrm{d}}$ increases for Strategies 2 and 3. In contrast, Strategy 1 leads to a continuously decreasing value of $c_{\mathrm{d\_opt}}$. Table 2 shows the multi-mode optimization parameters of the IMD obtained using the three optimization strategies. Notably, Strategy 3 achieves the maximum average cable damping ratio, followed by Strategy 2, while Strategy 1 performs the worst. Although the minimum cable damping ratio achieved by Strategy 3 is lower than that achieved by Strategy 2, Strategy 3 requires a smaller damping coefficient for the IMD than Strategy 1. Thus, Strategy 3 is preferred for designing IMD to suppress multi-mode cable vibrations.

Figure 6 compares the performance index J and optimal damping coefficient $c_{\mathrm{d\_opt}}$ of the NSD obtained using the three optimization strategies with respect to its dimensionless negative stiffness coefficient $\bar{k}_{\mathrm{d}}$. As illustrated in Figure 6a, J steadily increases as the increasing $\bar{k}_{\mathrm{d}}$ for the three optimization strategies. It can be found from Figure 6b that $c_{\mathrm{d\_opt}}$ gradually decreases with the increasing $\bar{k}_{\mathrm{d}}$ for the three optimization strategies. Table 3 summarizes the multi-mode optimization parameters of the NSD obtained using the three optimization strategies. Similarly to the IMD, a comprehensive comparison of the average and minimum cable damping ratios demonstrates that Strategy 3 outperforms Strategies 1 and 2, as it requires the least damping coefficient for the NSD to achieve relatively optimal performance in mitigating multi-mode cable vibrations. Therefore, Strategy 3 is recommended for the IMD or NSD to achieve better control effectiveness in suppressing multi-mode cable vibrations compared to Strategies 1 and 2. Notably, the NSD significantly improves the cable damping ratios for multiple vibration modes as its negative stiffness coefficient increases, while the IMD performs well only within a small inertia coefficient.

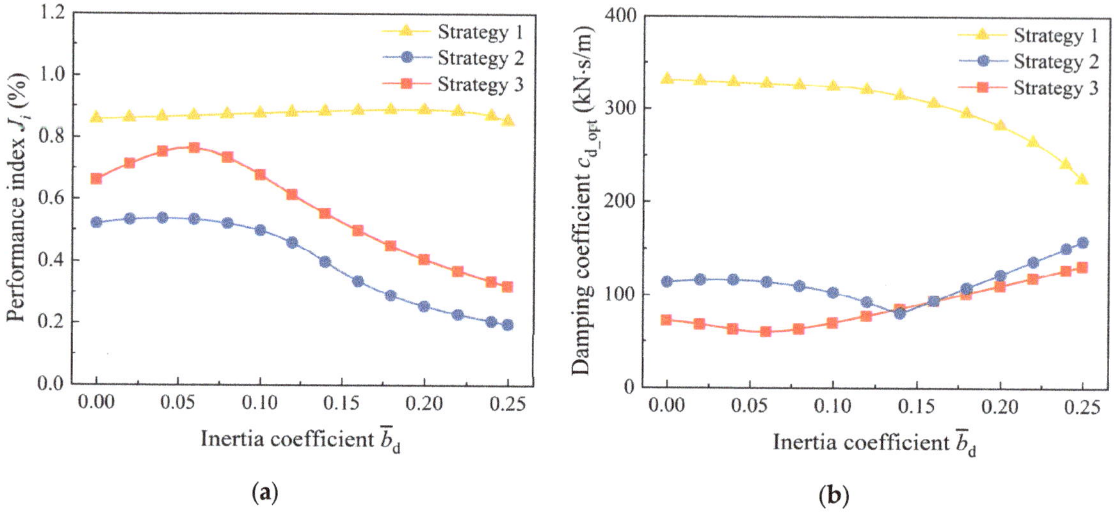

Figure 5. Comparisons of the performance index J and optimal damping coefficient c_{d_opt} of the IMD obtained using the three optimization strategies with respect to its dimensionless inertia coefficient $\overline{b}_d$. (**a**) Performance index J. (**b**) Optimal damping coefficient c_{d_opt}.

Table 2. Multi-mode optimization parameters of the IMD obtained using the three optimization strategies.

Strategy	Optimal Inertia Coefficient $\overline{b}_d$	Optimal Damping Coefficient c_{d_opt} (kN·s/m)	Average Value of Cable Damping Ratios (%)	Average Value of Cable Damping Ratios (%)
1	0.19	290.42	0.48	0.19
2	0.04	116.26	0.80	0.54
3	0.06	60.22	1.14	0.30

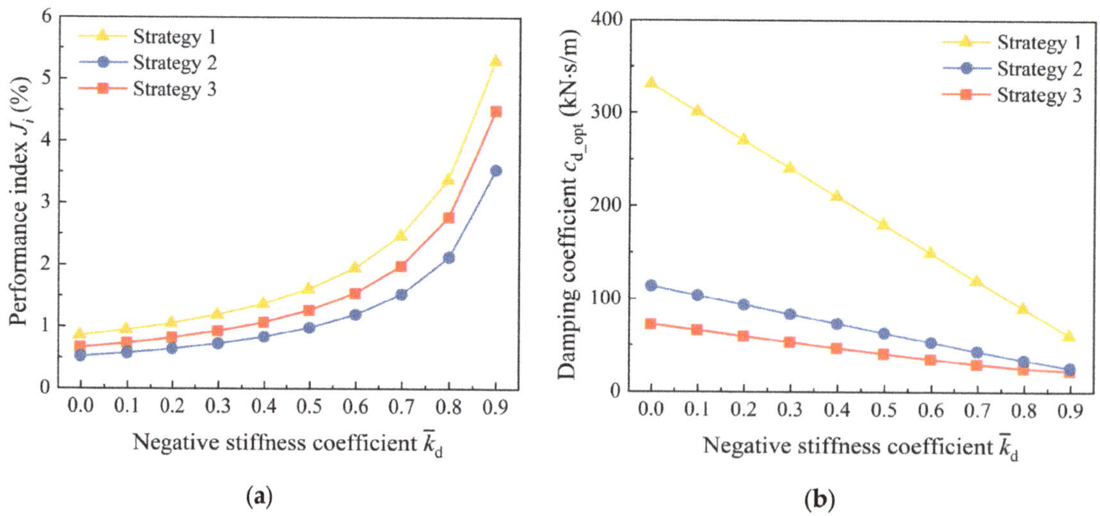

Figure 6. Comparisons of the performance index J and optimal damping coefficient c_{d_opt} of the NSD obtained using the three optimization strategies with respect to its dimensionless negative stiffness coefficient $\overline{k}_d$. (**a**) Performance index J. (**b**) Optimal damping coefficient c_{d_opt}.

Table 3. Multi-mode optimization parameters of the NSD obtained using the three optimization strategies.

Strategy	Negative Stiffness Coefficient $\bar{k}_d$	Optimal Damping Coefficient c_{d_opt} (kN·s/m)	Average Value of Cable Damping Ratios (%)	Average Value of Cable Damping Ratios (%)
1	−0.4 −0.6 −0.8	210.45 150.00 90.10	0.67 0.95 1.63	0.29 0.41 0.70
2	−0.4 −0.6 −0.8	73.23 53.22 34.04	1.20 1.71 2.96	0.84 1.20 2.13
3	−0.4 −0.6 −0.8	47.13 35.20 25.30	1.36 1.94 3.34	0.57 0.84 1.66

3.2. Design Parameters of IMD and NSD

The multi-mode optimization parameters of the IMD and NSD obtained using the most effective strategy, Strategy 3, are adopted as the final design parameters, as summarized in Table 4. The table shows that the NSD requires a smaller damping coefficient than the IMD, and its required damping coefficient decreases as the negative stiffness increases. Figure 7 compares the cable damping ratios provided by the designed IMD and NSD in the first eleven modes. As shown, the NSD offers higher modal damping ratios to the cable than the IMD, with the enhancement becoming more pronounced as the negative stiffness increases.

Table 4. Design parameters of the IMD and NSD for suppressing multi-mode cable vibrations.

Mode Order n	Damper Type	x_d	$\bar{b}_d$	$\bar{k}_d$	c_{d_opt} (kN·s/m)	J (%)
11	IMD	2%l	0.06	N/A	60.22	0.767
	NSD		N/A N/A N/A	−0.4 −0.6 −0.8	47.13 35.20 25.30	1.070 1.545 2.772

N/A: not applicable.

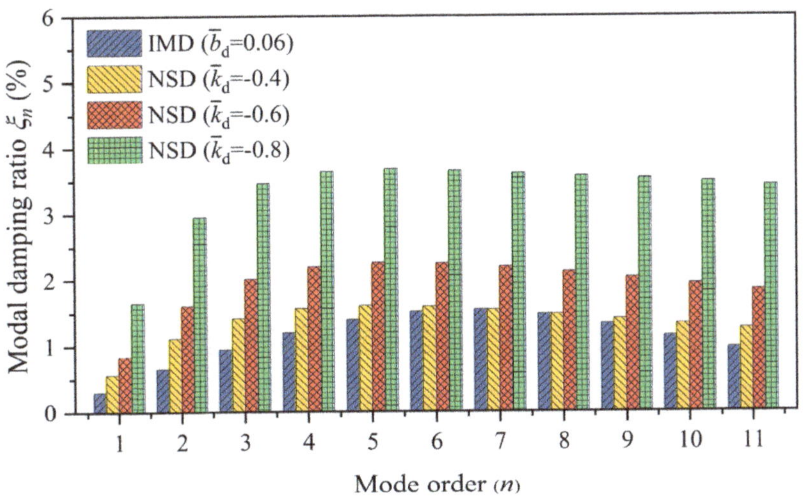

Figure 7. Comparisons of cable damping ratios in the first eleven modes.

Taking into account the cost-effectiveness and manufacturability of the NSD prototype, the configuration with $\bar{k}_d = -0.6$ is chosen for the subsequent performance comparison between the optimized IMD and the NSD. To assess their control performance from the perspective of robustness, a sensitivity analysis of the performance index J with respect to the damping coefficients of the IMD ($\bar{b}_d = 0.06$) and NSD ($\bar{k}_d = -0.6$) was performed, and the corresponding results are listed in Table 5. As can be seen, the performance index J progressively decreases as the damping coefficient deviates from its optimal value. When the damping coefficient deviates from the optimal value by +20% and −20%, the performance index of the IMD decreases by 1.83% and 2.87%, respectively, relative to its maximum value, whereas that of the NSD decreases by 3.17% and 4.66%, respectively. In comparison, the NSD exhibits greater differences than the IMD, while the overall discrepancy between the two remains relatively small.

Table 5. Sensitivity analysis of the performance index J with respect to the damping coefficients of the IMD and NSD in the first eleven modes.

Mode Order n	Damper Type	x_d	$\bar{b}_d$	$\bar{k}_d$	c_d (kN·s/m)	J (%)
11	IMD	2%l	0.06	N/A	$1.2c_{d\_opt}$	0.753
					$1.1c_{d\_opt}$	0.762
					$1.0c_{d\_opt}$	0.767
					$0.9c_{d\_opt}$	0.761
					$0.8c_{d\_opt}$	0.745
	NSD		N/A	−0.6	$1.2c_{d\_opt}$	1.496
					$1.1c_{d\_opt}$	1.531
					$1.0c_{d\_opt}$	1.545
					$0.9c_{d\_opt}$	1.528
					$0.8c_{d\_opt}$	1.473

N/A: not applicable.

Figure 8 compares the sensitivity of the cable damping ratios with respect to the damping coefficients of the IMD and NSD in the first eleven modes. As shown, for modes 1 to 4, deviations in the damping coefficient from its optimal value have little effect on the cable damping ratio. By contrast, for modes 5 to 11, the impact is considerably greater. This indicates that the cable damping ratios of low-order modes are less sensitive to deviations of the IMD or NSD damping coefficient from its optimal value, whereas those of high-order modes exhibit greater sensitivity. Overall, the sensitivities of the multi-mode cable damping with respect to the damping coefficients of the IMD and NSD are comparable, with no significant differences observed.

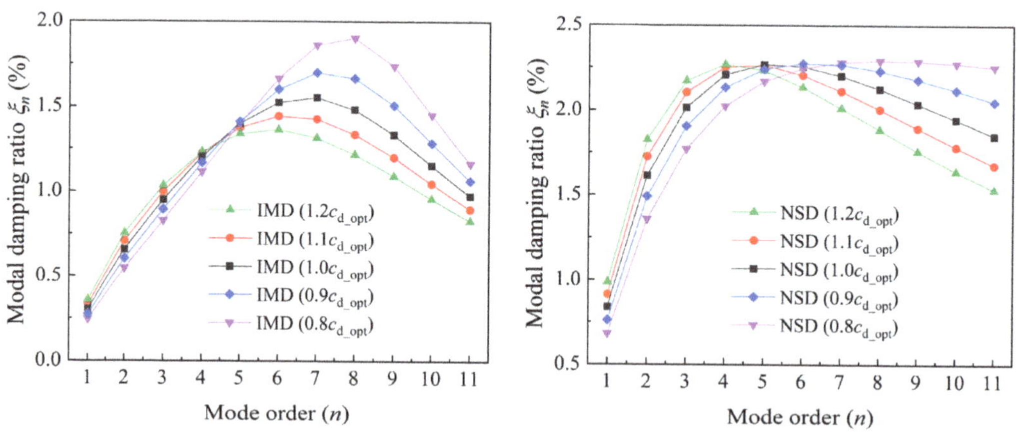

(**a**) (**b**)

Figure 8. Comparison of the sensitivity of the cable damping ratios with respect to the damping coefficients of the IMD and NSD in the first eleven modes. (**a**) IMD ($\bar{b}_d = 0.06$). (**b**) NSD ($\bar{k}_d = -0.6$).

4. Performance Comparison Between IMD and NSD

In Section 3, the optimal design parameters for the IMD and NSD are determined to mitigate multi-mode vibrations of a 493.72 (m) long ultra-long cable. On this basis, this section systematically evaluates the control effectiveness of the designed IMD and NSD in mitigating harmonic, white noise and wind-induced multi-mode vibrations of the ultra-long cable.

4.1. Harmonic Excitation

To evaluate the control effectiveness of the designed IMD and NSD, the ultra-long cable is first subjected to sinusoidal excitation, and Figure 9 shows the corresponding results. As shown, the mid-span cable vibrates predominantly in modes 1, 3, 5, 7, 9, and 11, while the one-tenth-span cable vibrates predominantly in modes 1 to 9 and 11. Overall, both the IMD and NSD can effectively reduce the cable dynamic responses. However, the NSD outperforms the IMD in reducing cable dynamic responses, highlighting its superior control effectiveness in suppressing multi-mode vibrations of stay cables subjected to harmonic excitation.

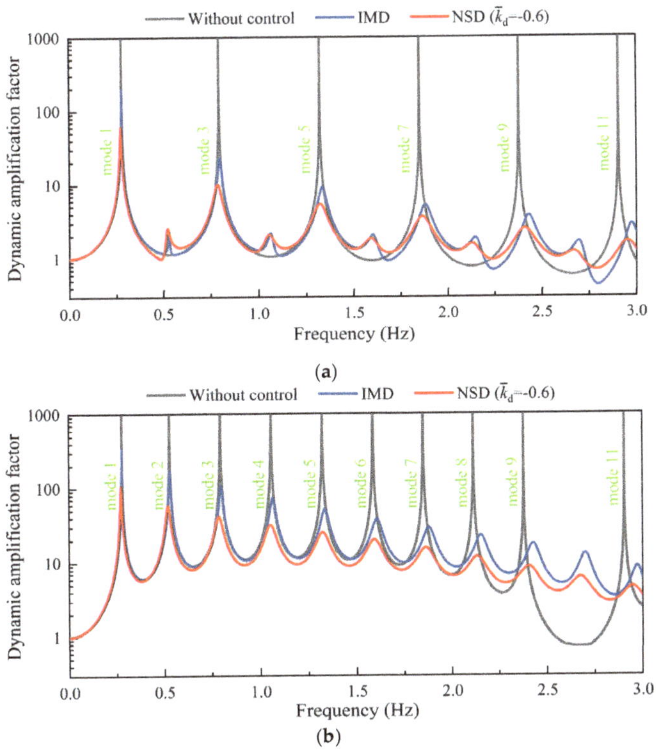

Figure 9. Dynamic amplification factor of cable displacements at various cable locations of the ultra-long cable subjected to sinusoidal excitation. (**a**) Mid-span. (**b**) One-tenth-span.

4.2. White Noise Excitation

Subsequently, white noise excitation is applied to the ultra-long cable. Figure 10 shows the displacement responses of the ultra-long stay cable subjected to the white noise excitation. As shown in Figure 10a, the displacement response of the cable equipped with the IMD or NSD is significantly less than that of the cable without control. The displacement response of the cable equipped with the NSD is dramatically reduced compared to that with the IMD. To quantitatively compare the cable displacement responses at multiple positions, the root-mean-square (RMS) of the cable displacements is calculated and compared in Figure 10b. For the cable equipped with the IMD, the RMS displacements of the cable at 0.02l (i.e., damper location) 0.25*l*, 0.5*l*, and 0.75*l* are 0.28 mm, 3.20 mm, 4.56 mm, and 3.20 mm, respectively, while in the case of cable equipped with the NSD, they are 0.48 mm, 2.49 mm, 3.34 mm, and 2.30 mm, respectively. Compared to the cable equipped with the IMD, the RMS displacements of the cable equipped with the NSD at 0.25*l*, 0.5*l*, and 0.75*l* are reduced by 22.19%, 26.75%, and 28.13%, respectively. In addition, the NSD exhibits greater displacement than the IMD under white noise excitation.

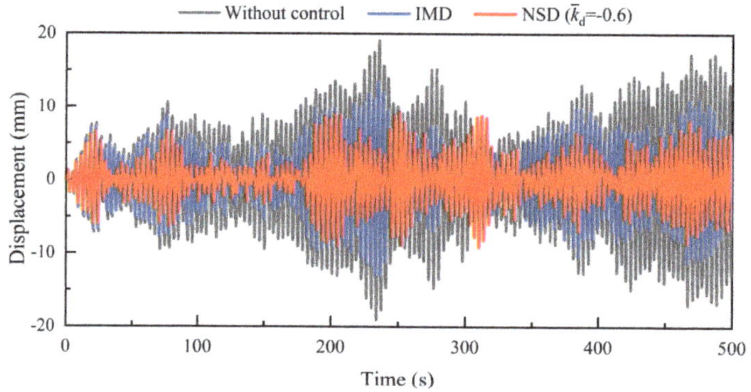

(**a**)

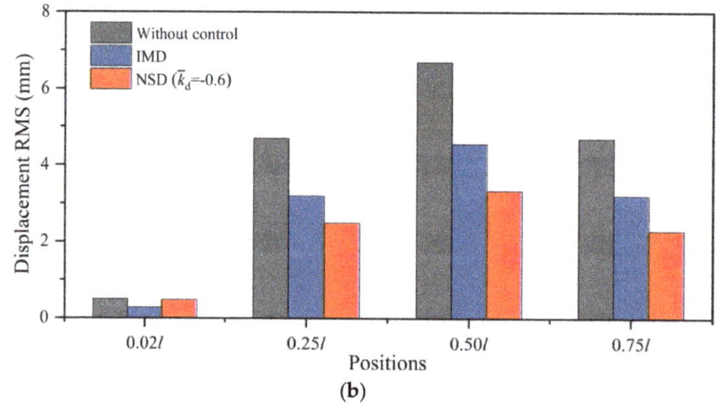

(**b**)

Figure 10. Displacement responses of the ultra-long stay cable subjected to the white noise excitation. (**a**) Displacement time-histories at cable mid-span. (**b**) RMS displacements at various cable positions.

4.3. Wind Excitation

Finally, wind excitation is applied to the ultra-long cable to further investigate the control effectiveness of the IMD and NSD. Figure 11 depicts the Fourier transform (FFT) spectrum of the wind-induced vibration responses of the uncontrolled cable with a one-tenth-span. As depicted, the wind-induced cable vibrations predominantly occur in modes 1 to 8. Figure 12 compares the displacement responses of the ultra-long cable under wind excitation. As illustrated in Figure 12a, both the IMD and NSD can reduce the displacement response of the ultra-long stay cable. Notably, the NSD shows better performance than the IMD in reducing the cable displacement response. Nevertheless, the reduction ratio varies in different time intervals, which can be attributed to the inherent randomness and complexity of wind loads. To further analyze the underlying reasons, the corresponding FFT spectrums during time intervals I (750 s~1050 s) and II (1150 s~1450 s) are compared in Figure 12b,c. During both time intervals, the IMD and NSD can effectively mitigate the first eight modal vibrations of the ultra-long cable. However, the IMD shows insufficient vibration mitigation performance in the first mode, whereas the NSD performs better. In comparison, the NSD is more applicable to mitigate wind-induced multi-mode cable vibrations.

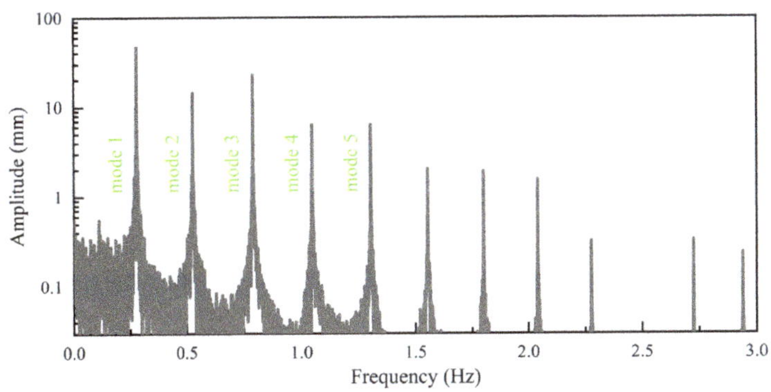

Figure 11. Dominant modes of wind-induced cable vibrations.

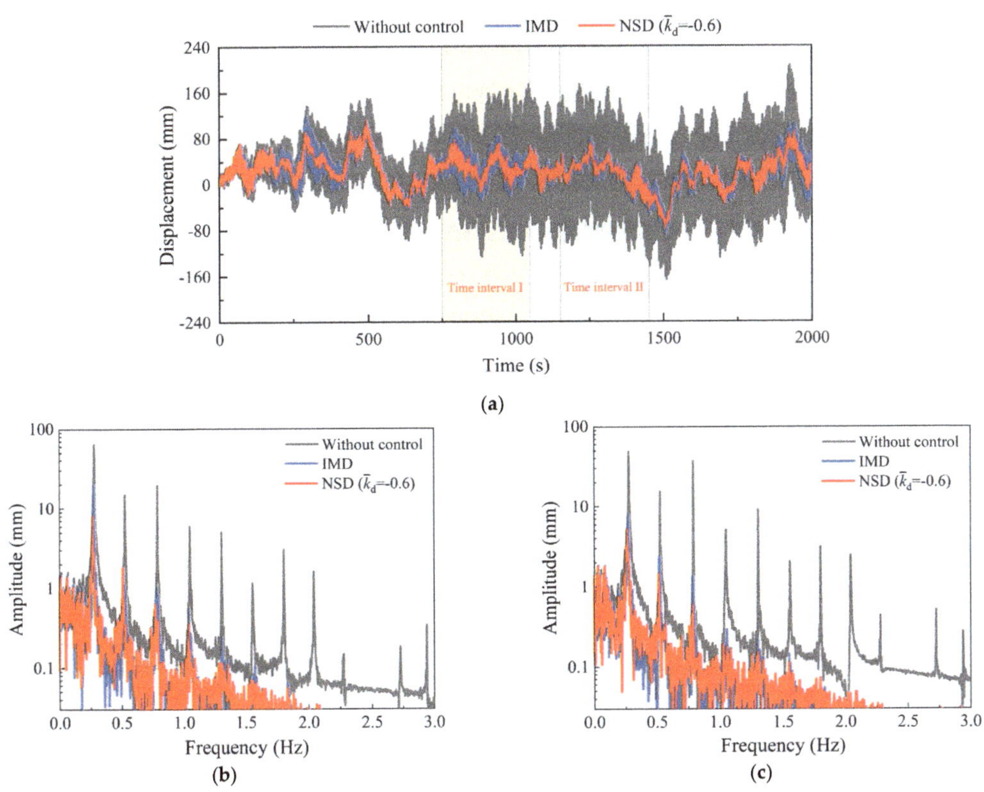

Figure 12. Displacement responses and corresponding FFT spectrums at one-tenth-span of the ultra-long cable subjected to wind excitation. (**a**) Displacement time-histories. (**b**) FFT spectrums in the time interval I. (**c**) FFT spectrums in the time interval II.

To elucidate the underlying reasons for the superior control effectiveness of the NSD over the IMD in multi-mode cable vibration mitigation, Figure 13 compares the hysteresis loops of the IMD and NSD in suppressing wind-induced multi-mode cable vibrations. As shown, the output force and displacement of the NSD are significantly greater than those of the IMD. Consequently, the negative stiffness of the NSD significantly amplifies the displacement of its damping element compared with the inertia of the IMD, resulting in a considerably greater energy dissipation capacity for the NSD in comparison to the IMD when applied to suppress multi-mode cable vibrations. Notably, the integration of artificial intelligence (AI)- and deep learning (DL)-based algorithms can facilitate more effective data analysis [61–63], which represents a key direction for our future work.

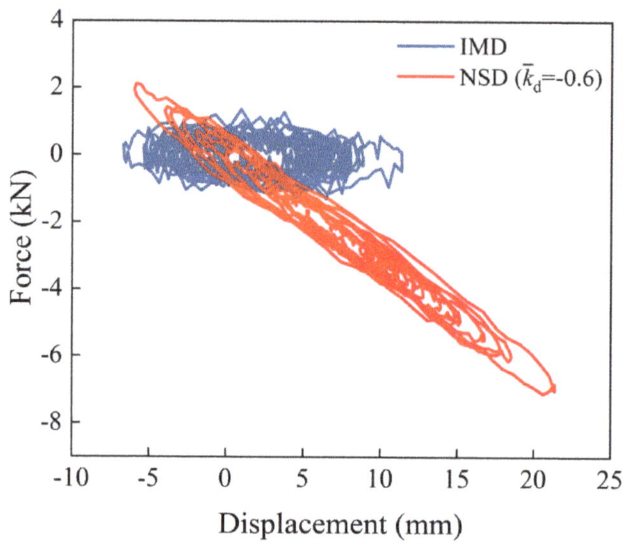

Figure 13. Comparisons of the hysteresis loops of the IMD and NSD in suppressing wind-induced multi-mode vibrations of the ultra-long cable in the time interval 950 s~1000 s.

5. Conclusions

In this paper, a comparative study is conducted to evaluate the control effectiveness of the IMDs and NSDs in suppressing multi-mode cable vibrations. The mechanical models of typical IMD and NSD are introduced in detail, and the numerical models of the two cable-damper systems are established utilizing the finite difference method. The effectiveness of three multi-mode optimization strategies is systematically evaluated, and the most effective strategy is chosen for the optimal design of the IMD and NSD. A performance comparison between the IMD and NSD is performed to assess their control effectiveness in suppressing harmonic, white noise, and wind-induced multi-mode vibrations of a 493.72 (m)-long ultra-long cable. According to the results, the key conclusions can be drawn:

(1) Strategy 3 is recommended for the IMD or NSD to achieve better control effectiveness in suppressing multi-mode cable vibrations compared to strategies 1 and 2.
(2) The NSD significantly improves the cable damping ratios for multiple vibration modes as its negative stiffness coefficient increases, while the IMD performs well only within a small inertia coefficient.
(3) The NSD requires a smaller damping coefficient than the IMD to achieve superior control performance, and its required damping coefficient decreases as the negative stiffness increases.

(4) The sensitivities of the multi-mode cable damping with respect to the damping coefficients of the IMD and NSD are comparable, with no significant differences observed.
(5) The NSD outperforms the IMD in suppressing multi-mode cable vibrations induced by harmonic, white noise and wind excitations. The dynamic responses of the ultra-long cable equipped with the NSD are significantly reduced compared to those equipped with the IMD under the three types of external excitations.
(6) The negative stiffness of the NSD significantly amplifies the displacement of its damping element compared to the inertia of the IMD, resulting in a considerably greater energy dissipation capacity for the NSD in comparison to the IMD when applied to suppress multi-mode cable vibrations.

In the present study, a relatively simple linear model is adopted to investigate the control effectiveness of IMD and NSD. In future studies, the influences of structural nonlinearities will be further investigated.

Author Contributions: Conceptualization, Z.S. and Z.C.; methodology, X.S. and Z.C.; software, Z.C.; validation, X.S., Z.C. and R.M.; formal analysis, Z.C.; investigation, Z.S.; resources, R.M.; data curation, Z.C.; writing—original draft preparation, Z.S.; writing—review and editing, Z.C., X.S. and R.M.; visualization, Z.C.; supervision, R.M.; project administration, Z.S.; funding acquisition, R.M. All authors have read and agreed to the published version of the manuscript.

Funding: The authors gratefully acknowledge the funding from the National Natural Science Foundation of China (Grant No. 52208452) and China Postdoctoral Science Foundation (Grant Nos. 2022M710283 and 2023T160033).

Data Availability Statement: Data will be made available on request.

Conflicts of Interest: Author Zhaowei Shen and Xiaohong Sun was employed by the company Shandong Electric Power Engineering Consulting Institute Corp., Ltd. The remaining authors declare that the research was conducted in the absence of any commercial or financial relationships that could be construed as a potential conflict of interest.

References

1. Chen, Z.; Wang, X.; Ko, J.; Ni, Y.; Spencer, B.; Yang, G.; Hu, J. MR damping system for mitigating wind-rain induced vibration on Dongting Lake Cable-Stayed Bridge. *Wind. Struct.* **2004**, *7*, 293–304. [CrossRef]
2. Liu, G.; Zhou, P.; Yu, T.; Li, Z. Optimal design of nonlinear negative-stiffness damper with flexible support for mitigating cable vibration. *Buildings* **2023**, *13*, 2620. [CrossRef]
3. Chen, L.; Di, F.; Xu, Y.; Sun, L.; Xu, Y.; Wang, L. Multimode cable vibration control using a viscous-shear damper: Case studies on the Sutong Bridge. *Struct. Control Health Monit.* **2020**, *27*, e2536. [CrossRef]
4. Di, F.; Sun, L.; Chen, L.; Zou, Y.; Qin, L.; Huang, Z. Frequency and damping of hybrid cable networks with cross-ties and external dampers: Full-scale experiments. *Mech. Syst. Signal Process.* **2023**, *197*, 110397. [CrossRef]
5. Calò, M.; Ruggieri, S.; Doglioni, A.; Morga, M.; Nettis, A.; Simeone, V.; Uva, G. Probabilistic-based assessment of subsidence phenomena on the existing built heritage by combining MTInSAR data and UAV photogrammetry. *Struct. Infrastruct. Eng.* **2024**, *2024*, 116. [CrossRef]
6. Li, H.; Zhou, P.; Li, Z. Using a rigid restraint with a built-in tuned mass damper to control the vibration of cables. *Buildings* **2024**, *14*, 3785. [CrossRef]
7. Li, J.; Shen, J.; Zhu, S. Adaptive self-powered active vibration control to cable structures. *Mech. Syst. Signal Process.* **2023**, *188*, 110050. [CrossRef]
8. Zhang, C. The active rotary inertia driver system for flutter vibration control of bridges and various promising applications. *Sci. China Technol. Sci.* **2023**, *66*, 390–405. [CrossRef]
9. Johnson, E.; Christenson, R.; Spencer, B. Semi-active damping of cables with sag. *Comput. Civ. Infrastruct. Eng.* **2003**, *18*, 132–146. [CrossRef]
10. Weber, F.; Distl, H. Semi-active damping with negative stiffness for multi-mode cable vibration mitigation: Approximate collocated control solution. *Smart. Mater. Struct.* **2015**, *24*, 115015. [CrossRef]
11. Li, J.; Zhu, S.; Shi, X.; Shen, W. Electromagnetic shunt damper for bridge cable vibration mitigation: Full-scale experimental study. *J. Struct. Eng.* **2020**, *146*, 04019175. [CrossRef]

12. He, D.; Xu, H.; Wang, M.; Wang, T. Transmission and dissipation of vibration in a dynamic vibration absorber-roller system based on particle damping technology. *Chin. J. Mech. Eng.* **2024**, *37*, 108. [CrossRef]
13. Zucca, M.; Longarini, N.; Simoncelli, M.; Aly, A. Tuned mass damper design for slender masonry structures: A framework for linear and nonlinear analysis. *Appl. Sci.* **2021**, *11*, 3425. [CrossRef]
14. Ma, R.; Han, C.; Jiang, S.; Bi, K.; Du, X. Concept, optimization and performance evaluation of a novel particle tuned inertial damper (PTID). *Eng. Struct.* **2024**, *317*, 118678. [CrossRef]
15. Ma, R.; Bi, K.; Zuo, H.; Song, J.; Han, Q. A novel tuned torsional mass damper (TTMD) for seismic protection of structures. *Eng. Struct.* **2024**, *300*, 117233. [CrossRef]
16. Kovacs, I. Zur frage der seilschwingungen und der seildampfung. *Die Bautechnik.* **1982**, *59*, 325–332. (In German)
17. Pacheco, B.; Fujino, Y.; Sulekh, A. Estimation curve for modal damping in stay cables with viscous damper. *J. Struct. Eng.* **1993**, *119*, 1961–1979. [CrossRef]
18. Krenk, S. Vibrations of a taut cable with an external damper. *J. Appl. Mech.* **2000**, *67*, 772–776. [CrossRef]
19. Fournier, J.; Cheng, S. Impact of damper stiffness and damper support stiffness on the efficiency of a linear viscous damper in controlling stay cable vibrations. *J. Bridg. Eng.* **2014**, *19*, 04013022. [CrossRef]
20. Fujino, Y.; Hoang, N. Design formulas for damping of a stay cable with a damper. *J. Struct. Eng.* **2008**, *134*, 269–278. [CrossRef]
21. Zhou, H.; Sun, L.; Xing, F. Damping of full-scale stay cable with viscous damper: Experiment and analysis. *Adv. Struct. Eng.* **2014**, *17*, 265–274. [CrossRef]
22. Main, J.; Jones, N. Free vibrations of taut cable with attached damper. II: Nonlinear damper. *J. Struct. Eng.* **2002**, *128*, 1072–1081. [CrossRef]
23. Krenk, S.; Høgsberg, J. Damping of cables by a transverse force. *J. Eng. Mech.* **2005**, *131*, 340–348. [CrossRef]
24. Chen, L.; Sun, L. Steady-state analysis of cable with nonlinear damper via harmonic balance method for maximizing damping. *J. Struct. Eng.* **2017**, *143*, 04016172. [CrossRef]
25. Zhou, Q.; Nielsen, S.; Qu, W. Semi-active control of three-dimensional vibrations of an inclined sag cable with magnetorheological dampers. *J. Sound Vib.* **2006**, *296*, 1–22. [CrossRef]
26. Xu, Y.; Yu, Z. Vibration of inclined sag cables with oil dampers in cable stayed bridges. *J. Bridg. Eng.* **1998**, *3*, 194–203. [CrossRef]
27. Hoang, N.; Fujino, Y. Analytical study on bending effects in a stay cable with a damper. *J. Eng. Mech.* **2007**, *133*, 1241–1246. [CrossRef]
28. Wang, Z.; Gao, H.; Fan, B.; Chen, Z. Inertial mass damper for vibration control of cable with sag. *J. Low Freq. Noise Vib. Act. Control* **2020**, *39*, 749–760. [CrossRef]
29. Wang, X.; Ni, Y.; Ko, J.; Chen, Z. Optimal design of viscous dampers for multi-mode vibration control of bridge cables. *Eng. Struct.* **2005**, *27*, 792–800. [CrossRef]
30. Weber, F.; Feltrin, G.; Malanka, M.; Fobo, W.; Distl, H. Design of viscous dampers targeting multiple cable modes. *Eng. Struct.* **2009**, *31*, 2797–2800. [CrossRef]
31. Yang, C.; Chen, Z.; Wang, W.; Hua, X.; Wang, Y. Optimal design of two viscous dampers for multi-mode control of a cable covering broad frequency range. *Eng. Struct.* **2021**, *245*, 112830. [CrossRef]
32. Cheng, S.; Darivandi, N.; Ghrib, F. The design of an optimal viscous damper for a bridge stay cable using energy-based approach. *J. Sound Vib.* **2010**, *329*, 4689–4704. [CrossRef]
33. Di, F.; Sun, L.; Chen, L. Cable vibration control with internal and external dampers: Theoretical analysis and field test. *Smart Struct. Syst.* **2020**, *26*, 575–589. [CrossRef]
34. An, M.; Li, S.; Sorokin, V.; Chen, Z.; Flay, R. Experimental study of vortex-induced vibration of stay cables installed with two types of perforated shroud light devices. *J. Wind Eng. Ind. Aerodyn.* **2024**, *253*, 105882. [CrossRef]
35. Li, H.; Liu, M.; Ou, J. Negative stiffness characteristics of active and semi-active control systems for stay cables. *Struct. Control Health Monit.* **2008**, *15*, 120–142. [CrossRef]
36. Xu, Y.; Xu, Z.; Guo, Y.; Zhou, M.; Zhao, Y.; Yang, Y.; Dai, J.; Zhang, J.; Zhu, C.; Ji, B.; et al. A programmable pseudo negative stiffness control device and its role in stay cable vibration control. *Mech. Syst. Signal Process.* **2022**, *173*, 109054. [CrossRef]
37. Li, H.; Bi, K.; Hao, H.; Yu, Y.; Xu, L. Experimental study of a novel quasi-active negative stiffness damper system for achieving optimal active control performance. *Eng. Struct.* **2024**, *299*, 117082. [CrossRef]
38. Ma, R.; Bi, K.; Hao, H. Inerter-based structural vibration control: A state-of-the-art review. *Eng. Struct.* **2021**, *243*, 112655. [CrossRef]
39. Wang, Z.; Song, L.; Cheng, Z.; Yang, H.; Wen, J.; Qi, M. Finite element model for vibration serviceability evaluation of a suspended floor with and without tuned mass dampers. *Buildings* **2023**, *13*, 309. [CrossRef]
40. Li, H.; Bi, K.; Han, Q.; Ma, R. A state-of-the-art review on negative stiffness-based structural vibration control. *Eng. Struct.* **2025**, *323*, 119247. [CrossRef]
41. Gao, H.; Wang, H.; Li, J.; Wang, Z.; Liang, R.; Xu, Z.; Ni, Y. Optimum design of viscous inerter damper targeting multi-mode vibration mitigation of stay cables. *Eng. Struct.* **2021**, *226*, 111375. [CrossRef]

42. Wang, Z.; Cheng, Z.; Wang, H.; Yue, F.; Gao, H.; Fan, B. Damping of a stay cable with two eddy-current inertial mass dampers: Theoretical analysis, experimental study, and parameter optimization. *Struct. Control Health Monit.* **2022**, *29*, e3085. [CrossRef]
43. Dong, S.; Duan, Y.; Xu, S.; Yun, C. Vibration mitigation of sagged cables using a viscous inertial mass damper: An experimental investigation. *Struct. Control Health Monit.* **2023**, *2023*, 4361740. [CrossRef]
44. Wang, Z.; Xu, Y.; Gao, H.; Cheng, Z.; Xu, K.; Zhao, S. Vibration control of a stay cable with a rotary electromagnetic inertial mass damper. *Smart Struct. Syst.* **2019**, *23*, 627–639. [CrossRef]
45. Li, Y.; Shen, W.; Zhu, H. Vibration mitigation of stay cables using electromagnetic inertial mass dampers: Full-scale experiment and analysis. *Eng. Struct.* **2019**, *200*, 109693. [CrossRef]
46. Shen, W.; Hao, J.; Zhu, H. Design approach and practical formulas of electromagnetic inertial mass dampers for cable multimode control. *J. Bridg. Eng.* **2022**, *27*, 04022121. [CrossRef]
47. Zhou, P.; Li, H. Modeling and control performance of a negative stiffness damper for suppressing stay cable vibrations. *Struct. Control Health Monit.* **2016**, *23*, 764–782. [CrossRef]
48. Chen, L.; Liu, Z.; Zou, Y.; Wang, M.; Nagarajaiah, S.; Sun, F.; Sun, L. Practical negative stiffness device with viscoelastic damper in parallel or series configuration for cable damping improvement. *J. Sound Vib.* **2023**, *560*, 117757. [CrossRef]
49. Shi, X.; Zhu, S.; Spencer, B. Experimental study on passive negative stiffness damper for cable vibration mitigation. *J. Eng. Mech.* **2017**, *143*, 04017070. [CrossRef]
50. Wang, Z.; Cheng, Z.; Yin, G.; Shen, W. A magnetic negative stiffness eddy-current inertial mass damper for cable vibration mitigation. *Mech. Syst. Signal Process.* **2023**, *188*, 110013. [CrossRef]
51. Cheng, Z.; Bi, K.; Wang, Z.; Ma, R.; Lin, W. Performance enhanced magnetic negative stiffness eddy-current damper: Numerical simulation and experimental investigation. *Mech. Syst. Signal Process.* **2025**, *223*, 111890. [CrossRef]
52. Chen, L.; Nagarajaiah, S.; Sun, L. A unified analysis of negative stiffness dampers and inerter-based absorbers for multimode cable vibration control. *J. Sound Vib.* **2021**, *494*, 115814. [CrossRef]
53. Wang, Y.; Chen, Z.; Yang, C.; Liu, Z.; He, J.; Feng, Z. A novel eddy current damper system for multi-mode high-order vibration control of ultra-long stay cables. *Eng. Struct.* **2022**, *262*, 114319. [CrossRef]
54. Mehrabi, A.; Tabatabai, H. Unified finite difference formulation for free vibration of cables. *J. Struct. Eng.* **1998**, *124*, 1313–1322. [CrossRef]
55. Kiani, K.; Efazati, M. Three-dimensional nonlocal-surface energy-based statics, dynamics, and divergence instability of movable cable-like nanostructures with arbitrary translational motion. *Arch. Appl. Mech.* **2021**, *91*, 3095–3123. [CrossRef]
56. Kiani, K.; Efazati, M. Nonlocal vibrations and instability of three-dimensionally accelerated moving nanocables. *Phys. Scr.* **2020**, *95*, 105005. [CrossRef]
57. Luo, J.; Jiang, J. Macdonald JHG. Cable vibration suppression with inerter-based absorbers. *J. Eng. Mech.* **2019**, *145*, 04018134. [CrossRef]
58. Huang, Z.; Hua, X.; Chen, Z.; Niu, H. Performance evaluation of inerter-based damping devices for structural vibration control of stay cables. *Smart. Struct. Syst.* **2019**, *23*, 615–626. [CrossRef]
59. Zhou, H.; Zhou, X.; Yao, G.; Sun, L.; Xing, F. Free vibration of two taut cables interconnected by a damper. *Struct. Control Health. Monit.* **2019**, *26*, e2423. [CrossRef]
60. *PTI-DC45.1*; Recommendations for Stay Cable Design, Testing and Installation. Post-Tensioning Institute: Farmington Hills, MI, USA, 2018.
61. Shen, J.; Wu, T.; Zou, J.; Wu, P.; Liao, Y. Development of deep ultraviolet LED packaging. *J. Opt. Photon. Res.* **2025**, *2*, 1–10. [CrossRef]
62. Lee, H.; Yoon, H.; Kim, S. Vibration detection of stay-cable from low-quality CCTV images using deep-learning-based dehazing and semantic segmentation algorithms. *Eng. Struct.* **2023**, *292*, 116567. [CrossRef]
63. Chin, S. Artificial perovskite multiple quantum well optoelectronics. *J. Opt. Photon. Res.* **2024**, *2024*, 1–9. [CrossRef]

buildings

MDPI

Article

Aseismic Experimental Research on Safety-Belt System of a Low-Gravity-Center Cable-Stayed Bridge

Qing Li [1,2], Xiangtao Lu [1,3,*], Zhen Wang [1] and Rong Fang [4]

1 Department of Civil Engineering, Beijing University of Technology, Beijing 100124, China; li602251892@emails.bjut.edu.cn (Q.L.); wangzhenskg@126.com (Z.W.)
2 Key Laboratory of Earthquake Engineering and Structural Retrofit of Beijing, Beijing 100124, China
3 School of Civil Engineering, Southeast University, Nanjing 211189, China
4 School of Civil and Transportation Engineering, Beijing University of Civil Engineering and Architecture, Beijing 102616, China; fangrong@bucea.edu.cn
* Correspondence: xiangtao.lu@outlook.com

Abstract: The seismic performance of low-gravity-center cable-stayed bridges is influenced by their structural system. This paper introduces a novel anti-seismic structural system designed to enhance the earthquake resistance of low-gravity-center cable-stayed bridges while reducing secondary internal forces during operation. A total of 47 groups of shaking table tests were conducted, considering four types of seismic waves and acceleration levels. In the shaking table tests, a safety-belt device was constructed and integrated into a scale model of a low-gravity-center cable-stayed bridge. The anti-seismic performance of the safety-belt system was validated by comparing the shaking table test results of the asymmetric restraint system and floating system reported in previous studies. The experimental findings revealed that the strain response at the tower bottom and the displacement response at the tower top were lower in the safety-belt system model. The safety-belt system shows potential for application in low-gravity-center cable-stayed bridges to reduce earthquake impacts. Furthermore, this study elucidates the damage mechanism of low-gravity-center cable-stayed bridges equipped with the safety-belt system under seismic loading, providing references for optimizing the anti-seismic design of such bridges.

Keywords: low-gravity-center cable-stayed bridge; shaking table test; anti-seismic structural system; safety-belt system; bridge engineering

1. Introduction

With the development of urban bridges, there are more and more cable-stayed bridges without navigation needs. Among them, the number of low-gravity-center cable-stayed bridges is also increasing, such as China's Jinan Third Bridge and Songhua River Bridge [1]. With the proposal of the new structural concept of the low-gravity-center cable-stayed bridge, it is urgent to carry out in-depth research on the differences between its structural characteristics and those of the conventional cable-stayed bridge. Based on the principle of structural mechanics, this study proposes determination criteria for the low-center-gravity cable-stayed bridge: Under the action of an earthquake, in the case that the ratio of the tower bottom moment in the floating system (FS) to that in the fixed articulated system (FAS) exceeds a certain threshold, the cable-stayed bridge can be identified as a low-gravity-center cable-stayed bridge [2].

The seismic analysis and design of cable-stayed bridges are critical areas of research [3]. In bridge engineering, the twin-tower suspension bridge is usually divided into three structural systems: asymmetric constraint system, FAS, and FS. The asymmetric constraint system is usually unfavorable to the seismic performance of a double-tower suspension bridge. In contrast, the FAS can improve the seismic resistance of the bridge through the coordinated force of the two towers. However, it cannot relieve the secondary internal force generated by the deformation of the girder under the action of hot load. Although the FS can effectively solve the problem of secondary internal force, it will lead to significant displacement response [4].

In recent years, numerous studies have investigated the seismic performance of conventional cable-stayed bridges under strong earthquakes through shaking table tests and even nonlinear analysis [5–9]. The seismic performance of these bridges is generally evaluated based on two aspects: internal forces and displacement [10]. A better seismic performance is characterized by lower internal forces and displacements. However, these two aspects often exhibit a trade-off under seismic loading; reducing internal force responses typically results in increased displacements and vice versa [11].

Many researchers have proposed passive energy dissipation devices for cable-stayed bridges, tailored to the dynamic characteristics of different structural systems. Generally, passive energy dissipation devices are widely applied in the seismic control and mitigation of cable-stayed bridges. Various types of passive control systems, such as friction dampers, metallic yield dampers, viscous fluid dampers, and viscoelastic dampers, have been employed [12,13]. Some studies have also explored anti-seismic systems that combine Transverse Steel Dampers (TSDs) with conventional sliding bearings [14–16]. However, these passive energy dissipation devices often involve complex installation processes and high fabrication costs [17–21].

Shaking table testing remains one of the most effective methods for assessing the performance of passive energy dissipation devices and studying the seismic response of scaled models under different seismic waves [21–25].

Later studies introduced the concept of the low-gravity-center cable-stayed bridge, focusing on the transmission path of the horizontal inertia force of the girder [2]. Criteria for identifying low-gravity-center cable-stayed bridges were established, based on simplified calculations of the fundamental frequency of cable-stayed bridges combined with response spectrum analysis [1]. These studies reported that the dynamic properties of low-gravity-center cable-stayed bridges differ from those of conventional cable-stayed bridges [26,27].

Furthermore, the nonlinear seismic response of cable-stayed bridges is a crucial issue in assessing structural safety [28,29]. For low-gravity-center cable-stayed bridges, there is a lack of real earthquake testing and nonlinear seismic response studies. Research on the structural damage of low-gravity-center cable-stayed bridges under strong earthquakes, particularly in terms of longitudinal seismic damage, is scarce [5].

In view of the above technical problems, this study proposes a system conversion scheme (safety-belt system, SBS) based on the belt connection device. This scheme effectively solves the problem of cooperative force under seismic load in an asymmetric constraint system (ARS) and eliminates the secondary internal force effect caused by deformation in FAS under temperature load. Additionally, the low-gravity-center cable-stayed bridge with the FAS is better than the FS in the control of internal force at the base of the tower [2]. In addition, it is not costly for the manufacturing of a safety-belt device fabricated by steel material, which has significant economic advantages.

To investigate the damage mechanism of the low-gravity-center cable-stayed bridges with a SBS under strong earthquakes, this study fabricated a safety-belt device, constructed a 1:75 scale model of a low-gravity-center cable-stayed bridge, and conducted 47 groups of shaking table tests. The safety-belt device was installed between tower and girder. The experiment considered four types of seismic waves (El Centro, TianJin, EMC_FAIRVIE, and site-specific artificial wave) and different acceleration levels (PGA ranging from 0.1 g to 1.4 g).

The safety-belt system's effectiveness was validated by comparing the shaking table test results from this study with those of the ARS and FS reported in previous research. The damage mechanism of low-gravity-center cable-stayed bridges under strong earthquakes was observed through the grading of seismic wave loading. This safety-belt system has potential in term of enhancing the aseismic performance of low-gravity-center cable-stayed bridges, and this study can provide a reference for optimizing the seismic design of such bridges.

2. Safety-Belt System

2.1. Design of Safety-Belt Device

The safety-belt system (SBS) is a vibration-reducing restraint system. It means a safety-belt device is installed at the tower–girder connection area of non-fixed pier on an asymmetric restraint system (ARS). The safety-belt device can be activated by dynamic acceleration [30]. This device consists of a safety-belt tie, a connecting block, a locking shaft, and a fixing platen, as illustrated in Figure 1a.

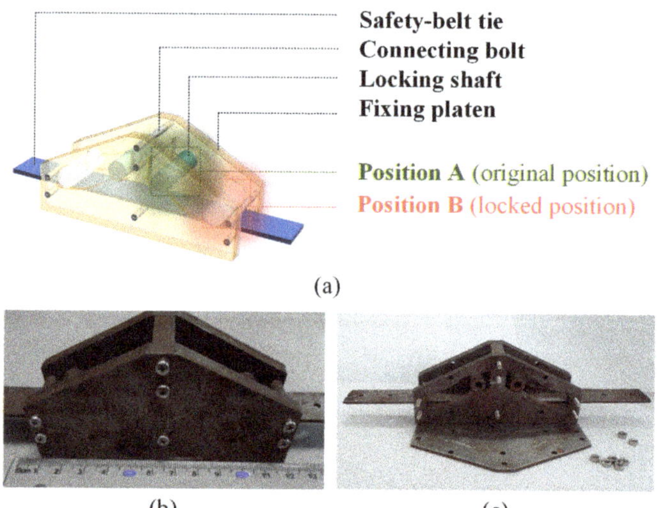

Figure 1. Structure of safety-belt device: (**a**) configuration of safety-belt device; (**b**) safety-belt device after installation; and (**c**) safety-belt device before installation.

During normal operation, the locking shaft remains in position A, allowing the safety-belt tie to move longitudinally with the beam. This design accommodates the relative displacement between the bridge tower and the main girder caused by daily temperature variations. At this stage, the locking shaft rolls within position A without dropping to position B, keeping the longitudinal connection between the tower and the girder unrestricted until the acceleration of the lower crossbeam reaches a predefined threshold for the first time. When an earthquake occurs, the ground motion acceleration causes the locking shaft to fall from position A to the locked position B. This action restricts the relative displacement between the girder and the bridge tower, thereby enhancing the seismic performance of the low-gravity-center cable-stayed bridge. The safety-belt device used in this study was fabricated as shown in Figure 1b,c.

2.2. Mechanical Principles

The mechanical principle of the safety-belt device is illustrated in Figure 2a. The relationship between the force and the relative displacement of the safety-belt device, corresponding to the locking process between the girder and the lower crossbeam, is depicted in Figure 2b.

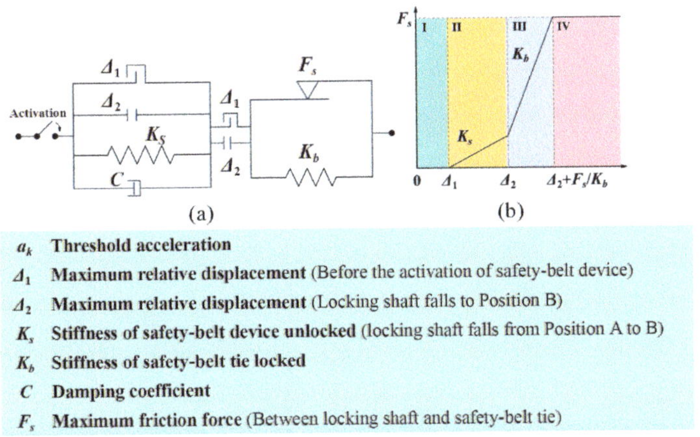

Figure 2. Mechanical principle and force–displacement relationship: (**a**) mechanical principle of safety-belt device; (**b**) force–displacement relationship of safety-belt device.

The connection between girder and lower crossbeam will not be fixed until the acceleration of the girder exceeds the acceleration threshold $\boldsymbol{a_k}$ (Stage I). Once this threshold is exceeded, the safety-belt device will be activated (Stage II), causing the locking shaft to roll from position A to position B.

If the relative displacement between girder and lower crossbeam exceeds $\boldsymbol{\Delta_1}$, an uncompleted locked connection is established (Stage II). At this moment, the restraint force $\boldsymbol{F_s}$ is mainly from the friction between the safety-belt tie and the locking shaft (at position B). When the relative displacement reaches $\boldsymbol{\Delta_2}$, a completed locked connection is established (Stage III). At this moment, the restraint force $\boldsymbol{F_s}$ is mainly offered by the deformation of the locked safety-belt tie. Once the force that the safety-belt tie is enduring exceeds the ultimate strength, the restraint force $\boldsymbol{F_s}$ reaches the maximum (Stage IV). The constitutive equations of the safety-belt device are shown in Equation (1):

$$F(t) = \begin{cases} 0 & ;t < t^* \\ 0 & ;t \geq t^*, \quad |d_t| < \Delta_1 \\ K_s \cdot (|d_t| - \Delta_1) & ;t \geq t^*, \quad \Delta_1 < |d_t| \leq \Delta_2 \\ K_b \cdot (|d_t| - \Delta_2) & ;t \geq t^*, \quad |d_t| < \Delta_2 \\ F_s & ;t \geq t^*, |d_t| \geq \Delta_2 + F_s/K_b \end{cases} \tag{1}$$

where $F(t)$ is the force exerted on the safety-belt device; t is the dynamic acceleration action time; t^* is the moment when the safety-belt device is activated; and *it* is the relative displacement between tower and girder.

3. Experiment Design

3.1. Design of Experimental Structure

3.1.1. Similarity Ratio

To ensure that the experimental structure exhibited the same dynamic characteristics as those observed in previous studies, the vertical compression–strain similarity ratio between the test model and the real bridge structure was set to 1.0, as referenced in earlier research [26]. Considering the dimensions and load capacity of the shaking table, the similarity ratio between the test model and the prototype structure was determined to be 1:75. The similarity ratio of the elastic modulus for the concrete main tower was set at 0.3, while the elastic modulus similarity ratio for the steel girder was 1.0. Additionally, the acceleration similarity ratio was maintained at 1.0, as shown in Table 1.

Table 1. Required scaling factors for the concrete towers.

Type	Physics Parameters	Symbol	Similarity Ratio (Model/Prototype)
Geometry property	Geometry	S_l	0.0133
Material properties	Strain	$S\varepsilon$	1.0
	Elastic modulus	$S_E = S_\sigma$	0.3
	Stress	S_σ	0.3
	Mass density	$S_\rho = S_E/S_a S_L$	22.5
	Poisson's ratio	S_μ	1.0
	Mass	$S_m = S_\rho {S_l}^3$	0.000053
Dynamic properties	Time	$S_t = (S_m/S_k)\ 0.5$	0.1155
	Frequency	$S_f = 1/S_t$	8.66
	Damping ratio	$S_c = S_m/S_t$	0.00046
	Velocity	$S_v = S_l/S_t$	0.1155
	Acceleration	S_a	1.0

3.1.2. Experimental Structure

The dynamic characteristics of a low-gravity-center cable-stayed bridge are significantly influenced by the different connection types between the tower and the girder. Additionally, the stress distribution within the main tower is heavily impacted by the type of restraint used between the tower and the girder.

To enable a meaningful comparison between the experimental results in this paper and those from previous studies, and to validate the advantages of the SBS, the experimental structure was designed to closely resemble the structure used in earlier research [26]. The seismic systems in the earlier research are the ARS and the FS.

It facilitates a comprehensive demonstration of the SBS's advantages by comparing SBS with ARS and FS. The structural system of the SBS is similar to that of the ARS before the safety-belt device is activated. Comparison with the ARS can intuitively reflect the advantages of the SBS. FS as a traditional structural system, has limitations in terms of seismic performance, and has a potential risk of girder collision resulting from excessive displacement response under great earthquake action. The comparison between SBS and FS could provide an experimental reference for expanding the experiment record of the low-gravity-center cable stayed bridges with FS.

The types of restraints and their corresponding designations for towers in different structural systems are detailed in Table 2.

Table 2. Constraint types of different structural systems.

Classification	1# Tower–Girder Connection	2# Tower–Girder Connection
FS	Free sliding	Free sliding
ARS	Fixed constraint	Free sliding
SBS	Fixed constraint	Safety-belt device

The connection of the safety-belt device is presented in Figure 3a. The fixed constraint type of the SBS and ARS is illustrated in Figure 3b, and the free-sliding constraint of FS is illustrated in Figure 3c.

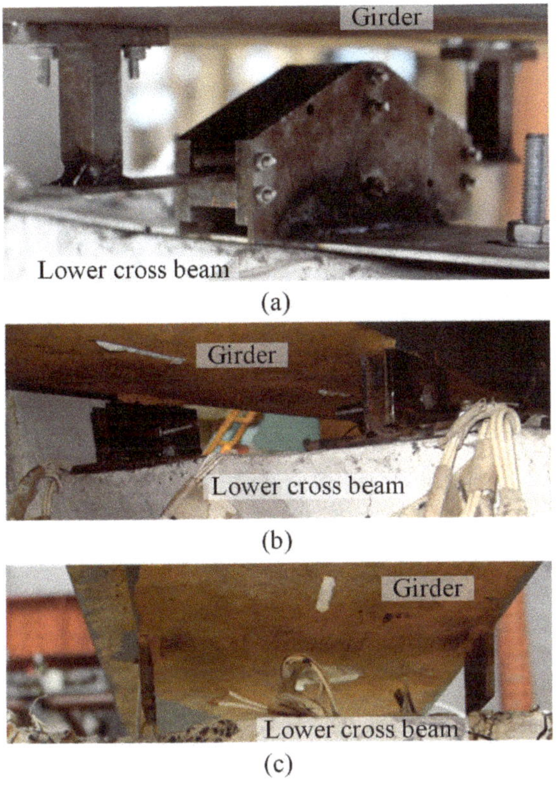

Figure 3. Support types: (**a**) Connection of safety-belt device; (**b**) Fixed support constraint and (**c**) Free-sliding constraint.

The cross-sectional design of the main tower, main girder, and auxiliary pier in the experimental structure was guided by the principle of flexural stiffness equivalence. In the reinforcement design for the main concrete tower, the bearing capacity of the normal section was determined using the principle of equivalent bending capacity, while the oblique section's bearing capacity was designed based on the principle of equivalent shear capacity.

The H-shaped main tower of the experimental structure was constructed using C15 self-compacting concrete (SCC), reinforced with six HRB335 round steel bars and 14# galvanized iron wires for stirrups. The dimensions of the main tower structure are illustrated in Figure 4a, and the cross-section of the main girder is depicted in Figure 4b.

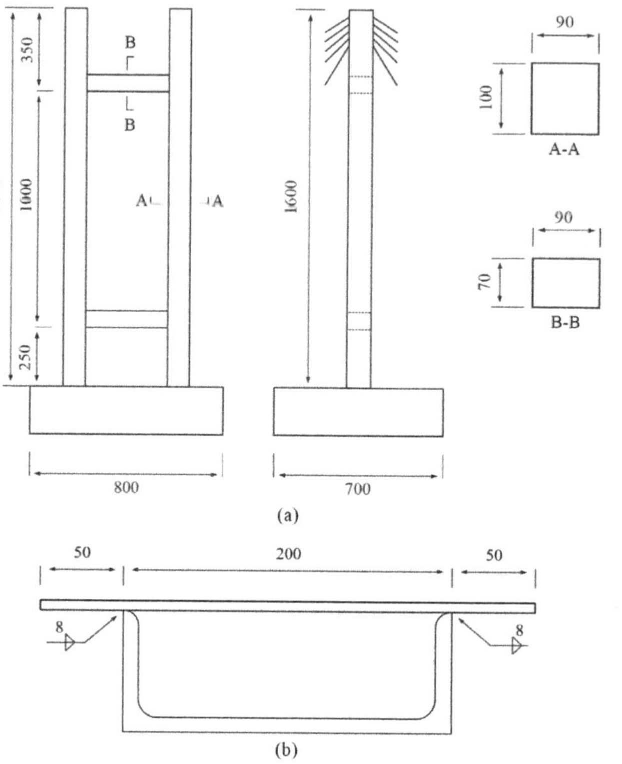

Figure 4. Structural design: (**a**) tower (unit: mm) and (**b**) girder (unit: mm).

The main girder of the experimental structure was constructed using Q345D H-shaped steel, consistent with the materials used in previous studies [26]. The mechanical properties of the selected materials were tested, and the results are summarized in Table 3.

Table 3. Test result of micro concrete and steel bar specimens.

Index	Compressive Strength of Concrete (MPa)	Elastic Modulus of Concrete (GPa)	Yield Stress of Steel Bar (MPa)
Measured value	17.57 16.82 15.64	12.1 10.1 14.14	367 362 376
Average value	16.68	12.11	368

3.1.3. Counterweight

In this experiment, standard iron blocks identical to those used in previous studies were employed to simulate the weight of a low-gravity-center cable-stayed bridge. The counterweight configuration of the experimental structure is detailed in Table 4.

Table 4. Additional masses of the single tower and steel girder.

Component	Self-Weight (kg)	Additional Mass (kg)	Total Mass (kg)	Theoretical Mass (kg)
Steel girder	193.36	320	513.36	514.13
Single tower	82.71	140	222.71	224.64

3.2. Design of Ground Motions

To investigate the damage characteristics and seismic response of a low-gravity-center cable-stayed bridge under great earthquake, three actual seismic records, including El Centro (ELC), TianJin (TJ), and EMC_FAIRVIE (EMC), and one site-specific artificial (SA) wave were selected as input seismic waves for the shaking table test.

The selected ground motions are representative of different site conditions. Specifically, the ELC record is typical of Site Classes II–III, while the TJ record corresponds to Site Classes III–IV. The EMC record is representative of a Site Class III condition, and the site category of the AS wave also falls within Site Class III. All site classifications are defined in accordance with the Chinese Seismic Design Code [26,30].

To achieve consistent earthquake loading, the duration of each seismic wave was compressed to 1/5 of its original length. Following the normalization of peak acceleration and the adjustment of input time compression, the time-history curves of ground motion acceleration for the four waves are presented in Figure 5. Additionally, the initial acceleration peaks of these four seismic waves were uniformly adjusted to 0.1 g.

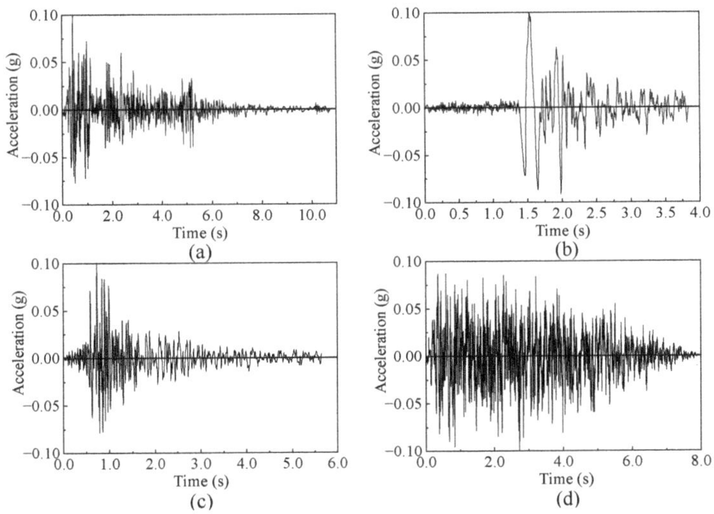

Figure 5. Acceleration time-history curves of ground motions: (**a**) ELC wave; (**b**) TJ wave; (**c**) EMC wave; and (**d**) SA wave.

Each seismic wave was applied along the longitudinal direction of the experimental structure. During each loading stage, the seismic waves were applied in the following order: ELC, TJ, EMC, and SA. The peak ground acceleration (PGA) of each seismic wave was incrementally increased from 0.1 g to 1.1 g, with 0.1 g increments at each loading stage. Only the PGA of the TJ wave was subsequently increased to 1.4. The changes in the dynamic characteristics of the test structure were recorded before and after the initiation of each loading stage.

3.3. Experimental Arrangement

During the shaking table test, displacement, acceleration, and strain were recorded at each loading step. The study evaluated the seismic performance of various anti-seismic structural systems from these three aspects [31,32]. The arrangements of the accelerometer, displacement sensor, and strain sensor are illustrated in Figure 6a, and the experimental structure is depicted in Figure 6b.

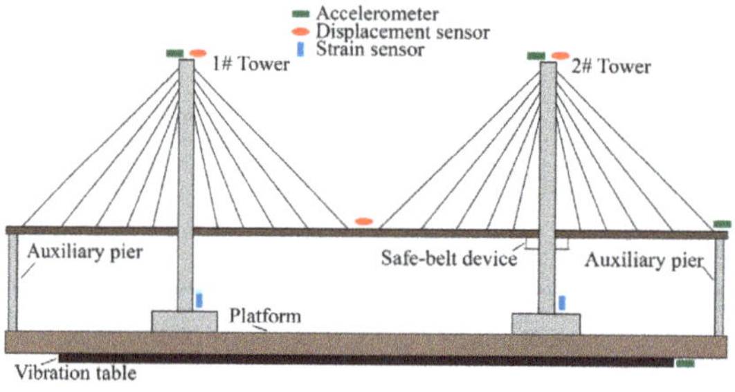

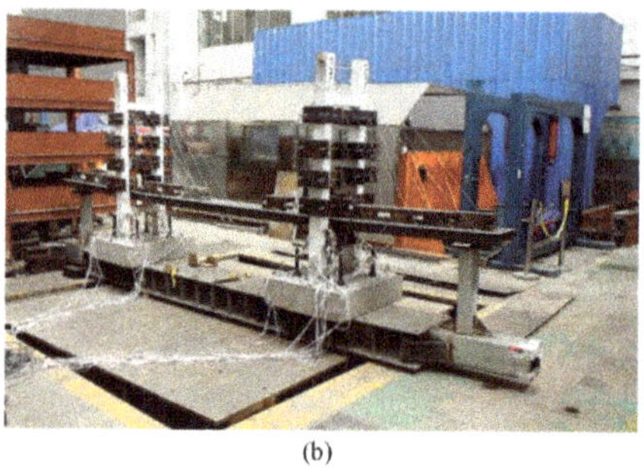

Figure 6. Arrangement of shaking table test: (**a**) arrangement of sensors and (**b**) experimental structure.

Before each loading step, the self-vibration characteristics of the experimental structure were assessed using a white noise vibration test. This test was conducted to evaluate the dynamic response capabilities of the structure and to detect any structural faults. During each loading step, responses at monitoring sites (displacement, acceleration, and strain) were recorded, and the locations and extent of any damage were observed throughout the experiment.

4. Results

4.1. Self-Vibration Frequency

In the previous study, the self-vibration frequencies of two anti-seismic structural systems were evaluated: the ARS and the FS. The acceleration time-history curve at the top of the main tower, recorded before each loading step, was Fourier transformed to obtain the corresponding spectral curve [26]. For this study, the dynamic characteristics of the SBS needed to be comparable to those of the ARS. The self-vibration frequencies of the three anti-seismic structural systems are shown in Figure 7.

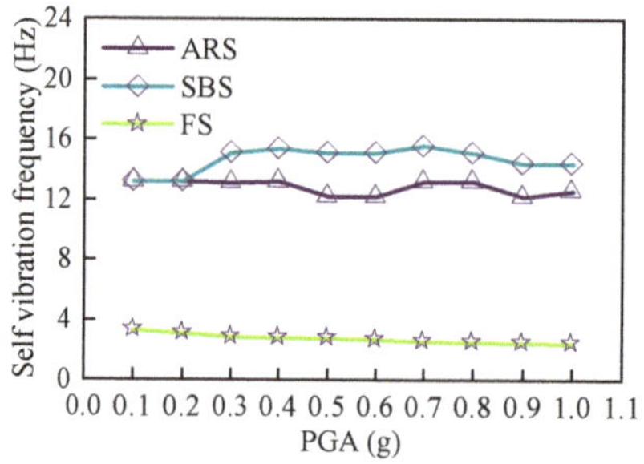

Figure 7. Self-vibration frequencies of the three anti-seismic structural systems.

Before the activation of the safety-belt device, the self-vibration frequency of the SBS was almost similar to that of the ARS. This facilitated the comparison of anti-seismic capabilities between experimental structures with the ARS and the SBS. After the activation of the safety-belt device, the self-vibration frequencies of the SBS were slightly higher than those of the ARS. This conformed to the expectation of the dynamic characteristic of the SBS.

4.2. Experimental Phenomenon

The threshold accelerations were recorded when the safety-belt device was activated under the four seismic waves, as shown in Table 5.

Table 5. PGA when the safety-belt device was activated and functioned.

Seismic Waves	ELC	TJ	EMC	SA
PGA (Activated)	0.4 g	0.3 g	1.0 g	0.5 g
PGA (Functioned)	0.8 g	0.5 g	1.0 g	0.8 g

The first activation of the safety-belt device occurred under the action of the TJ wave when the PGA was 0.3 g, but the locking shaft had not fallen into the locked position; at this moment, the relative displacement between the girder and tower was small, and the tower and girder were not locked, thus the safety-belt device has not yet fully functioned.

The safety-belt device was activated for the first time under the 0.4 g PGA of the ELC wave but it has not yet fully functioned. It was activated for the first time under the 0.5 g PGA of the SA wave, but it has not yet fully functioned as well, while it fully functioned under the action of the TJ wave with 0.5 g PGA. When PGA reached 0.8 g, it functioned under the action of both the EL Centro wave and the SA wave. When PGA reached 1.0 g, it was activated and functioned under the action of the EMC wave.

The damage observed with the gradual increase of PGA is shown in Table 6. It can be seen that the first crack appeared at the location of the lower crossbeam of 1# tower with some concrete surface peeling off when PGA was 0.8 g. When PGA was 1.4 g, the safety-belt ties buckled and the bolts at the connection between the safety-belt device and the main girder broke due to the shear force.

Table 6. Phenomena of the SBS under different PGA action.

PGA	Phenomena
0.1 g~0.7 g	No damage
0.8 g	Obvious cracks with some concrete surface peeling appeared on 1# tower crossbeam, as Figure 8a,b show.
1 g	Horizontal cracks appeared on the two tower bottoms of 1# tower, as Figure 8g,h show. The original cracks on the lower crossbeam of 1# tower extended and concrete fragmentation occurred, as Figure 8c,d show.
1.1 g	The original cracks on 1# tower bottom extended to an around-tower-bottom circle crack and aggravated concrete fragmentation occurred. Reinforcement at the column corner of the tower bottom was exposed. The lower crossbeam had further concrete damage, as Figure 8f shows.
1.2 g	New oblique cracks appeared on the 1# tower nearing the lower crossbeam, as Figure 8e shows.
1.4 g	On the lower crossbeam of the 2# tower, the bolt used in fixing the safety-belt device and main girder broke under shearing, and the safety-belt device was broken, as Figure 8i shows.

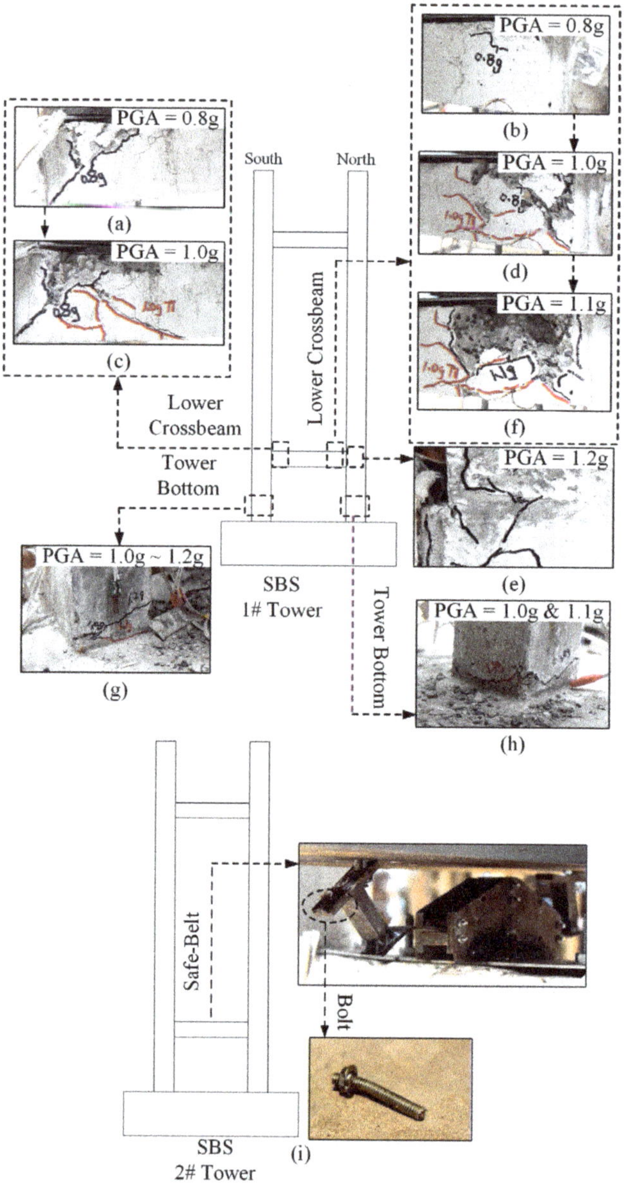

Figure 8. The damage observed of the SBS 1# tower: (**a**) the southern lower crossbeam when PGA was 0.8 g; (**b**) the northern lower crossbeam when PGA was 0.8 g; (**c**) the southern lower crossbeam when PGA was 1.0 g; (**d**) the northern lower crossbeam when PGA was 1.0 g; (**e**) the northern tower column when PGA was 1.2 g; (**f**) the northern lower crossbeam when PGA was 1.1 g; (**g**) the southern tower bottom when PGA was 1.0~1.2 g; (**h**) the northern tower bottom when PGA was 1.0 and 1.1 g; and (**i**) the safety-belt system 2# tower damage state.

4.3. Experimental Phenomena Comparison

The damage information of the ARS and the FS are summarized, based on the previous study [26]. The damage of the ARS occurred earlier than that of the SBS. The first appearance of cracks was at the moment when PGA reached 0.7 g, as Table 7 shows. The damage of the ARS is shown in Figure 8.

Table 7. Phenomena of the ARS experimental structure under different PGA action.

PGA	Phenomena
0.1 g~0.6 g	No damage
0.7 g	A crack located at the lower tower column of 1# tower appeared as Figure 9c shows. A diagonal crack at the location of the fixed support of 1# tower appeared as Figure 9f shows.
0.8 g	The extension of original cracks of 1# lower crossbeam beam occurred with local concrete fragmentation, as Figure 9e shows.
1.1 g	A through crack appeared at 1# tower bottom with a large amount of concrete fragmentation, and the main reinforcements of the column corner of the 1# tower bottom were exposed, as Figure 9d shows.
1.4 g	A large amount of concrete fragmentation occurred at the lower crossbeam of 1# tower, and the reinforcements were exposed. Meanwhile, many cracks appeared at the bottom of the 1# tower, as Figure 9a,b show.

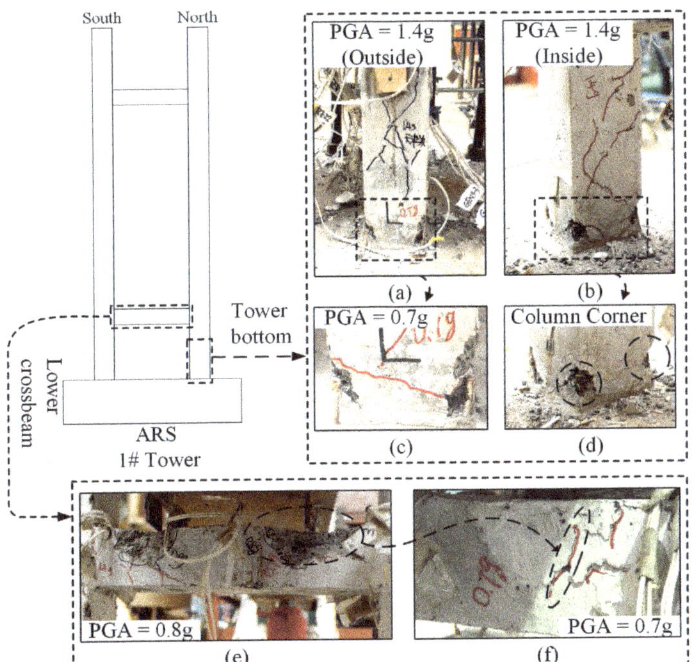

Figure 9. The damage observed of the ARS 1# tower: (**a**) the outside of the northern lower tower column when PGA was 1.4 g; (**b**) the inside of the northern lower tower column when PGA was 1.4 g; (**c**) the outside of northern lower tower column when PGA was 0.7 g; (**d**) the inside of northern lower tower column corner when PGA was 0.7 g; (**e**) the eastern side of lower crossbeam when PGA was 0.8 g; and (**f**) the eastern side of lower crossbeam when PGA was 0.7 g.

The damage of the FS occurred earlier than the ARS or the SBS. The first appearance of cracks in the FS was at the moment when PGA was 0.5 g, as Table 8 shows. The damage of the FS is shown in Figure 9.

Table 8. The phenomenon of FS experimental structure under different PGA action.

PGA	Phenomenon
0.1 g~0.4 g	No damage
0.5 g	Two obvious cracks appeared at the location near the lower crossbeam of 1# tower, as Figure 10a shows. One crack appeared at the location near the lower crossbeam of the 2# tower, as Figure 10c shows.
0.6 g	A new crack appeared at the location near the lower crossbeam of the 2# tower, as Figure 10c shows. The obvious collision between the girder and the auxiliary pier occurred under the action of the TJ wave.
0.8 g	Four parallel cracks appeared at the central part of the 2# tower column, as Figure 10b shows. Two cracks appeared at 2# tower bottom, as Figure 10d shows.
1 g	The collisions between the girder and auxiliary pier occurred under the action of the ELC wave, TJ wave, and SA wave.
1.1 g	The outermost diagonal cable of 2# tower lost the anchored state.

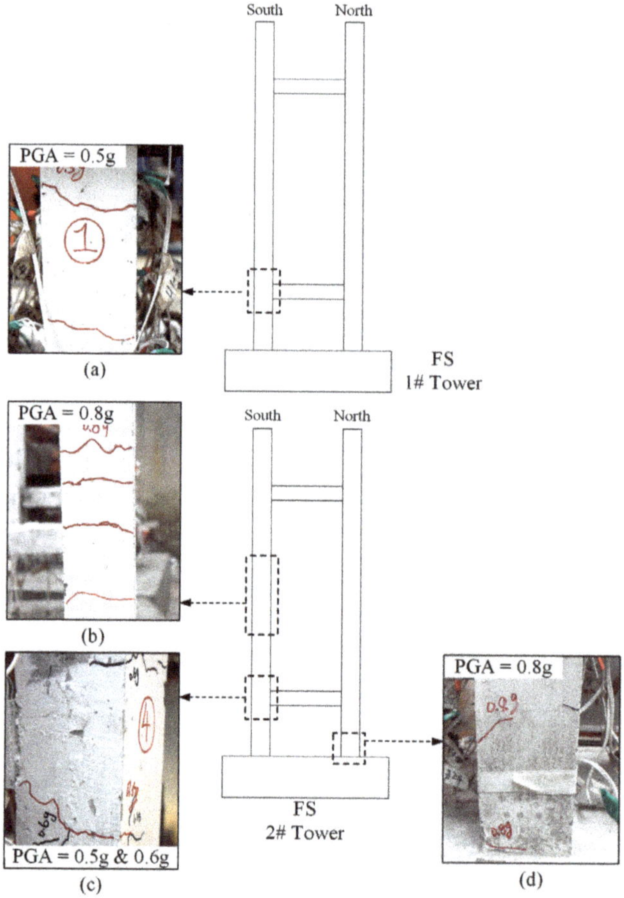

Figure 10. The damage observed of FS 1# and 2# tower: (**a**) the location of 1# tower southern side near the lower crossbeam when PGA was 0.5 g; (**b**) the location of 2# tower southern tower column when PGA was 0.8 g; (**c**) the location of 2# tower column southern side near the lower crossbeam when PGA was 0.5 g and 0.6 g; and (**d**) the northern side of 2# tower bottom.

The damage details for different structural systems can be summarized by combining the results from previously reported studies. Obviously, the PGA (0.8 g), first rendering the SBS to be damaged, is higher than those PGAs (0.7 g and 0.5 g) first rendering the ARS and the FS to be damaged. Additionally, the cracks of the SBS were less. The damage situation of experimental structures with different anti-seismic systems is shown in Table 9.

Table 9. Observed damage details.

PGA	FS	ARS	SBS
0.5 g	2 cracks occurred (1# tower) 1 crack occurred (2# tower)	No damage	No damage
0.6 g	1 new crack occurred (2# tower) Pier-girder collision	No damage	No damage
0.7 g	—	1 crack occurred (1# tower) 1 crack occurred (1# tower fixed pier)	No damage
0.8 g	4 parallel cracks occurred (2# tower) 2 cracks occurred (2# tower)	Original cracks extended (1# tower) Concrete fragmentation (1# tower)	Concrete fragmentation (of 1# tower crossbeam)
1.0 g	Intense pier-girder collision (ELC, TJ, and SA wave)	—	Horizontal cracks occurred (1# tower column) Original cracks extended (1# tower bottom)
1.1 g	Outmost stayed cable unanchored (2# tower)	Penetrating cracks occurred (1# tower bottom) Intense fragmentation occurred (1# tower column corner)	Original cracks extended (1# tower bottom)
1.2 g	—	—	1 new crack occurred (1# tower) 1 oblique crack occurred (1# tower)
1.4 g	—	Intense fragmentation (1# tower lower crossbeam) Plenty of cracks occurred (1# tower)	Bolt was broken (safety belt device)

5. Analysis of Seismic Response

5.1. Displacement Response

The displacement response of the three structural systems under different seismic waves is shown in Figure 11a,b. After the activation of the safety-belt device, the displacement response of the tower top and girder of the SBS is slightly smaller than that of the ARS under the action of ELC, EMC, and SA waves.

However, under the action of the TJ wave, the displacement response of the tower top and girder of the SBS is larger than that of the ARS. This might result for the following reasons: (1) Although the safety belt was activated, due to the pulse effect of TJ waves, the maximum structural response occurred before the safety-belt device could fully function; (2) the SBS has a larger dynamic magnification factor under the effect of TJ wave. This indicates that the seismic damage characteristics of low-gravity-center cable-stayed bridges are not only related to the type of structural system but also related to the characteristics of the ground vibration suffered.

Compared with the FS, the displacement response of the tower top and girder of the SBS is smaller under the action of different seismic waves. This indicated that the SBS could effectively control the displacement response of the structure compared with the FS.

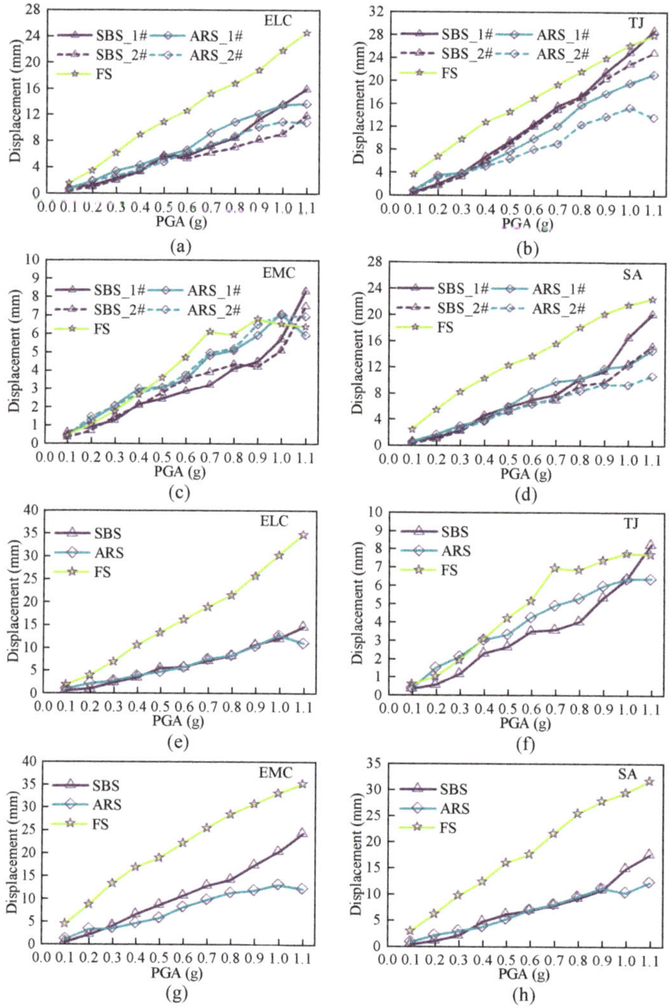

Figure 11. Displacement response of tower and girder under different waves: (**a**) tower response under ELC; (**b**) tower response under TJ; (**c**) tower response under EMC; (**d**) tower response under SA; (**e**) girder response under ELC; (**f**) girder response under TJ; (**g**) girder response under EMC; and (**h**) girder response under SA.

5.2. Acceleration Response

The acceleration responses of different structural systems were compared. An acceleration amplification factor λ (the ratio of the acceleration to the inputted PGA) was used in normalizing the acceleration response.

After the safety-belt device was activated, the maximum acceleration response of the SBS tower top is lower than that of the ARS experimental structure 1# tower, under the action of the ELC wave, and SA wave, as Figure 12a shows. However, the acceleration response at the tower top of the SBS is larger than that of the ARS under the action of the TJ wave and EMC wave. This might be caused by the additional restraint on the tower of the free pier that resulted from activation of the safety-belt device.

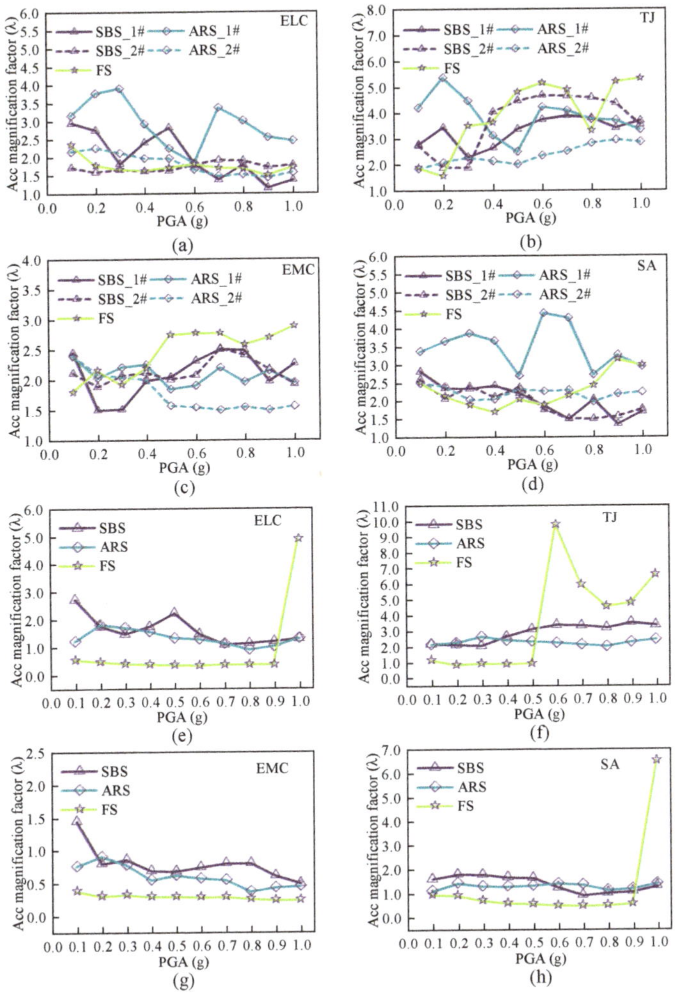

Figure 12. Acceleration response of tower and girder under different waves: (**a**) tower response under ELC; (**b**) tower response under TJ; (**c**) tower response under EMC; (**d**) tower response under SA; (**e**) girder response under ELC; (**f**) girder response under TJ; (**g**) girder response under EMC; and (**h**) girder response under SA.

From Figure 12b, the acceleration response of the girder of the FS increases significantly when the PGA effect is large, under the action of the ELC wave, TJ wave, and SA wave. This is due to the sudden increase in the acceleration response caused by the excessive displacement of the girder, which is lifted by the tension cable and collides with the auxiliary pier when it falls back. Although the FS has a smaller acceleration response, excessive displacement response may lead to unforeseen collisions.

5.3. Strain Response

The internal force response and damage mechanism of the SBS under seismic action were analyzed by analyzing the strain of reinforcement in the tower. The strain response at the bottom of 1# tower and 2# tower was selected for analysis.

Compared with the 1# tower of the ARS and the FS, after activation of the safety-belt device, the strain response of the 1# tower bottom of the SBS is smaller under the action of the four seismic waves, as shown in Figure 13a.

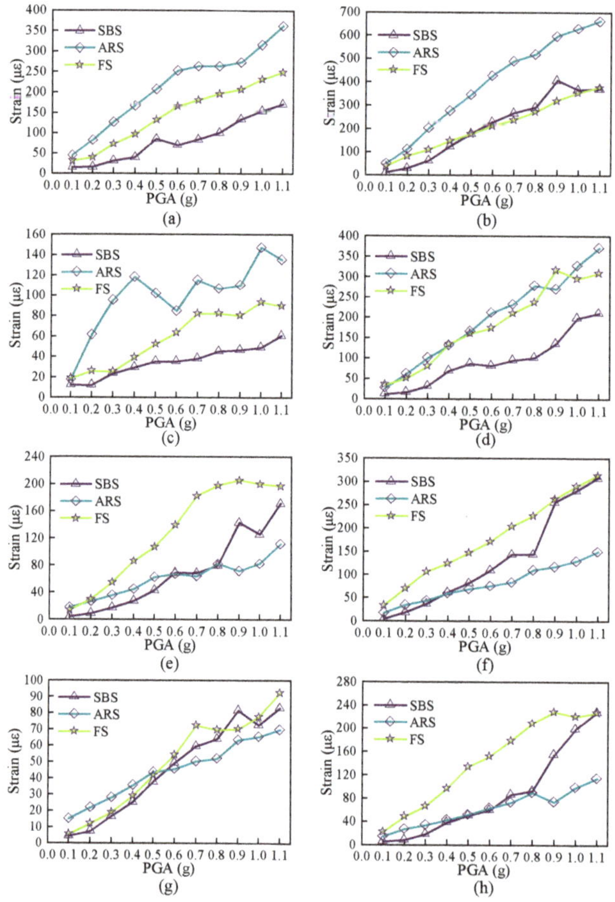

Figure 13. Strain response of 1# and 2# tower: (**a**) 1# tower response under ELC; (**b**) 1# tower response under TJ; (**c**) 1# tower response under EMC; (**d**) 1# tower response under SA; (**e**) 2# tower response under ELC; (**f**) 2# tower response under TJ; (**g**) 2# tower response under EMC; and (**h**) 2# tower response under SA.

Compared with the 2# tower of the ARS, after activation of the safety-belt device, the strain response of the 2# tower bottom of the SBS is larger than that of the ARS under the action of the four seismic waves, as Figure 13b shows. This might result from the load-sharing effect of the safety-belt device set at the connection between the 2# tower and the girder. It confirms the expectation of the effect of a safety-belt device.

5.4. Evaluation of Decreasing Amplitude Effect

The decreasing amplitude effect of the SBS in displacement and strain compared with the ARS and the FS is used to present the seismic performance of the SBS. The decreasing amplitude effect is defined by the decreasing amplitude ratio, as Equation (2):

$$\lambda = \frac{R_{SBS} - R_0}{R_0} \times 100\% \tag{2}$$

where λ denotes the decreasing amplitude ratio, R_{SBS} denotes the seismic response indicator of SBS, and R_0 denotes the seismic response indicator of the ARS or the FS.

In terms of displacement response, the SBS and the ARS both had a better vibration damping effect than the FS, which was been reported in Section 5.1.

Under some specific seismic waves, like ELC and EMC waves, the SBS performed better than the ARS with an effect of 10–30%, in term of the tower displacement, as shown in Figure 14a. Notably, the decreasing amplitude ratio degraded since the PGA of the TJ wave exceeded 0.4 g. This was likely caused by the locked safety-belt device being prematurely broken, resulting from a suddenly increased acceleration of the TJ wave. Additionally, the decreasing amplitude ratio degraded since the PGA of the SA wave exceeded 1.0 g. It is resulted from that the locked safety-belt device was broken and failed to offer a restraint for displacement.

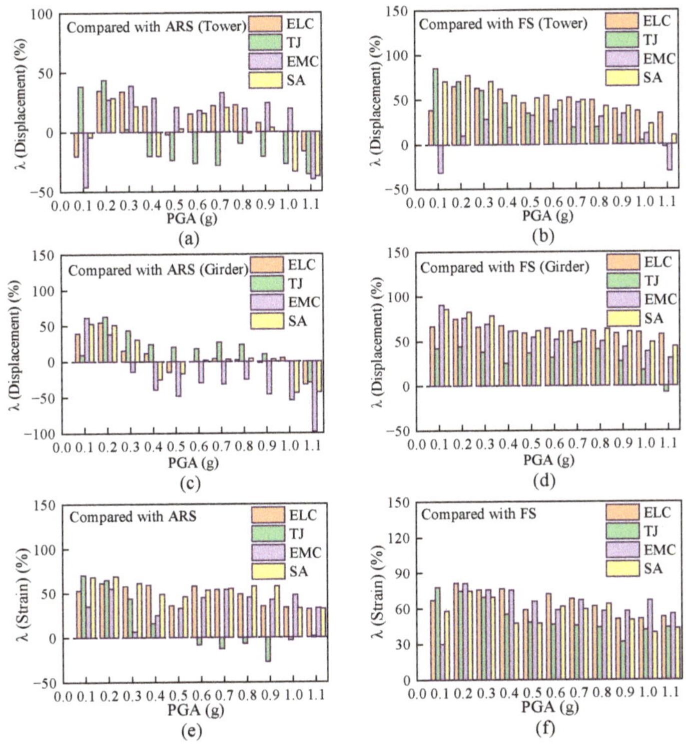

Figure 14. Decreasing amplitude effect of displacement and strain: (**a**) Decreasing amplitude ratio of tower displacement compared with ARS, (**b**) Decreasing amplitude ratio of tower displacement compared with FS; (**c**) Decreasing amplitude ratio of girder displacement compared with ARS; (**d**) Decreasing amplitude ratio of girder displacement compared with FS; (**e**) Decreasing amplitude ratio of strain compared with ARS; (**f**) Decreasing amplitude ratio of strain compared with FS.

Under the actions of ELC and TJ waves, the SBS performed better than ARS in decreasing girder displacement, as Figure 14c shows. However, the performance of the SBS apparently displayed a degradation in decreasing displacement amplitude under EMC wave action. This was likely resulted from that the safety-belt device was not being activated and did not function until the PGA of EMC wave reached 1.0 g. Additionally, the decreasing amplitude ratio degraded under the SA wave of 1.0–1.1 g PGA, which was caused by the reason that the locked safety-belt device had been broken.

The vibration damping effect of the SBS is overall better than the ARS in terms of strain response, as Figure 14e shows. This suggests that the functioning safety-belt device allowed the free tower to participate in bearing seismic action and reduced the strain response. Compared with the FS, the SBS performs better entirely, as Figure 14b,d,f show.

6. Conclusions

To improve the anti-seismic capability of a low-gravity-center cable-stayed bridge, a seismic structural system named the SBS was proposed. An experimental structure was constructed for shaking table tests. The anti-seismic performance of SBS was obtained. The advantages of the SBS were validated by comparison with the seismic performances of the ARS and the FS. The main research conclusions were as follows:

(1) The installation of the safety-belt device has no obvious impact on the dynamic characteristics (self-vibration frequency) of the low-gravity-center cable-stayed bridge, before the activation of the safety-belt device. It will not influence the daily operation of the bridge. In the aspect of alleviating the influence of secondary internal force caused by the deformation of a girder, the SBS is the same as the ARS.
(2) The seismic damage characteristics of low-gravity-center cable-stayed bridges are related to the type of structural system and the characteristics of seismic sites.
(3) The safety-belt device can improve the seismic performance of low-gravity-center cable-stayed bridges. The functioning safety-belt device may reduce the internal force response of the restrained tower and the displacement response of the girder by letting the free tower with the safety-belt device bear the seismic action together.
(4) The SBS has potential value in the seismic resistance of low-gravity-center cable-stayed bridges. In this paper, preliminary research was conducted on the seismic performance of the SBS low-gravity-center cable-stayed bridge by shaking table tests. The study on the seismic design theory and seismic fortification level division of the SBS low-gravity-center cable-stayed bridge should be carried out in the future.

Additionally, the core function of the SBS is to achieve a system conversion (from sliding support to fixed support) under great earthquake action. It is theoretically applicable to all bridge structures that require such systemic conversion. For instance, it can be installed at the free pier of continuous beam bridges to replace the conventional expansion joint and convert the sliding support into a bending resistant node. In addition, the modular design of the SBS allows convenient installation and prevents traffic from interruption.

7. Limitations and Recommendations

In this research, the SBS experimentally performed better than the ARS and the FS, but a more meticulous comparison between the SBS and the FAS is needed, so as to achieve a more precise verification for the superiority of the SBS. The constraint type of which the safety-belt device is installed changes after the activation and functioning. In this case, the SBS will be similar to the FAS. Thus, the differences between their seismic performance need to be clarified. Notably, the significant difference between them is the ductility of safety-belt device when it is fully locked. Under great earthquake action, the fully locked safety-belt device can be sacrificed to protect the concrete components. Meanwhile, the broken safety-belt device is easy to replace with a new one.

Additionally, the SBS has not been applied in real bridges. Further studies concentrating on the application of the SBS are needed, such as refined numerical and experimental research on the SBS application on different low-gravity-center cable-stayed bridges.

In this paper, only longitudinal direction earthquake action was considered. In fact, vertical earthquake action is also an important factor resulting in bridge damage. It is necessary to study the effect of vertical vibration on the SBS for systemically optimizing the design of the SBS. Furthermore, it is possible to quantify the vertical seismic response of the SBS by a computer-vision-based modal identification method for a more systemic assessment of the SBS seismic performance [33].

In addition, the temperature is also a factor affecting the durability and sensitivity of passive devices. It is necessary to study the effect of temperature on the performance degradation of the safety-belt device.

Author Contributions: Methodology, Q.L. and X.L.; Validation, Q.L. and R.F.; Resources, Z.W.; Writing—original draft, Q.L. and X.L.; Writing—review & editing, Q.L. and X.L. All authors have read and agreed to the published version of the manuscript.

Funding: This research received no external funding.

Data Availability Statement: Data are contained within the article.

Conflicts of Interest: The authors declare no conflict of interest.

References

1. Chen, Y.; Wang, Y.; Zhang, W.; Liang, K.; Du, X. Simplified Criterion for Low Gravity Center Cable-Stayed Bridge Based on Response Spectrum Analyses. *Sustainability* **2022**, *14*, 10570. [CrossRef]
2. Zhang, W.; Li, J.; Li, H. Study of Seismic Response Characteristics of Low Gravity Center Cable-Stayed Bridge. *Bridge Constr.* **2007**, *5*, 21–23+41.
3. Franchini, A.; Sebastian, W.; D'Ayala, D. Surrogate-based fragility analysis and probabilistic optimisation of cable-stayed bridges subject to seismic loads. *Eng. Struct.* **2022**, *256*, 113949. [CrossRef]
4. Li, H.; Liu, J.; Ou, J. Investigation of Seismic Damage of Cable-Stayed Bridges with Different Connection Configuration. *J. Earthq. Tsunami* **2009**, *03*, 227–247. [CrossRef]
5. Xu, Y.; Zeng, S.; Duan, X.; Ji, D. Seismic experimental study on a concrete pylon from a typical medium span cable-stayed bridge. *Front. Struct. Civ. Eng.* **2018**, *12*, 401–411. [CrossRef]
6. Javanmardi, A.; Ibrahim, Z.; Ghaedi, K.; Jameel, M.; Khatibi, H.; Suhatril, M. Seismic response characteristics of a base isolated cable-stayed bridge under moderate and strong ground motions. *Arch. Civ. Mech. Eng.* **2017**, *17*, 419–432. [CrossRef]
7. Yoo, H.; Na, H.S.; Choi, D.H. Approximate method for estimation of collapse loads of steel cable-stayed bridges. *Steel Constr.* **2012**, *72*, 143–154. [CrossRef]
8. Song, W.K.; Kim, S.E. Analysis of the overall collapse mechanism of cable-stayed bridges with different cable layouts. *Eng. Struct.* **2007**, *29*, 2133–2142. [CrossRef]
9. Raheem, S.E.A.; Hayashikawa, T. Parametric Study on Steel Tower Seismic Response of Cable-Stayed Bridges Under Great Earthquake Ground Motion. *Doboku Gakkai Ronbunshu* **2003**, *2003*, 25–41. [CrossRef]
10. Guo, R. Research on Static and Dynamic Performance of Different Structural Systems of 140 m+240m+140m Single Cable-Stayed Bridge. Master's Thesis, Shijiazhuang Tiedao University, Shijiazhuang, China, 2015.
11. Ye, A.; Hu, S.; Fan, L. Research on Aseismatic Structural System of Cable-stayed Bridge. *Bridge Constr.* **2002**, *4*, 1–4.
12. Soong, T.T.; Spencer, B.F., Jr. Supplemental energy dissipation: State-of-the-art and state-of-the-practice. *Eng. Struct.* **2002**, *24*, 243–259. [CrossRef]
13. Soong, T.T. Supplemental Energy Dissipation: State-of-the-Art and State-of-the-Practice. In Proceedings of the International Conference on Advances in Structural Dynamics Vol. I, Hong Kong, China, 13–15 December 2000.
14. Shen, X.; Wang, X.; Ye, Q.; Ye, A. Seismic performance of Transverse Steel Damper seismic system for long span bridges. *Eng. Struct.* **2017**, *141*, 14–28. [CrossRef]
15. Ahmad, F.; Rawat, S.; Yang, R.C.; Zhang, L.; Zhang, Y.X. Fire resistance and thermal performance of hybrid fibre-reinforced magnesium oxychloride cement-based composites. *Constr. Build. Mater.* **2025**, *472*, 140867. [CrossRef]
16. Nasir, A.; Butt, F.; Ahmad, F. Enhanced mechanical and axial resilience of recycled plastic aggregate concrete reinforced with silica fume and fibers. *Innov. Infrastruct. Solut.* **2025**, *10*, 4. [CrossRef]
17. Niu, J.; Ding, Y.; Shi, Y.; Li, Z. A simplified design method for metallic dampers used in the transverse direction of cable-stayed bridges. *Earthq. Eng. Eng. Vib.* **2020**, *19*, 15. [CrossRef]
18. Raza, M.U.; Butt, F.; Ahmad, F.; Waqas, R.M. Seismic safety assessment of buildings and perceptions of earthquake risk among communities in Mingora, Swat, Pakistan. *Innov. Infrastruct. Solut.* **2025**, *10*, 124. [CrossRef]
19. Ahmad, F.; Rawat, S.; Yang, R.; Zhang, L.; Fanna, D.J.; Soe, K.; Zhang, Y.X. Effect of metakaolin and ground granulated blast furnace slag on the performance of hybrid fibre-reinforced magnesium oxychloride cement-based composites. *Int. J. Civ. Eng.* **2025**, *23*, 853–868. [CrossRef]

20. Ahmad, F.; Qureshi, M.I.; Rawat, S.; Alkharisi, M.K.; Alturki, M. E-waste in concrete construction: Recycling, applications, and impact on mechanical, durability, and thermal properties—A review. *Innov. Infrastruct. Solut.* **2025**, *10*, 246. [CrossRef]
21. Rawat, S.; Saliba, P.; Estephan, P.C.; Ahmad, F.; Zhang, Y. Mechanical performance of hybrid fibre reinforced magnesium oxychloride cement-based composites at ambient and elevated temperature. *Buildings* **2024**, *14*, 270. [CrossRef]
22. Pang, Y. Seismic Fragility Assessment of an Isolated Multipylon Cable-Stayed Bridge Using Shaking Table Tests. *Adv. Civ. Eng.* **2017**, *2017*, 9514086. [CrossRef]
23. Li, J.; Yan, J.; Peng, T.; Han, L. Shake Table Studies of Seismic Structural Systems of a Taizhou Changjiang Highway Bridge Model. *J. Bridge Eng.* **2015**, *20*, 04014065. [CrossRef]
24. Zhou, R.; Zong, Z.H.; Huang, X.Y.; Xia, Z.H. Seismic response study on a multi-span cable-stayed bridge scale model under multi-support excitations. Part II: Numerical analysis. *J. Zhejiang Univ. Sci. A Appl. Phys. Eng.* **2014**, *15*, 405–418. [CrossRef]
25. Zong, Z.H.; Zhou, R.; Huang, X.Y.; Xia, Z.H. Seismic response study on a multi-span cable-stayed bridge scale model under multi-support excitations. Part I: Shaking table tests. *J. Zhejiang Univ. Sci. A Appl. Phys. Eng.* **2014**, *15*, 351–363. [CrossRef]
26. Wang, Z.; Zhang, W.; Fang, R.; Zhao, H. Dynamic Model Testing of Low-Gravity-Center Cable-Stayed Bridges with Different Girder-to-Tower Connections. *J. Bridge Eng.* **2021**, *26*, 04020112. [CrossRef]
27. Zhao, L.; Huang, J.; Zhang, W. Influence of Height of Gravity Center on Tower-girder Elastic Connecting Device's Damping Effect in Cable-stayed Bridge. *Highway* **2015**, *60*, 79–82.
28. Yi, R.; Li, X.; Zhu, S.; Li, Y.; Xu, X. A deep leaning method for dynamic vibration analysis of bridges subjected to uniform seismic excitation. *Soil Dyn. Earthq. Eng.* **2023**, *168*, 107830. [CrossRef]
29. Zhang, J.; Yang, Z.Y.; Yuan, D.P.; Zheng, S.X.; Hu, Y.R. Seismic Response of Two Site Models and Their Effects on the Railway Cable-Stayed Bridge. *KSCE J. Civ. Eng.* **2021**, *25*, 4724–4741. [CrossRef]
30. Fang, R.; Zhang, W.; Chen, Y.; Zhang, L. Aseismic effect and parametric analysis of the safe-belt device for a continuous bridge with equal height piers. *Eng. Struct.* **2019**, *199*, 109553. [CrossRef]
31. Ahamed, I.Z.A.K.; Magar, R.B. Dynamic Response of Cable-Stayed Bridge Under Seismic Vulnerability Analysis: State of the Art. In *Emerging Trends in Composite Structures Select Proceedings of ICC-IDEA 2023*; Lecture Notes in Civil Engineering; Springer: Singapore, 2024; Volume 387.
32. Yan, X.; Cunyu, C.; Zeng, Z.; Shijie, Z. Shake table experimental study of cable-stayed bridges with two different design strategies of H-shaped towers. *Earthq. Eng. Eng. Vib.* **2021**, *20*, 483–493. [CrossRef]
33. Liao, R.; Zhang, Y.; Wang, H.; Lu, L.; Chen, Z.; Wang, X.; Zuo, W. An effective ship detection approach combining lightweight networks with supervised simulation-to-reality domain adaptation. *Comput.-Aided Civ. Infrastruct. Eng.* **2025**, 1–26. [CrossRef]

MDPI AG
Grosspeteranlage 5
4052 Basel
Switzerland
Tel.: +41 61 683 77 34

Buildings Editorial Office
E-mail: buildings@mdpi.com
www.mdpi.com/journal/buildings

www.ingramcontent.com/pod-product-compliance
Lightning Source LLC
Chambersburg PA
CBHW061816210926
51850CB00006B/717

9783725857432